Neurotoxicity of Nanomaterials and Nanomedicine

Neurotoxicity of Nanomaterials and Nanomedicine

Edited by

Xinguo Jiang
School of Pharmacy, Fudan University, Shanghai, China

Huile Gao
West China School of Pharmacy, Sichuan University, Chengdu, China

ELSEVIER

AMSTERDAM • BOSTON • HEIDELBERG • LONDON
NEW YORK • OXFORD • PARIS • SAN DIEGO
SAN FRANCISCO • SINGAPORE • SYDNEY • TOKYO

Academic Press is an imprint of Elsevier

Academic Press is an imprint of Elsevier
125 London Wall, London EC2Y 5AS, United Kingdom
525 B Street, Suite 1800, San Diego, CA 92101-4495, United States
50 Hampshire Street, 5th Floor, Cambridge, MA 02139, United States
The Boulevard, Langford Lane, Kidlington, Oxford OX5 1GB, United Kingdom

Library of Congress Cataloging-in-Publication Data
A catalog record for this book is available from the Library of Congress

British Library Cataloguing-in-Publication Data
A catalogue record for this book is available from the British Library

ISBN: 978-0-12-804598-5

For information on all Academic Press publications
visit our website at https://www.elsevier.com/

Publisher: Mica Haley
Acquisition Editor: Erin Hill-Parks
Editorial Project Manager: Tracy Tufaga
Production Project Manager: Chris Wortley
Designer: Matthew Limbert

Typeset by TNQ Books and Journals

Contents

4. Current Perspective on Nanomaterial-Induced Adverse Effects: Neurotoxicity as a Case Example

J. Li and F.L. Martin

5. Toxicity of Titanium Dioxide Nanoparticles on Brain

L. Shao and B. Song

8. The Applications, Neurotoxicity, and Related Mechanism of Gold Nanoparticles

S.J.S. Flora

9. The Applications, Neurotoxicity, and Related Mechanisms of Manganese-Containing Nanoparticles

K. Ikeda, T. Tachibana and Y. Manome

10. The Application, Neurotoxicity, and Related Mechanism of Silica Nanoparticles

C. Zhong, M. He, K. Lou and F. Gao

11. The Synthesis, Application, and Related Neurotoxicity of Carbon Nanotubes

D. Shi, G. Mi and T.J. Webster

12. The Application, Neurotoxicity, and Related Mechanism of Cationic Polymers

Y. Li and D. Ju

List of Contributors

F. Brandão Portuguese National Institute of Health, Porto, Portugal; University of Porto, Porto, Portugal

H. Chen Shanghai Jiao Tong University School of Medicine, Shanghai, China

C. Costa Portuguese National Institute of Health, Porto, Portugal; University of Porto, Porto, Portugal

N. Fernández-Bertólez Universidade da Coruña, A Coruña, Spain

S.J.S. Flora Defence Research and Developmental Establishment, Gwalior, India

F. Gao East China University of Science and Technology, Shanghai, China

H. Gao Sichuan University, Chengdu, China

X. Gao Shanghai Jiao Tong University School of Medicine, Shanghai, China

X. Gu Shanghai Jiao Tong University School of Medicine, Shanghai, China

M. He East China University of Science and Technology, Shanghai, China

Q. He Sichuan University, Chengdu, China

K. Ikeda The Jikei University School of Medicine, Tokyo, Japan

X. Jiang Fudan University, Shanghai, China

D. Ju Fudan University, Shanghai, China

G. Kiliç Karolinska Institutet, Stockholm, Sweden

B. Laffon Universidade da Coruña, A Coruña, Spain

J. Li Lancaster University, Lancaster, United Kingdom

Y. Liu Sichuan University, Chengdu, China

Y. Li Fudan University, Shanghai, China; University of Pennsylvania, Philadelphia, PA, United States

K. Lou East China University of Science and Technology, Shanghai, China

Y. Manome The Jikei University School of Medicine, Tokyo, Japan

F.L. Martin Lancaster University, Lancaster, United Kingdom; University of Central Lancashire, Preston, United Kingdom

G. Mi Northeastern University, Boston, MA, United States

E. Pásaro Universidade da Coruña, A Coruña, Spain

L.Q. Shao Southern Medical University, Guangzhou, China

D. Shi Northeastern University, Boston, MA, United States

B. Song Southern Medical University, Guangzhou, China; Guizhou Provincial People's Hospital, Guiyang, China

L. Strużyńska Mossakowski Medical Research Centre, Polish Academy of Sciences, Warsaw, Poland

T. Tachibana The Jikei University School of Medicine, Tokyo, Japan

J.P. Teixeira Portuguese National Institute of Health, Porto, Portugal; University of Porto, Porto, Portugal

V. Valdiglesias Universidade da Coruña, A Coruña, Spain

T.J. Webster Northeastern University, Boston, MA, United States; King Abdulaziz University, Jeddah, Saudi Arabia

C. Zhong East China University of Science and Technology, Shanghai, China

Biography

Xinguo Jiang, BS

Professor, Key Laboratory of Smart Drug Delivery (Ministry of Education), School of Pharmacy, Fudan University, Shanghai, China.

Prof. Xinguo Jiang obtained BS in Pharmacy from Shanghai First Medical College in July 1982. Then he worked in the Department of Pharmaceutics in Fudan University (originally Shanghai First Medical College) since October 1986. Between January and July 1996, he was a guest professor in the Department of Pharmacy in Nagasaki University (Japan). He also served as the vice editor for *Chinese Journal of Clinical Pharmacy* and as editorial board member for eight other journals including *Acta Pharmaceutica Sinica* and *Chinese Pharmaceutical Journal*. Prof. Jiang's research interests focus on the development of novel drug delivery systems especially for brain-targeting drug delivery. Having served as the Principal Scientist, Prof. Jiang received a research grant from the State Plan for Development of Basic Research in Key Areas. He also has won eight research grants from National Natural Science Foundation of China and five research grants from Shanghai government. He has accomplished over 70 R&D projects on novel drug delivery systems under cooperation with pharmaceutical companies. Prof. Jiang has published over 120 research papers in peer-reviewed journals and has 15 patents. He has developed three marked formulations. He served as guest editor and edited two theme issues for *Pharmaceutical Research* and *Current Pharmaceutical Biotechnology*. He also wrote several book chapters and edited three books (in Chinese).

Huile Gao, PhD

Associate Professor, Key Laboratory of Drug Targeting and Drug Delivery Systems, West China School of Pharmacy, Sichuan University, Chengdu, Sichuan, China.

Dr. Huile Gao received his PhD in Pharmaceutics from School of Pharmacy, Fudan University, in 2013 under the supervision of Prof. Xinguo Jiang. Then he joined the West China School of Pharmacy, Sichuan University, as Instructor in July 2013 and was promoted to Associate Professor in July 2014. Dr. Gao's research interests focus on the design, synthesis, characterization, and evaluation of stimuli-responsive nanomaterials for drug and imaging probe delivery to improve treatment and diagnosis of human diseases especially tumor and brain diseases. He has published over 50 peer-reviewed articles. His research is supported by the National Natural Science Foundation of China (81373337, 81402866), Excellent Young Scientist Foundation of Sichuan University (2015SCU04A14), and four other grants. He has been given the Excellent Doctoral Dissertation of Shanghai award and the Young Excellent Pharmaceutics Scientist award, both in 2015.

Preface

Development of nanomaterials and nanomedicines is an impressive endeavor by researchers from chemistry, physics, engineering, biology, and medicine to improve the quality of our life. Many kinds of nanomaterials and nanomedicines are approved for human use or are under evaluation. The extensive application considerably elevates the exposure of nanomaterials and toxicity potential to human beings. Even as we enjoy the benefit from nanomaterials and nanomedicines, the toxicity potential should not be ignored.

The central nervous system (CNS) is the most important part of the human body, and should be paid serious attention when we talk about toxicity. The blood–brain barrier (BBB) has contributed to the protection of the CNS from harm by toxicants. Knowledge about BBB and the interaction between BBB and compounds has greatly expanded in the past several decades especially the past few years. The depth of knowledge about BBB enables researchers to develop nanomaterials and nanomedicines for delivering drugs or imaging probes to brain that display engaging potential for CNS disorders. However, neurotoxicity is emerging as a critical concern. The knowledge about neurotoxicity urgently needs to expand.

The chapters in this book were written by active researchers who dedicate their effort to enlarge the brain application of nanomaterials, develop brain nanomedicines, and improve the understanding of interaction of nanomaterials and the CNS. We appreciate these authors for sharing their knowledge and insights about this topic, which provide all of us, especially who work toward nanomaterials and nanomedicines, with an overview of the field and spark to think more about our research.

The early chapters provide an overview of the application of nanomaterials and nanomedicines in brain. The main routes by which nanomaterials enter brain are described. Then the excretion routes are discussed, although there are few studies related to this important topic.

A general review about neurotoxicity is provided in Chapter 4, with emphasis on the neurotoxicity mechanism. Then contributors describe in detail the application and neurotoxicity of many kinds of nanomaterials (Chapters 5–12) that are widely used, including titanium dioxide nanoparticles, iron oxide nanoparticles, silver nanoparticles, gold nanoparticles, manganese-containing nanoparticles, silica nanoparticles, carbon nanotubes, and cationic polymers. These kinds of nanoparticles represent the most commonly used nanomaterials in this field. Finally, an overview about neurotoxicity is provided with discussion of potential strategies to reduce the neurotoxicity of nanomaterials and nanomedicines. Researchers will benefit from this knowledge to design novel nanomaterials and nanomedicines with minimum neurotoxicity.

The editors greatly thank the individual chapter contributors for kindly sharing their knowledge and ideas. It is a great pleasure to collaborate with them to develop this book. We admire the outstanding work they contributed so that researchers from nanomaterials and nanomedicines can benefit from this book and make greater success in their own work.

Xinguo Jiang
Huile Gao
April 2016

Introduction and Overview

H. Gao[1], X. Jiang[2]
[1]*Sichuan University, Chengdu, China;* [2]*Fudan University, Shanghai, China*

Nanoparticles (NPs) are generally described as particles of size about 0.1–100 nm, but particles that are several hundred nanometers in size are also called NPs. Thus in this book, we expand the definition of NPs to all particles around 0.1–1000 nm. The materials of NPs are called nanomaterials. One of the most important applications of NPs is delivering drugs and imaging probes to human body for disease diagnosis and treatment. These NPs are called nanomedicines.

Accompanied with the development of chemistry, biology, and materials science, many kinds of nanomaterials are synthesized with impressive characteristics. Nanomedicines are also extensively designed for both peripheral diseases and central nervous system (CNS) disorders. Despite great achievements, the toxicity of nanomaterials has been emerging as an increasingly important concern. The CNS is the most important part of human being, thus the toxicity to CNS, named neurotoxicity, should be paid particular attention.

In this book, the application and neurotoxicity of nanomaterials and nanomedicines are reviewed and discussed. The book includes the following:

- The application of nanomaterials and nanomedicines in brain targeting drug delivery
- The routes by which nanomaterials and nanomedicines enter into brain
- The excretion of nanomaterials and nanomedicines from brain
- The neurotoxicity of many kinds of nanomaterials, including titanium dioxide nanoparticles, iron oxide nanoparticles, silver nanoparticles, gold nanoparticles, manganese-containing nanoparticles, silica nanoparticles, carbon nanotubes, and cationic polymers
- The general mechanism of neurotoxicity and strategies to reduce the neurotoxicity.

HOW DO NANOMATERIALS AND NANOMEDICINES ENTER INTO AND GET EXCRETED FROM BRAIN?

In the CNS, the blood–brain barrier (BBB) considerably restricts the brain distribution of nanomaterials and nanomedicines. In healthy condition, the BBB protects the CNS from harm by toxicants in blood. But in CNS disorders, the

BBB also restricts the brain access of drugs, making CNS disorders the most difficult diseases to treat. Many researchers dedicate their efforts to enhance brain drug delivery using various methods. In these methods, nanomaterials play a critical role. In Chapter 1, we discuss the general application of nanomaterials and nanomedicines in brain targeting drug delivery and then in Chapter 2, Dr. Liu and Dr. He further summarize the routes that the nanomaterials enter brain, including penetrating the BBB through receptor-mediated endocytosis, transporter-mediated endocytosis, adsorptive-mediated endocytosis, bypassing BBB through intranasal delivery, inhibiting the function of BBB by inhibition of efflux pumps, and disturbing the structure of BBB. The distribution of nanomaterials and nanomedicines in brain is influenced by many factors, such as size, shape, surface charge, and ligand modification of NPs; administration routes; chronobiology; and disease condition, which is reviewed by Dr. Gao in Chapter 3. After entering into brain, the nanomaterials could be degraded or excreted from brain, which is affected by the deformability, biodegradability, size, shape, and ligand modification of the nanomaterials and nanomedicines. The conscious state and disease condition also affect this procedure, which is discussed in Chapter 3.

WHAT IS THE NEUROTOXICITY OF NANOMATERIALS?

Because nanomaterials could enter into brain through various routes, the contact of nanomaterials and nanomedicines with CNS may cause some neurotoxicity. The common neurotoxicity includes oxidative stress, inducing cell apoptosis and autophagy, immune response and inflammation, activating specific signaling pathway, affecting BBB function, and so on, which are reviewed in Chapter 4. Many kinds of nanomaterials have been developed with potential for CNS exposure, among which the widely used nanomaterials are selected for detailed description, such as metal NPs, carbon nanotubes, and cationic polymers. The application and neurotoxicity of these nanomaterials are reviewed in Chapters 5–12. The potential mechanism and influence factors are also reviewed, such as size, shape, crystal type, charge, surface property, release of ions, and administration route. Because neurotoxicity is a critical concern in the application of nanomaterials and nanomedicines, strategies to reduce neurotoxicity are important for researchers. Based on the reasons involved in the neurotoxicity, we conclude several strategies in Chapter 13, including reducing brain exposure and decreasing inherent toxicity of nanomaterials and nanomedicines.

CONCLUSION

This book is prepared with the purpose of benefiting nanomaterial and nanomedicine researchers in the following areas:

- Fundamental knowledge about nanomaterials and nanomedicines application in CNS disorders

- Routes by which nanomaterials and nanomedicines enter into and get excreted from brain
- Common neurotoxicity of nanomaterials and the influence factors
- Mechanism of neurotoxicity and strategies to reduce neurotoxicity.

The valuable insights shared by chapter authors are intended to expand the fundamental knowledge about brain application of nanomaterials and nanomedicines and the corresponding neurotoxicity. Researchers from all related fields may further develop nanomaterials and nanomedicines with lower neurotoxicity for more satisfying application in humans.

Chapter 1

The Medical Applications of Nanomaterials in the Central Nervous System

H. Gao[1], X. Jiang[2]
[1]*Sichuan University, Chengdu, China;* [2]*Fudan University, Shanghai, China*

Chapter Outline

1. INTRODUCTION

The development of nanotechnology has been influencing many aspects of human life. Various nanomaterials have been developed to address critical requirements including disease diagnosis and therapy, energy, and environmental protection. Especially in medicine, the application of nanomaterials has gained increasing attention from not only academics but also industry.

The central nervous system (CNS) is one of the most important systems in the human body. To protect and keep a stable microenvironment for the normal activities of neurons, several barriers play a key role in guarding the brain parenchyma against invading substances or organisms (Neuwelt et al., 2008; Tuladhar et al., 2015). Over a hundred years ago, Goldman proved the existence of a physiological barrier that separated the brain from the circulation system

Neurotoxicity of Nanomaterials and Nanomedicine. http://dx.doi.org/10.1016/B978-0-12-804598-5.00001-5

using trypan blue. Till 1976, the modern concept of the blood–brain barrier (BBB) was proposed by Hugh Davson (1976). A series of research confirmed that a barrier system did exist between brain, cerebrospinal fluid, and blood aiming at limiting the exchange of substances (Zlokovic et al., 1990; Zloković et al., 1987). This system is a dynamic structure that adjusts and regulates the balance between blood and brain to maintain the normal function of the CNS (Abbott, 2005).

Owing to the aging population, the incidence of CNS disorders increases gradually, such as brain tumor, Alzheimer disease (AD), Parkinson disease (PD), and stroke. Because most of the CNS disorders gravely decrease the life quality of patients, they have been considered as the most serious threats to humans. Unfortunately, in contrast to peripheral diseases, the diagnosis and treatment of CNS disorders are restricted by the BBB (Wohlfart et al., 2012). The BBB is a tight barrier that is constituted by several kinds of cells, such as brain microvessel endothelial cells, astrocytes, microglial cells, and pericytes (Begley, 2004). The BBB has a high transendothelial electrical resistance, which considerably restricts the diffusion and transportation of molecules from blood to brain. Actually, almost 98% of small molecules (<500 Da) and 100% of large molecules are not able to penetrate through the BBB (Pardridge, 2007). Generally, the integrity of the BBB protects the brain from harm by toxins and other substances, which is important for keeping the normal function of brain. Unfortunately, the BBB also restricts the access of drugs used to treat diseases involving the brain. Therefore, it is important to enhance brain delivery of drugs to improve the diagnosis and treatment of CNS disorders.

Nanomaterials are widely used as carriers of drugs and probes because they can act as Trojan horse to deliver the drugs and probes to certain tissues and cells after effective disguise (Gao et al., 2013; Gao and Jiang, 2015). Many kinds of nanomaterials have been utilized for CNS disorders. In this chapter, we focus on the application of various nanomaterials in delivering different kinds of drugs and probes to treat CNS disorders.

2. SMALL CHEMICAL DRUGS DELIVERY

Small chemical drugs are still the most commonly used in disease treatment. Except certain lipophilic drugs, such as temozolomide and L-dopa, most of the chemical drugs could not penetrate the BBB. Although some researchers argued the BBB is disrupted in some conditions, such as brain tumor, the BBB was indeed complete in at least the filtrated region and the diseased brain tissue is still hard to be accessed by drugs (Gao et al., 2013). For example, the distribution of lapatinib in brain metastasis is only 19% as that in lung metastasis (Taskar et al., 2012). To address this issue, nanomaterials are often used to deliver the chemical drugs into brain. Owing to the small size and good stability, the chemical drugs could be encapsulated into or anchored onto most nanomaterials.

2.1 Natural Nanomaterials

Natural nanomaterials, such as lipid and natural proteins, are widely used for constructing drug delivery systems because of their safety. Several kinds of natural nanomaterials-based/derived nanodrugs are available clinically, such as doxorubicin (DOX)-loaded liposomes (Doxil), daunorobicin-loaded liposomes (DaunoXome), and paclitaxel (PTX)-loaded albumin nanoparticles (NPs) (Abraxane) (Weissig et al., 2014). However, these nanodrugs were not decorated with specific ligand, thus they could be used for the treatment of peripheral diseases rather than CNS disorders because of the restriction by BBB. Decorating with BBB-specific ligands could improve the distribution in brain (Gao et al., 2013), thus many studies developed various brain targeting drug delivery systems based on the natural nanomaterials.

Liposomes, consisting of lipid and cholesterol, are the most commonly used drug delivery systems not only because of their safety but also because of their wide application in almost all kinds of drugs. For chemical drugs, both hydrophilic and hydrophobic drugs could be loaded into liposomes because they have an aqueous solution core and a hydrophobic membrane. There are many kinds of active targeting ligand-modified liposomes to deliver small molecules to the brain (Lai et al., 2013). Transferrin (Tf) receptor is highly expressed on the BBB, thus the corresponding ligand, Tf, was modified onto liposomes to deliver 5-florouracil to brain tumor (Soni et al., 2008). The Tf-liposomes caused 17-fold higher 5-florouracil brain delivery compared with free 5-florouracil. Modification with two ligands can further elevate the BBB penetration and disease targeting. Ying et al. (2010) comodified *p*-aminophenyl-α-D-mannopyranoside (MAN) and Tf onto liposomes. MAN is a mannose analog that has a high affinity for BBB overexpressed glucose transporter 1, which can facilitate the penetration through BBB of the constructed systems. Tf receptors are highly expressed on BBB and glioma cells, which could further enhance the BBB transportation and glioma targeting of liposomes. In combination, the dual-modified liposomes showed higher BBB model penetration and C6 glioma cells uptake compared with MAN and Tf solo-modified liposomes. Except the dual modification, it is promising to combine the motifs of different ligands into one, which could combine the different functions from two ligands (Gao et al., 2014b). Liu et al. (2014) decorated liposomes with R8-RGD, a tandem peptide of octaarginine and RGD peptide, to deliver DOX into brain tumor, which may benefit from the active targeting effect of RGD and cell penetrating effect of octaarginine. Results showed the uptake of R8-RGD-liposomes by both brain endothelial cells and brain tumor cells was much higher than that of unmodified liposomes. In vivo, R8-RGD-liposomes showed considerably higher accumulation in brain tumor than that of unmodified liposomes. As a result, DOX-loaded R8-RGD-liposomes could considerably prolong the median survival time of brain tumor–bearing mice from 26 to 48 days. Replacing traditional cell penetrating peptide octaarginine with tumor microenvironment activatable TH peptide also showed efficient brain tumor targeting accompanied with long blood circulation time (Shi et al., 2015). More importantly, the blood circulation time was extended because the activatable cell penetrating peptide TH was negative in neutral pH.

Owing to the safety of liposomes, several active targeting ligand-modified liposomes were under preclinical and clinical evaluation (van der Meel et al., 2013). Glutathione (GSH) can penetrate through the BBB mediated by GSH transporter, thus GSH-liposomes were developed as a brain targeting drug delivery platform (To-BBB technology). In rats receiving 7 mg/kg of either DOX-loaded GSH-liposomes (2B3-101) or PEGylated liposomal DOX, the brain-to-blood ratio of DOX was 4.8-fold higher after administration of 2B3-101 than that of PEGylated liposomal DOX (Birngruber et al., 2014). In the mice model, 2B3-101 successfully prolonged the survival of mice (van der Meel et al., 2013), suggesting this nanomedicine was useful in brain tumor treatment. At present, it is under phase I/II clinical evaluation, and some positive data have been obtained, which showed that 2B3-101 is well tolerated.

As albumin is an endogenous protein, albumin-based NPs are impressive systems for drug delivery because of their good biocompatibility, and Abraxane has been approved by the US Food and Drug Administration in 2007 (Fu et al., 2009). The encapsulation of drugs mainly depends on the highly affinity between the drugs and albumin. Our group prepared docetaxel-loaded albumin NPs with a drug-loading capacity of 7.5% (Gao et al., 2015). The NPs could effectively distribute to brain tumor mainly attributing to the weak enhanced permeability and retention (EPR) effect of brain tumor and the interaction between albumin and the secreted protein, acidic and rich in cysteine, which is highly expressed on many kinds of tumor cells (Fu et al., 2009). Similarly, lapatinib-loaded albumin NPs (LTNPs) effectively target to glioma. The concentration achieved in the glioma was about 0.08%ID/g tissue, which was 25-fold higher than the commercial tablet (Fig. 1.1) (Gao et al., 2014a). In addition, the glioma/normal brain ratio was as high as 30, demonstrating the LTNPs could selectively target to glioma. Regarding the superiority of albumin-based NPs, there are several albumin nanodrugs under evaluation, including docetaxel, lapatinib, and pirarubicin (Gao et al., 2014a, 2015; Zhou et al., 2013).

To further enhance the targeting delivery efficiency, albumin NPs could be modified with targeting ligands. Su et al. (2014) decorated lactoferrin (Lf) onto albumin NPs to deliver DOX into brain tumor. The Lf decoration considerably increased the uptake by both brain capillary endothelial cells and C6 glioma cells. Consequently, the Lf-decorated albumin NPs delivered about three times higher DOX into tumor-bearing brain compared with unmodified albumin NPs. Dual ligand modification may further improve the targeting efficiency. A kind of RGD and KALA peptide dual-modified albumin NPs was constructed to deliver DOX (Chen et al., 2015a). In this system, RGD could recognize integrin

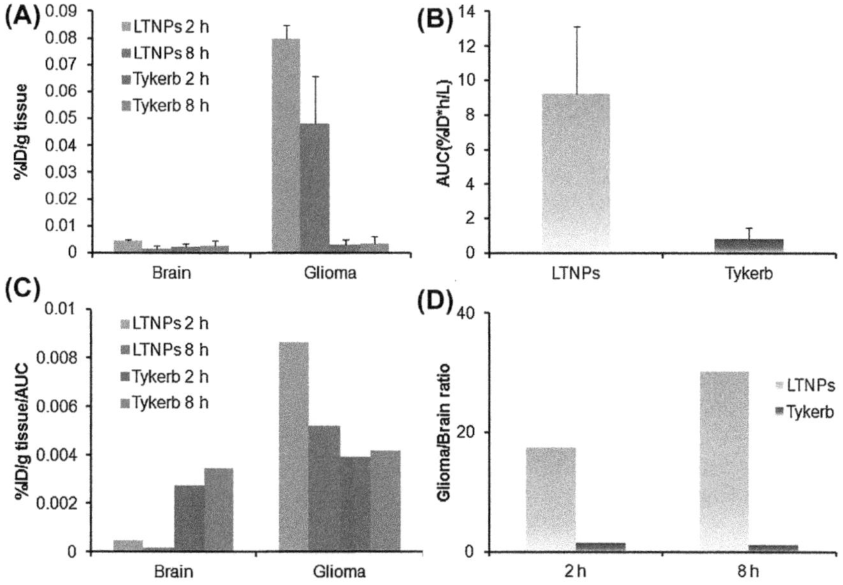

FIGURE 1.1 Pharmacokinetics and glioma distribution of lapatinib-loaded albumin NPs (LTNPs) and the commercial tablet (Tykerb). (A) Lapatinib concentrations in normal brain and glioma of LTNPs and Tykerb group at 2 and 8 h (B) $AUC_{0-\infty}$of LTNPs and Tykerb group. (C) Concentration in brain and glioma deducted $AUC_{0-\infty}$ of LTNPs and Tykerb group. (D) Glioma/brain ratio of LTNPs and Tykerb group, $n=5$. *Reprinted from Gao, H., Wang, Y., Chen, C., Chen, J., Wei, Y., Cao, S., Jiang, X., 2014a. Incorporation of lapatinib into core-shell nanoparticles improves both the solubility and anti-glioma effects of the drug. Int. J. Pharm. 461 (1–2), 478–488 with permission of the copyright holder, Elsevier, Amsterdam.*

receptors on U87 glioma cells and thus mediate the active targeting of the system, whereas KALA peptide, a kind of cell-penetrating peptide, further elevated the internalization of particles. After internalization, the particles would distribute in endosomes with low pH. At such a low pH value (about 5), the negative charge of albumin decreased significantly because the endosome pH was near the isoionic point of albumin (pI=4.7). The low charge density of albumin decreased the electronic interaction between albumin and cationic KALA peptide, leading to the disassembly of the particles and release of DOX. Consequently, the DOX-loaded RGD and KALA dual-modified albumin NPs showed high toxicity to U87 cells. The half inhibitory concentration (IC_{50}) of the dual targeting system was 2.6 μg/mL, which was significantly lower than that of free DOX (9.4 μg/mL).

2.2 Anionic and Neutral Polymers

Biodegradable polymers are the most commonly used nanomaterials, such as poly (ethylene glycol)-block-poly(epsiloncaprolactone) (PEG-PCL), poly(ethylene glycol)-block-poly(lactic acid) (PEG-PLA), and poly(ethylene glycol)-block-poly(lactide-co-glycolide) (PEG-PLGA), mainly because of the safety and controllable properties of these polymers. PEG-PLA and PEG-PLGA have been approved for injection, and a PTX-loaded PEG-PLA-based micelle has been approved in 2001 for breast cancer and pancreatic cancer treatment (Weissig et al., 2014). Although no nanomedicine has been approved for the treatment of CNS disorders, PEG-PLA and PEG-PLGA still attracted many researchers' attention, making them the most widely used nanomaterials in CNS disorders.

There are several studies that suggested unmodified PLGA NPs could distribute into brain after oral administration, but no mechanism was reported (Semete et al., 2010). In most studies, specific ligand modification was required to penetrate the BBB and target to the diseased cells. Several peptides, such as tLyp-1 (Hu et al., 2013), iNGR (Kang et al., 2014), APTEDB peptide (Gu et al., 2014), EGFP-EGF1 (Zhang et al., 2014b), and CLT1 peptide (Zhang et al., 2014a), were used to modify PEG-PLA NPs and then deliver various chemical drugs into brain. All these studies showed promising treatment effect, demonstrating that ligand modification is an effective method for improving brain targeting delivery. For example, ALMP peptide-modified PEG-PLA NPs delivered approximately two-fold higher PTX to brain tumor than unmodified PEG-PLA NPs, resulting in 13.5 days (39.1%) longer median survival time of brain tumor–bearing mice (Gu et al., 2013). In addition, the drug encapsulation efficiency could be considerably elevated using the nanoemulsion templating method to prepare the brain targeting delivery system. For example, PLGA NPs prepared using nanoemulsion templating showed an encapsulation efficiency higher than 90% (Fornaguera et al., 2015), which is useful for improving the brain drug delivery efficiency.

PEG-PCL is another kind of biodegradable polymer that has been extensively used in recent years. Our laboratory used PEG-PCL to construct NPs of size about 150 nm, which could load both hydrophobic and hydrophilic drugs. Similarly, PEG-PCL NPs are also modified with specific ligands for brain targeting drug delivery. For example, we modified PEG-PCL NPs with both TGN peptide and AS1411 aptamer to penetrate the BBB and then target to brain tumor cells (Gao et al., 2012). Using near infrared dye as a probe, it was shown that the dual-modified PEG-PCL NPs could specifically accumulate in brain tumor site with high tumor/normal brain ratio. After loading with docetaxel, a common chemotherapeutic drug, the dual-modified NPs prolonged the median survival time of brain tumor–bearing mice from 17 to 32 days. More importantly, the PEG-PCL NPs did not display obvious side effects to most normal tissues after administering several times.

Other kinds of biodegradable polymers are also used in brain targeting delivery of chemical drugs in a similar way, such as poly(butyl cyanoacrylate) (PBCA), poly(isohexyl cyanoacrylate), and poly(alkyl cyanoacrylates) (Kreuter, 2014). For example, RGD-modified (PEG)-b-poly-(L-glutamic acid) micelles were capable of delivering platinum to brain tumor with rapid accumulation and high permeability from vessels into the tumor parenchyma (Miura et al., 2013).

2.3 Dendrimers

Different from other kinds of polymers, dendrimers are well defined with precise size, shape, surface groups, and architectures. There are several kinds of dendrimers, such as poly(amidoamine) (PAMAM), poly(etherhydroxylamine), and poly(propyleneimine) (PPI), which are all exploited extensively for drug and gene delivery (Menjoge et al., 2010). PAMAM is the first commercial dendrimer. Because of the fruitful amino groups on the surface, PAMAM often directly conjugates with drugs through linkers. Zhu et al. (2010) conjugated DOX onto PAMAM through *cis*-aconitic anhydride, and 1 molecule of PAMAM could be loaded with 14 molecules of DOX. Zhang et al. (2011) further modified the DOX-PAMAM with RGD for brain tumor targeting delivery because RGD can interact with integrin receptors that overexpressed on both brain tumor cells and neovasculatures. The distribution of RGD-modified DOX-PAMAM in brain tumor was 21.6-fold and 1.3-fold higher than that of free drug and unmodified DOX-PAMAM, respectively. Consequently, the RGD-modified DOX-PAMAM significantly prolonged the median survival time of brain tumor–bearing mice. Dual targeting strategy could further improve the drug delivery efficiency due to the complex microenvironment of brain tumor (Gao et al., 2013). Tf and wheat germ agglutinin (WGA) dual-modified PAMAM was constructed to deliver DOX because these two ligands could both enhance the BBB penetration and elevate brain tumor cell uptake (He et al., 2010). The transportation ratio across BBB of the dual targeting system achieved 13.5% of DOX after 2 h incubation, which was significantly higher than that of Tf-modified PAMAM (7%) and WGA-modified PAMAM (8%). Regarding the multidrug resistance proteins are responsible for exocytosis drugs by BBB and tumor cells, tamoxifen, an estrogen receptor antagonist that could inhibit the multidrug resistance proteins thus improve the BBB penetration, was comodified with Tf onto DOX-PAMAM (Li et al., 2012). The transportation ratio across the BBB model of this dual-modified system was 6.1%, which was higher than single-ligand-modified DOX-PAMAM. As a result, in the BBB model and brain tumor cell coculture system, the dual-modified DOX-PAMAM led to 31% apoptosis of brain tumor cells, whereas the number for Tf-modified DOX-PAMAM was only 24%.

Dendrigraft poly-lysine (DGL) is a kind of dendrimer that is similar to PAMAM. Because of the lower toxicity compared with PAMAM, DGL has gained increasing attention in drug delivery. Although there are several studies that evaluated the chemical drug delivery efficiency of DGL (Hu et al., 2015a,b), no study has evaluated the delivery to treat CNS disorders. Most related studies focused on gene delivery, which would be discussed in the next section.

2.4 Metal Nanoparticles

Several kinds of metal-based NPs were established for biological application, such as gold NPs (AuNPs), silver NPs, iron NPs, copper NPs, quantum dots

(QDs), upconversion NPs, and metal organic framework. These nanomaterials showed distinguished properties compared with natural materials and traditional polymers, and some preliminary studies were performed to evaluate the potential of these nanomaterials in brain targeting delivery. However, the toxicity of these materials is a major barrier in biological application. In this section, we focus on the application of AuNPs. Other metal NPs showed similar application as AuNPs in brain targeting delivery.

The common application of AuNPs is delivery of drugs to brain tumor. Our group decorated AuNPs with angiopep-2, a ligand for low-density lipoprotein-related protein, which overexpressed on both BBB and brain tumor cells (Demeule et al., 2008; Ruan et al., 2015b). The model drug, DOX, was anchored onto the AuNP surface through hydrozone, a pH-sensitive linker. The drug-loading capacity was as high as 9.7%, which was higher than for many other kinds of NPs. Owing to the hydrozone between AuNPs and DOX, the release of DOX was obviously pH sensitive. The 48-h cumulative release in pH 7.4 was 21.9%, whereas the number in pH 5.0 was elevated to 88.3%. Combining BBB and brain tumor targeting of anigopep-2 and tumor-specific release of DOX, DOX-loaded angiopep-2-modified AuNPs significantly prolonged the median survival time of brain tumor–bearing mice from 19 to 55 days. The Tf peptide-modified AuNPs also could elevate the delivery of photodynamic prodrug Pc4 into brain tumor, which was sixfold higher than that of unmodified AuNPs (Dixit et al., 2015). However, due to the heterogenous microenvironment, the targeting delivery efficiency of NPs with fixed size was still modest because high tumor retention requires large particle size, whereas high tumor penetration requires small size (Cabra et al., 2011; Kibria et al., 2013). To satisfy the controversial requirement of tumor retention and penetration, our group fabricated AuNPs with gelatin NPs (Ruan et al., 2015a), because the gelatin could be degraded by tumor overexpressed matrix metalloproteinase-2 (MMP-2), which resulting size reduction from large to small (Wong et al., 2011). After 24-h incubation with MMP-2, the size of fabricated NPs considerably reduced from 188.2 to 55.9 nm. Thus this kind of NPs showed better tumor penetration and retention, resulting in higher brain tumor accumulation of DOX.

2.5 Carbon-Based Inorganic Nanomaterials

There are several kinds of carbon-based inorganic nanomaterials involved in the diagnosis and treatment of brain diseases. Carbon nanotubes (CNTs) received much attention because they could easily load drugs such as DOX and small interfering RNA (siRNA) onto the polyaromatic surface of nanotubes through π-π stacking (Wang and Al-Jamal, 2015). The needle-like shape of CNTs provides them a unique mechanism to facilitate membrane penetration and cell internalization (Wong et al., 2013). There was direct evidence that CNTs could penetrate through in vitro microvascular cerebral endothelial monolayers (Shityakov et al., 2015). In vivo, acetylcholine-loaded CNTs without further modification successfully reverse the kainic acid-induced AD symptom, suggesting CNTs could directly penetrate through the BBB (Yang et al., 2010). Our group modified oxidized multiwalled CNTs (OMWNTs) with anogiopep-2 to construct brain tumor targeting delivery systems (Ren et al., 2012). The DOX loading efficiency of this system was as high as 80%. In vitro, the DOX-loaded angiopep-2-modified OMWNTs showed significantly higher cytotoxicity (IC_{50} was 2.98 μg/mL against C6 glioma cells) than the unmodified OMWNTs (IC_{50} was 88.98 μg/mL). In vivo, angiopep-2-modified OMWNTs accumulated much more in brain tumor than unmodified OMWNTs. As a result, the median survival time of brain tumor–bearing mice prolonged to 43 days after treatment by DOX-loaded angiopep-2-modified OMWNTs, which was significantly better than free DOX (33.5 days) and DOX-loaded unmodified OMWNTs (36 days). This study clearly demonstrated that CNTs could be used for brain tumor targeting delivery.

Graphene is a two-dimensional nanomaterial with a single layer. The large surface of graphene makes it easy to load or chemically conjugate various drugs with high loading capacity (Zhang et al., 2012). To improve biocompatibility, graphene is often coated with PEG. Chowdhury et al. (2015) conjugated lucanthone onto PEGylated graphene. It was shown that the uptake by glioma cells was much higher than that of control cells, suggesting the PEGylated graphene could deliver drugs into glioma cells. Polyacrylic acid (PAA) was also used to improve the solubility and biocompatibility of graphene oxide. Then the PAA-modified graphene oxide was conjugated with 1,3-bis(2-chloroethyl)-1-nitrosourea (BCNU), a chemotherapeutic drug, to treat brain tumor because this drug itself could penetrate the BBB (Lu et al., 2012). A drug-loading capacity of 19.8% could be achieved, and 70% of the drug activity was retained after conjugation onto the graphene oxide. In vitro antitumor study demonstrated that PAA-modified graphene oxide significantly decreased the IC_{50} of BCNU.

Carbon dots (CDs) are also suitable for drug delivery and cancer treatment because of their small size and potential photothermo effect. Graphene QDs showed ability in inhibition of β-amyloid aggregation (Liu et al., 2015). As the dots may penetrate the BBB without modification (Qian et al., 2014a), the graphene QDs may be used for AD treatment. Our group decorated angiopep-2-modified CDs with DOX using a disulfide bond to deliver drugs into brain tumor cells (Chen et al., 2015b). The angiopep-2 could effectively mediate the internalization of the system by C6 glioma cells; then the DOX is released into the cells due to the high concentration of GSH in C6 cells (Ballatori et al., 2009). The DOX-loaded angiopep-2-modified CDs showed higher C6 cellular uptake and cytotoxicity compared with unmodified CDs and free DOX, suggesting that angiopep-2-modified CDs could serve as a drug delivery system for brain tumor. However, no in vivo study was reported.

3. PEPTIDE AND PROTEIN DELIVERY

Peptide and protein have gained increasing attention in pharmaceutical industry because of their high specificity and therapeutic potential (Lalatsa et al., 2014). Many peptides and proteins display high therapeutic efficiency to CNS disorders especially neurodegenerative diseases. Neurotrophins are particularly impressive in the treatment of neurodegenerative diseases, such as AD and PD, because these diseases are strongly related to decreased neurotrophic support, loss of neurons, and low activity of certain enzymes (Brasnjevic et al., 2009). However, even worse than chemical drugs, almost 100% of peptides and proteins cannot transport through the BBB, restricting their application in brain disorders (Pardridge, 2007). Nanomaterials can act as Trojan horse and deliver the encapsulated peptides and proteins to brain. Owing to the poor stability, it is a great challenge to retain the activity of peptides and proteins during the preparation, delivery, and release process.

3.1 Natural Nanomaterials

As discussed earlier, liposomes contain an aqueous solution core, which could load peptides and proteins. VHH-pa2H is a heavy-chain antibody fragment to detect Aβ depositions with high affinity. Rotman et al. (2015) loaded it into GSH-liposomes to enhance the accumulation of VHH-pa2H in the brain of mice bearing AD. To track the distribution of VHH-pa2H, ^{111}In was used as label. The overall labeling yield of VHH-pa2H-loaded GSH-liposomes was 43.1%. In vivo, GSH-liposomes delivered significantly more VHH-pad2H into brain. Importantly, the distribution in brain of AD-bearing transgenic mice was twofold higher than that in brain of wild-type mice, demonstrating the system could effectively deliver proteins to brain. The GSH-liposomes could also load opioid peptide DAMGO (H-Tyr-d-Ala-Gly-MePhe-Gly-ol) (Lindqvist et al., 2013). The blood circulation half-time largely increased from 9.2 to 417 min after encapsulation into GSH-liposomes, which was important for enhancing specific target delivery. To further evaluate the brain targeting efficiency, $K_{p,uu}$, the ratio of unbound drug in brain to blood at equilibrium, was introduced. For free DAMGO, the $K_{p,uu}$ was only 0.09, whereas the GSH-liposomes significantly increased the $K_{p,uu}$ to 0.21. These quantitative data demonstrated GSH-liposomes were promising carriers for delivering peptides into brain.

Intranasal administration is a promising route for effective brain delivery even for NPs without specific ligand modification. H102 (HKQLPFFEED) peptide is a β-sheet breaker that shows activity in improving memory of amyloid precursor protein transgenic mice (Lin et al., 2014). It can be loaded into liposomes at a high encapsulation efficiency of 71.35% (Zheng et al., 2015b). When loaded into the liposomes, the β-sheet conformation of H102 peptide increased from 9.16% to 14.08% because of the interaction between the peptide and liposomes. After intranasal administration, the highest concentration was reached at

30 min, and the absolute bioavailability of H102-loaded liposomes was 30.2%, which was 2.69-fold higher than that of intranasal free H102 peptide. This system not only significantly improved the spatial memory impairment of AD model rats, but also showed no toxicity to nasal mucosa, demonstrating the liposomes could considerably enhance the brain delivery of peptides.

Chitosan is a cationic polysaccharide obtained by deacetylation of the natural polysaccharide chitin. It is an attractive nanomaterial in pharmaceutical applications because it is water soluble, does not cause allergic reaction, and can be degraded into nontoxic amino sugars. Z-DEVD-FMK peptide is a specific caspase inhibitor to reduce the neuronal cell death of caspase. The Z-DEVD-FMK could be loaded into chitosan NPs at a drug-loading capacity of 1.9% and OX26 (an antibody that could bind with the Tf receptor) was introduced to the surface of chitosan NPs to increase affinity with brain cells (Aktas et al., 2005). Intravenous injection of OX26-modified chitosan NPs led to considerably higher accumulation in brain than that of unmodified chitosan NPs.

3.2 Polymers

Polymers are widely used in delivering peptides and proteins. The peptides and proteins could be directly incorporated into the polymers during NP formation. Pang et al. modified OX26 onto PEG-PCL NPs to deliver neuroprotective peptide NC1900 to brain for AD treatment. The NC1900 loading capacity was 1.31% with an encapsulation efficiency of 20.6%. Importantly, 2-h cumulative release of NC1900 was lower than 25%, indicating the system could protect the NC1900 peptide from degradation by proteinases in blood. In the scopolamine-induced learning and memory deficit rat model, Morris water maze test showed that NC1900-loaded OX26 NP treatment could effectively improve the cognitive memory of mice to near-normal level (Pang et al., 2008). NAP (NAPVSIPQ) peptide shows neuroprotection effect at very low concentrations (10^{-17}–10^{-10} M), which is a candidate drug for AD treatment (Gozes et al., 2005). PEG-PCL NPs could load it with a drug-loading capacity of 0.71% and encapsulation efficiency of 54% (Liu et al., 2013b). To enable brain delivery, Lf was modified onto NPs for intranasal administration because the Lf receptor was overexpressed on the apical surface of respiratory epithelial cells, brain endothelial cells, and neurons (Elfinger et al., 2007; Suzuki et al., 2005). In the $A\beta_{1\text{-}40}$ and ibotenic acid coinjection-induced AD mice model, both behavior test and acetylcholinesterase and choline acetyltransferase activity evaluation determined that the NAP-loaded Lf-NPs improved the memory and neuroprotection effect (Liu et al., 2013b).

PEG-PLA was also used for encapsulation of peptides and proteins. Liu et al. (2013a) encapsulated NAP in PEG-PLA NPs and decorated the NPs with B6 peptide (CGHKAKGPRK), which can bind with the Tf receptor and mediate the brain targeting delivery (Nie et al., 2011). Similar to the NAP-loaded Lf-NPs, the NAP-loaded B6-NPs could considerably attenuate learning impairment, cholinergic disruption, and loss of hippocampal neurons, which was better than unmodified NAP-loaded NPs.

The dual targeting delivery strategy can provide better delivery efficiency than the BBB targeting systems (Gao et al., 2013). Zhang et al. (2014d) loaded H102 peptide into a dual targeting delivery system: TGN (BBB targeting) and QSH (Aβ targeting) comodified PEG-PLA NPs (TQNPs). Although a lower drug-loading capacity (about 0.5%) was obtained, the encapsulation efficiency was as high as about 60%. After intravenous injection into AD-bearing mice, the blood half-life of TQNPs was 1.57 h, whereas the peak in brain appeared at approximately 1 h (Fig. 1.2). The maximum concentrations in cerebrum and cerebellum of H102-loaded TQNPs was 1.8-fold higher than that of unmodified NPs, and the area under the time–concentration curve (AUC) was about 1.4-fold higher. In the hippocampus, the H102 concentration of TQNPs-treated mice reached as high as 785 ng/g at 1 h after injection, demonstrating this system could effectively enhance the brain delivery of peptide.

Intranasal administration could also improve the brain delivery of peptides. The same research group loaded basic fibroblast growth factor (bFGF) to solanum tuberosum lectin PEG-PLA NP (STLNPs), in which the solanum tuberosum lectin can selectively bind to N-acetylglucosamine on the nasal epithelial membrane to improve brain delivery (Zhang et al., 2014c). The AUC in olfactory bulb, cerebrum, and cerebellum of rats that were administered STL-NPs intranasally was 1.79–5.17 times higher than that of rats with intravenous administration, and 0.61–2.21 times higher than that of intranasal solution. As a result, the spatial learning and memory of AD rats treated with bFGF-loaded STLNPs was considerably improved compared with intranasal solution, unmodified NP, and intravenous STLNPs. Other ligands, such as WGA, lectin, and ulex europaeus agglutinin I, can also improve the brain delivery of peptide through intranasal administration because the ligands showed high affinity for receptors on the olfactory epithelium (Gao et al., 2006, 2007a,b). For example, the AUC of intact ^{125}I-vasoactive intestinal peptide in brain of mice following the intranasal administration of ^{125}I-vasoactive intestinal peptide-loaded WGA-modified NPs was significantly increased by about 3.5- to 4.7-fold and about 5.6- to 7.7-fold, respectively, compared with that after intranasal application of ^{125}I-vasoactive intestinal peptide solution (Gao et al., 2007b).

Similar to PEG-PLA, PEG-PLGA could effectively deliver peptides to brain. Our group developed TGN-conjugated PEG-PLGA NPs (TGN-NPs) to deliver neuroprotective NAP peptide to brain (Li et al., 2013b). The TGN-NPs showed similar peptide-loading capacity and encapsulation efficiency compared with PEG-PLA NPs. In the AD mice model, the NAP-loaded TGN-NPs displayed better learning and memory improvement compared with unmodified NPs.

Except these polymers, many other polymers showed promising application in peptide and protein delivery, including poly(methyl)methacrylate, poly(alkyl) cyanoacrylate, pluronic block polymers, PBCA, poly(hexylcyanoacrylate monomers), PEG-poly(α,β-aspartic acid), PEG–poly(ethyleneimine) (PEG–PEI),

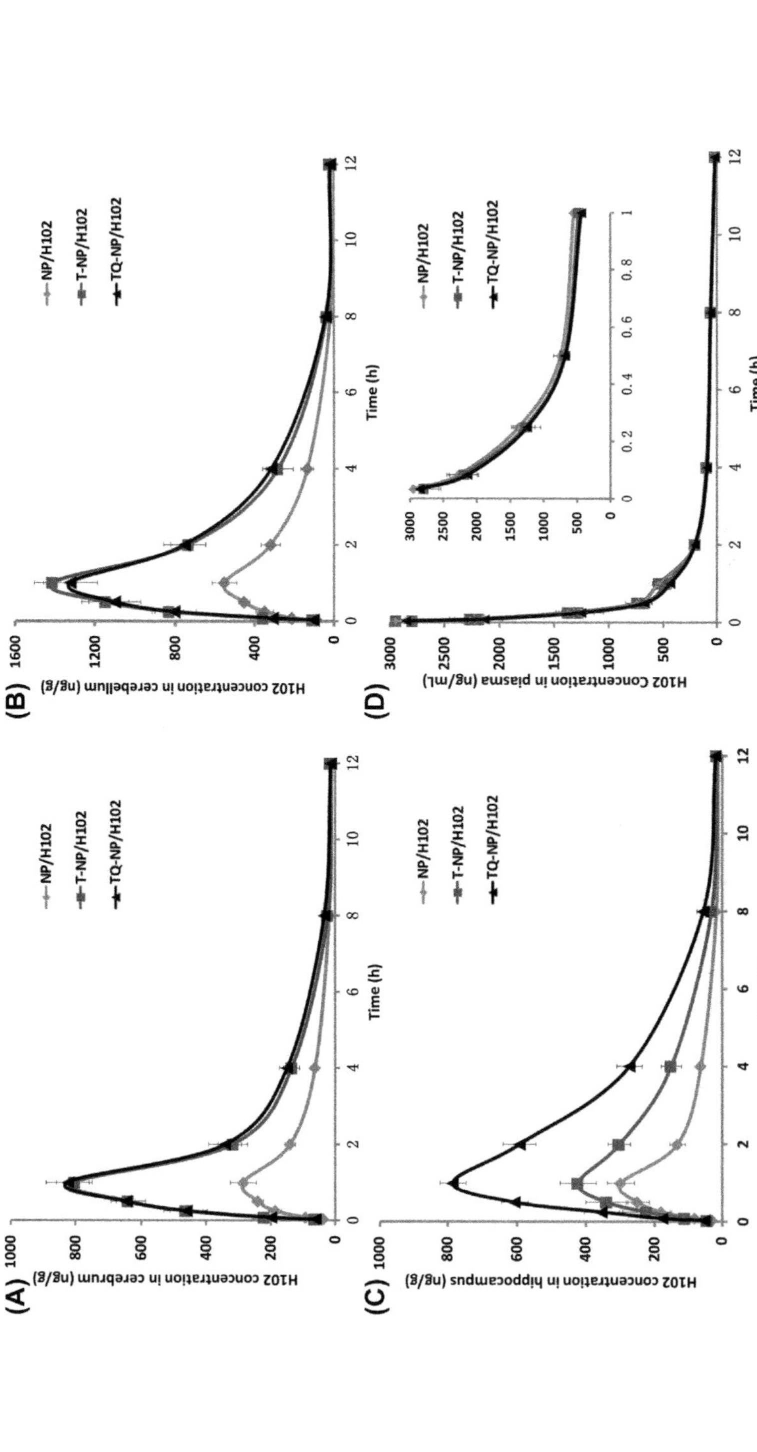

FIGURE 1.2 Concentration versus time curves of H102 in the cerebrum (A), cerebellum (B), hippocampus (C), and plasma (D). AD model mice were treated with NP/H102, TNP/H102, and TQNP/H102. Data are represented as mean ± S.D., $n = 4$. *Reprinted from Zhang, C., Zheng, X., Wan, X., Shao, X., Liu, Q., Zhang, Z., Zhang, Q., 2014d. The potential use of H102 peptide-loaded dual-functional nanoparticles in the treatment of Alzheimer's disease. J. Control. Release 192, 317–324 with permission of the copyright holder, Elsevier, Amsterdam.*

poly(carboxyl ethylene glycol-g-glutamate)-co-poly(distearin-g-glutamate), and PEG–poly(L-lysine) (Brasnjevic et al., 2009; Chen et al., 2014; Yi et al., 2014). These nanomaterials, decorated with specific ligands, could improve the distribution of peptides and proteins in brain, resulting in elevated therapeutic outcomes.

4. GENE DELIVERY

Gene therapy is considered as powerful and ultimate cure for many diseases. Commonly, gene delivery vectors are divided into viral vectors and nonviral vectors (Guo and Huang, 2012). Viral vectors are still the most efficient gene delivery vectors because of their high gene transfection. There are several gene drugs under clinical evaluation, which have all utilized viral vectors. However, because of safety concern, nonviral vectors gained great attention, and various nanomaterials are synthesized to deliver gene into brain.

4.1 Lipid Nanomaterials

Cationic liposome is one of the most commonly used nonviral gene delivery vectors and it is widely used as a standard gene vector especially for in vitro transfection (Ozpolat et al., 2014). The low toxicity, low immunogenicity, high loading capacity, and easy preparation contributed to the wide application of cationic liposomes. Our group modified the cationic liposomes with angiopep-2 to deliver therapeutic gene tumor necrosis factor-related apoptosis-inducing ligand (TRAIL) into brain glioma (Sun et al., 2012). The TRAIL plasmid-loaded angiopep-2-modified cationic liposomes could considerably prolong the median survival time from 23 (saline group) to 45.5 days, which was significantly higher than that of TRAIL plasmid-loaded cationic liposomes group, suggesting the angiopep-2-modified cationic liposomes could effectively deliver the DNA to brain glioma with high transfection efficiency. Cationic liposomes modified with anti-Tf receptor single chain antibody fragment also could effectively deliver p53 gene to brain tumor and cancer stem cells, enhancing the antitumor effect in combination with temozolomide (Kim et al., 2014). The p53 gene-loaded system also could sensitize the cancer stem cells to temozolomide and increase tumor cell apoptosis.

Immunoliposomes with neutral charge showed potential in delivery of various genes into brain, and the neutral charge provided the liposomes with long blood circulation time, which was superior to cationic liposomes. Shi and Pardridge (2000) first described the use of OX26 mAb-modified liposomes to deliver luciferase and β-galactosidase gene to brain. Commonly, cationic liposomes tend to nonspecifically deliver gene to lung, resulting in high gene expression in lung. However, the OX26-liposomes showed much lower distribution and gene expression in lung. After a single dose of intravenous injection of 10 μg of DNA plasmid, the expression in brain achieved peak at 48 h, and the gene expression could be observed in the whole brain, such as neurons, choroid plexus epithelium, and microvasculature. These results demonstrate that the immunoliposomes are powerful vectors for brain targeting gene delivery. Following this report, several therapeutic genes were delivered to brain using the same vectors (Lai et al., 2013). For example, tyrosine hydroxylase expression plasmid was loaded to OX26-liposomes for gene therapy of PD induced by 6-hydroxydopamine (Zhang et al., 2003). After a single injection, the tyrosine hydroxylase activity normalized both in the dorsal striatum ipsilateral to the 6-hydroxydopamine injection and in the contralateral striatum.

4.2 Cationic Polymers

Cationic polymers are used for gene delivery because they could effectively condense the anionic DNA and RNA and show low immune response. PEI is the most widely used cationic polymer for gene delivery. A kind of microRNA (miRNA), miR-124a, can activate neuronal differentiation for functional improvement. Hwang et al. (2011) encapsulated it into rabies virus glycoprotein (RVG) peptide-conjugated PEI because RVG is a substrate of neuronal cells overexpressed acetylcholine receptor (Kumar et al., 2007). After intravenous injection, the distribution in brain of cy5.5-miR-124a-loaded RVG-PEI was significantly higher than that of PEI. Poly(mannitol-co-PEI) composed of branched PEI and mannitol, showed elevated gene delivery through stimulation of caveolae-mediated endocytosis. Park et al. (2015) modified the RVG peptide onto the materials to deliver therapeutic gene to brain. β-Secretase 1 (BACE1) plays an early role in amyloid plaque generation and knockdown of BACE1 by siBACE1 may have therapeutic potential for AD (Kobayashi et al., 2008). In vitro, the RVG peptide modified complex showed a twofold higher gene silencing effect on neuronal cell lines. In vivo, after labeling with fluorescent Cy5.5, it was shown that the RVG peptide modified complex delivered more siRNA to brain. The BACE1 silencing efficiency of RVG peptide modified complexes was mostly seen in hippocampus and cortex parts in AD, whereas the gene silencing showed about threefold reduction of BACE1 expression, and the silencing efficiency was also demonstrated by Western blotting. These results suggested this system was an impressive gene delivery system. Several other peptides could also mediate the brain targeting gene delivery of PEI, such as myristic acid (Meng et al., 2010), cyclic RGD peptide (Zhan et al., 2011), *p*-hydroxybenzoic acid (Li et al., 2011), fatty acid (Shen et al., 2013), C-end rule RGERPPR peptide (Wang et al., 2013), monoclonal anti-CD11b antibody OX42 (Smolny et al., 2014), and retroinverso C-end rule peptide $_D$(RPPREGR) (Wang et al., 2014).

Although PEI-loaded gene showed high transfection efficiency, the application of PEI was restricted by the poor stability of the PEI/gene complex and rapid blood clearance (Fischer et al., 2004). Surface modification with PEG could avoid unintended interaction with negative blood component and cell membrane, thus improving the blood circulation time. However, this modification also decreased the gene transfection efficiency (Song et al., 2002). To solve this problem, cleavable PEG was established. Lei et al. (2013) modified PEI with PEG through disulfide bond and RGD was utilized for brain tumor targeting (RGD-PEG-SS-PEI). Because the concentration of GSH in the intracellular environment was much higher than in the extracellular environment (Ballatori et al., 2009), the PEG could be detached in cells, improving gene transfection efficiency. Transfection of the RGD-PEG-SS-PEI/DNA complex in U87 glioma cells was 5.5-fold higher than that of PEG-PEI/DNA complex. In vivo, the system led to increased gene expression in brain tumor. Encapsulation in liposomes also could avoid the problems faced by the PEI/DNA complex (Ko et al., 2009). The PEI/oligodeoxynucleotides (ODN) complex showed 95% encapsulation efficiency in liposomes. After intravenous injection, less than 10% of injected PEI/ODN remained in blood after 10 min, whereas more than 10% of the injected liposomal PEI/ODN was still in blood. After labeling the liposomes with anti-Tf receptor antibody, the distribution of liposomal PEI/ODN in brain achieved nearly 10-fold increase compared with unmodified liposomal PEI/ODN. Gene transfection in bEnd5 cells also was considerably elevated.

Apart from exploring strategies to attenuate the toxicity of PEI, researchers were interested in developing other cationic polymers with high gene transfection efficiency accompanied with low toxicity. Poly(2-(N,N-dimethylamino) ethyl methacrylate) (PDMAEMA) is a gene delivery carrier with promising potential because of its relatively high transfection efficiency, good safety, and versatile synthesis strategies (van de Wetering et al., 1998). After surface modification with brain targeting peptide TGN, it could be used for brain targeting gene delivery (Qian et al., 2013). When the N/P ratio was higher than five, the TGN-PEG-PDMAEMA completely condenses the DNA without release in gel retardation assay. Compared with PEI, the cytotoxicity of TGN-PEG-PDMAEMA/DNA complexes was obviously lower at 48 h after transfection when $N/P < 24$, whereas the highest transfection efficiency was achieved at $N/P = 10$, suggesting this kind of nanomaterials indeed had low toxicity. After labeling the DNA with fluorescent EMA, there was considerably higher fluorescence accumulation in brain at 1 h after intravenous injection, demonstrating this system effectively delivered the DNA into brain. Using GFP pDNA as a model gene, it was shown that the GFP expressed in many parts of brain, such as third ventricle and hypothalamus. At 48 h after injection, the brain gene expression level was 5800 U/mg protein, which was threefold higher than that of unmodified PEG-PDMAEMA/DNA, demonstrating the TGN-PEG-PDMAEMA was a promising vector for brain targeting gene delivery. To enable codelivery of gene and chemotherapeutics, star-branched amphiphilic copolymer PLA-PDMAEMA was synthesized (Qian et al., 2014b). The star copolymer displayed low critical micelle concentration of 320 ng/mL. The optimized structure of copolymer showed 2.5-fold higher gene transfection efficiency compared with PEI. After coloading with miR-21i and DOX, the complex exhibited a considerably better antiproliferation effect against glioma cells compared with DOX and miR-21i treatment alone.

Cleavable poly-(β-amino ester) (PBAE) is a superior gene vector because it can release cargoes under certain microenvironment. Kozielski et al. (2014) synthesized a serial of disulfide bonds-containing PBAE polymers to make the polymers degrade in brain tumor cells due to the high level of GSH in tumor cells. Through tuning the properties, such as bioreducibility and hydrophobicity, the gene knockdown efficiency of optimized PBAE/siRNA complex was as high as 91% in brain glioblastoma cells, whereas the cytotoxicity was not significant. Even at a dose of as low as 1 nM of siRNA, the polymer achieved 63% gene knockdown, demonstrating that the cleavable PBAE was highly efficient in siRNA delivery.

4.3 Dendrimers

PAMAM is a dendritic cationic polymer that has unified size at several nanometers (depending on the generation). The cationic PAMAM can directly bind with anionic DNA and RNA, condense them, and form gene-loaded NPs. Using plasmid EGFP as model DNA, the gene expression of plasmid EGFP-loaded NPs in brain capillary endothelial cells increased accompanied with the increase of the weight ratio of PAMAM to DNA (Huang et al., 2008). Lf modification considerably enhanced the gene expression both in vitro and in vivo, suggesting that the Lf-modified PAMAM was a successful brain gene delivery carrier. Therapeutic genes encoding glial cell line-derived neurotrophic factor (GDNF) have been widely used in gene therapy of PD. Huang et al. (2010) loaded the human GDNF gene (*hGDNF*) into Lf-modified PAMAM to evaluate the treatment outcome to PD rats. A single injection of *hGDNF*-loaded Lf-modified PAMAM expressed 653 ng GDNF/mg protein in brain, which was over twofold higher than the unmodified PAMAM. Consequently, the *hGDNF*-loaded Lf-modified PAMAM showed the best treatment effect of rotenone-induced PD-bearing rats using both behavior observation and quantitative neurotransmitter analysis as indexes.

Similar to PAMAM, DGL could be used for gene delivery due to the cationic surface. Liu et al. (2010) modified DGL with a peptide leptin30 derived from leptin, which could be specifically taken up by brain (Barrett et al., 2009). NPs of size 140 nm were formed when the leptin30-modified DGL was loaded with red fluorescence protein plasmid DNA. Electrophoresis demonstrated that the DNA could be completely encapsulated in the NP when the DGL to DNA weight ratio was 6:1. And the NPs could protect DNA from degradation by DNase (Fig. 1.3). In vitro, this system obviously enhanced the expression in brain capillary endothelial cells, with intensity fourfold higher than that of unmodified DGL. In vivo, the leptin30-modified DNA-loaded DGL NPs caused twofold higher gene expression in brain compared with that of unmodified DGL, demonstrating this system could successfully

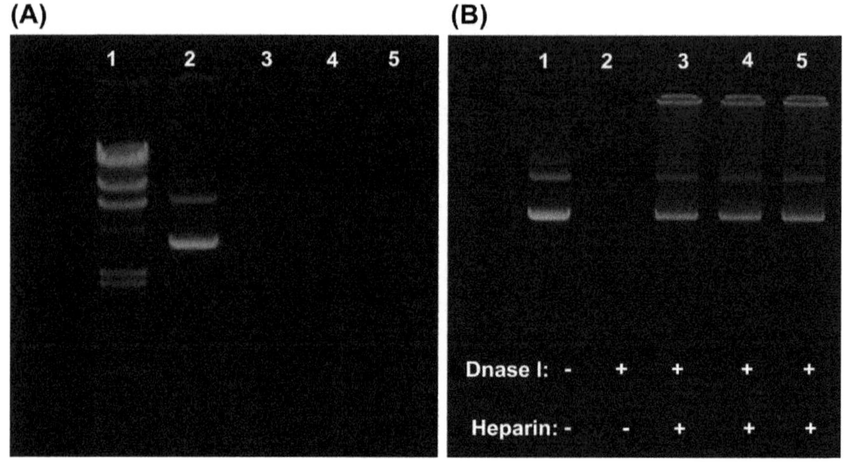

FIGURE 1.3 (A) Agarose gel electrophoresis was performed. Lane 1: marker; lane 2: naked DNA; lane 3: DGL/DNA NPs; lane 4: DGL-PEG/DNA NPs; lane 5: DGL-PEG-Leptin30/DNA NPs. (B) The stability of DNA-loaded NPs against DNase I digestion. Plasmid DNA was released from the NPs by the addition of sodium heparin separated by agarose gel electrophoresis after DNase I incubation. Lane 1: naked plasmid DNA; lane 2: naked plasmid DNA treated with DNase I; lanes 3–5: DGL/DNA, DGL-PEG/DNA, and DGL-PEG-leptin30/DNA NPs with treatment of heparin after DNase I incubation. *DGL*, dendrigraft poly-lysine; *NP*, nanoparticle; *PEG*, poly(ethylene glycol). *Reprinted from Liu, Y., Li, J., Shao, K., Huang, R., Ye, L., Lou, J., Jiang, C., 2010. A leptin derived 30-amino-acid peptide modified pegylated poly-L-lysine dendrigraft for brain targeted gene delivery. Biomaterials 31 (19), 5246–5257 with permission of the copyright holder, Elsevier, Amsterdam.*

deliver gene to brain. Li et al. (2013c) modified the DGL with choline derivate for brain targeting delivery mediated by BBB overexpressed choline transporter. In this system, Li et al. (2013a) coloaded with DOX and human tumor necrosis factor-related apoptosis-inducing ligand (pORF-hTRAIL, Trail) gene, which showed a synergistic effect in brain tumor treatment. The gene for PD treatment, *hGDNF*, also could be loaded into DGL NPs for brain delivery after surface modification with angiopep, which can recognize the BBB overexpressed low-density lipoprotein-related protein (Demeule et al., 2008; Huang et al., 2013). In behavior test, rotenone-induced PD rats treated with multidoses of *hGDNF*-loaded angiopep-modified DGL NPs displayed an increased number of line crossing and decreased inactive time compared with the control group, suggesting the treatment improved locomotor activity of rats. In addition, the dopaminergic neuron loss attenuated considerably after the treatment. These results demonstrated that DGL was a promising carrier for brain targeting gene delivery.

PPI is a promising alternative of PAMAM for gene delivery to brain (Somani and Dufes, 2014). Interestingly, the electrostatic interaction between PPI with DNA only involved the peripheral amine groups of PPI, thus the tertiary amino groups inside the dendrimer acted as proton sponge in endosome (Kabanov et al., 1998), allowing the quick escape of DNA from endosome, which was critical for effective gene delivery. Tf-decorated PPI/DNA complex was able to induce twofold higher gene expression in brain compared with unmodified PPI/DNA complex, and decreased the nonspecific gene expression in lung (Somani et al., 2014). More importantly, at 24 h the gene expression of Tf-decorated PPI/DNA complex in brain was the highest among all the organs and at least threefold higher than in other organs, suggesting that Tf-modified PPI was a promising brain gene delivery system.

4.4 Other Materials

Some cationic peptide-based oligomers or polymers also showed potential in brain targeting gene delivery. Kumar et al. (2007) directly conjugated 29-amino-acid peptide RVG peptide with nonamer arginine residues (RVG-9R) for delivering of siRNA. The RVG-9R chimeric peptide could effectively bind with and deliver siRNA to neuronal cells, leading to apparent gene silencing. In vivo, the specific gene silencing of the RVG-9R/siRNA complex provided robust protection against fatal viral encephalitis in mice. There were several peptides able to deliver genes to brain, including R7L10 (composed of a 7-arginine stretch and a 10-leucine stretch) (Oh et al., 2015), reducible poly(oligo-D-arginines) (connected 9-arginine oligopeptides internal disulfide linkages) (Hyun et al., 2010), and arginine-rich polyplexes (Morris and Labhasetwar, 2015).

RNA-based NPs could act as vectors for gene delivery. A multivalent naked RNA nanoparticle based on pRNA three-way junction from bacteriophage phi29 was used to target glioblastoma cells with folate ligand (Lee et al., 2015). Systemic injection of this system efficiently delivered siRNA into brain tumor cells of mice and reduced luciferase reporter gene expression, which was fourfold lower than in the control, suggesting that the RNA NP was a useful vector for siRNA delivery.

Hybridized materials were also used for brain targeting gene delivery. A hybridized nanovector consisted of an iron oxide NP core and a surface coated with copolymer of chitosan, PEG, and PEI (Kievit et al., 2010). In this system, chlorotoxin was used to enhance the glioma targeting effect, which could effectively improve the glioma cell uptake of the system and elevate gene transfection as monitored by a fluorescent imaging system. PAMAM grafted graphene oxide was capable of codelivering the anticancer drug epirubicin and Let-7g miRNA to U87 glioma cells, resulting in higher gene transfection efficiency and better antiproliferation effect compared with graphene oxide (Yang et al., 2014). Many other nanomaterials also successfully deliver genes to brain, including alkylglyceryl-modified dextran-graft-PLA (Toman et al., 2015) and RGD-modified and disulfide cross-linked short PEI functionalized gold nanorod (Wang et al., 2015).

5. NANOMATERIALS AS IMAGING PROBE

Many kinds of nanomaterials could deliver probes to brain for imaging. Because most of the loaded imaging probes are small molecular compounds or metal ions with chelating agents, and the delivery of these probes is similar to the delivery of small chemical drugs, it would not be discussed in this chapter. In addition to this application, there are several kinds of nanomaterials that have intrinsic properties enabling them to directly serve as imaging probes, such as iron oxide NPs for magnetic resonance imaging (MRI) and QDs and CDs for fluorescent imaging.

5.1 Iron Oxide Nanoparticles

Iron oxide NPs are good contrast for MRI and one of the products has been approved for clinical use in 1996. To enable them specifically target to brain tumor, Lf was anchored onto the superparamagnetic iron oxide NPs (SPIONs) (Xie et al., 2011). The size of Lf-conjugated SPIONs was 75 nm, the saturation magnetization was 51 emu/g Fe and T_2 relaxivity was 75.6 $mM^{-1} s^{-1}$, suggesting that the Lf-conjugated SPIONs were a good contrast agent for MRI. After intravenous injection at the dose of 12 mg/kg, Lf-modified SPIONs considerably increased the contrast of brain tumor, whereas unmodified SPIONs showed no obvious change (Fig. 1.4). The distribution of contrast agent in brain tumor was further evaluated by Prussian staining. Many of the Lf-modified SPIONs could be observed in the brain tumor site, whereas little unmodified SPIONs

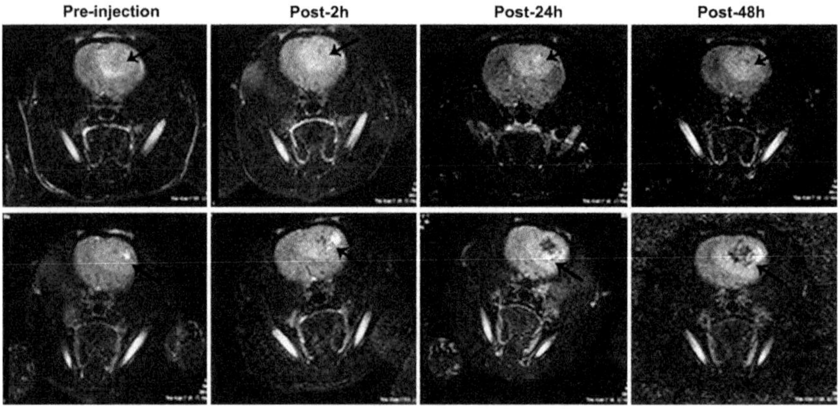

FIGURE 1.4 In vivo T_2-weighted magnetic resonance images of rats' brain bearing C6 gliomas. Upper row: acquired after administration of SPIONs (12 mg/kg); lower row: acquired after administration of lactoferrin-conjugated SPIONs (12 mg/kg). *Reprinted from Xie, H., Zhu, Y., Jiang, W., Zhou, Q., Yang, H., Gu, N., Zhang, Y., Xu, H., Xu, H., Yang, X., 2011. Lactoferrin-conjugated superparamagnetic iron oxide nanoparticles as a specific MRI contrast agent for detection of brain glioma in vivo. Biomaterials 32 (2), 495–502 with permission of the copyright holder, Elsevier, Amsterdam.*

were observed, demonstrating the Lf-modified SPIONs were a successful brain tumor MRI contrast agent. Modification with other kinds of ligands, such as epidermal growth factor (EGF) (Shevtsov et al., 2014), EGF receptor monoclonal antibody (Mu et al., 2015), heat shock protein Hsp70-specific antibody (Shevtsov et al., 2015a), and interleukin-1 receptor antagonist (Shevtsov et al., 2015b), could also improve the brain tumor targeting of SPIONs, making them promising MRI contrast probes.

After conjugation with anti-AβPP antibody, SPIONs could be used for imaging Aβ plaques in AD mice (Sillerud et al., 2013). Without anti-AβPP-conjugated SPIONs, the number of MRI visible plaque in AD brain was only 347. After treatment with the contrast, the number increased to 668, indicating the anti-AβPP-conjugated SPIONs could improve the detection sensitivity.

5.2 Quantum Dots

QDs are nanocrystals with tunable size and multicolor fluorescence (Jaiswal et al., 2003), making them attractive in labeling and tracking particles and cells (Onoshima et al., 2015). In addition to being used as a tracking probe, QDs, with modifications, could be directly used for diseases or makers imaging. Intracerebroventricular injection of anti-Aβ antibody-functionalized ZnS/CdSe QDs into *APP717 V-F* transgenic mice clearly imaged the distribution of Aβ, which mainly located in hippocampus, cerebral cortex, sagittal septum, and striatum of the brain (Feng et al., 2013). After quantitative analysis, the fluorescence intensity in healthy brain was only 12.3, whereas the number increased to 46.6 in 16-month-old transgenic mice, demonstrating this probe could be used for Aβ imaging. After labeling with EGF or anti-EGF receptor antibody, QDs could specifically and sensitively image glial tumor cells in cell culture, glioma mouse models and human brain-tumor biopsies (Arndt-Jovin et al., 2009). Clear demarcation could be observed by fluorescence imaging between brain and tumor tissue at the macroscopic as well as the cellular level, which was useful for surgery resection of brain tumor.

5.3 Carbon Dots

CDs could be directly used for disease imaging because of their unique fluorescent properties, including no photobleaching, good biocompatibility, and absence of optical blinking. Using glycine as precursor, our laboratory prepared CDs with size about 3–4 nm (Ruan et al., 2014b). The highest emission wavelength of the CDs was 500 nm, and strong emission could be observed at as long as 600 nm, making the CDs suitable for in vivo tumor imaging. In vivo, the CDs could accumulate in brain tumor site owing to the weak EPR effect. To further improve the brain tumor imaging effect, we decorated angiopep-2 onto the CDs (Ruan et al., 2014a). The angiopep-2-modified CDs showed higher uptake by

C6 glioma cells than that of unmodified CDs, demonstrating the angiopep-2 modification could improve the brain tumor targeting effect of nanomaterials. Consequently, the angiopep-2-modified CDs displayed twofold higher intensity in brain tumor than unmodified CDs. More importantly, the glioma-to-normal brain ratio (G/N ratio) of angiopep-2-modified CDs was 1.73, which was considerably higher than that of unmodified CDs (1.12), demonstrating ligand is necessary for brain tumor targeting imaging. Alternatively, using some kinds of ligands as precursor is also useful for enhancing brain tumor targeting efficiency. Zheng et al. (2015a) prepared CD-Asp using glucose and L-aspartic acid as precursors. Compared with other kinds of CDs, such as CD-Glu (using glucose and glutamic acid as precursors), CD-G (using glucose as the only precursor), and CD-A (using L-aspartic as the only precursor), CD-Asp showed considerable C6 glioma cell selectivity. The cellular uptake of CD-Asp by C6 cells was significantly higher than that by noncancerous L929 cells. As a result, CD-Asp displayed much higher G/N ratio that other CDs (1.42, 0.91, 0.86, and 0.88 for CD-Asp, CD-G, CD-A, and CD-Glu, respectively). The self-targeting effect of CD-Asp may be contributed by aspartic acid and glucose on the surface of CD-Asp, because both glucose transporter and ASCT2 are highly expressed on BBB and/or brain tumor cells (Luciani et al., 2004; McAllister et al., 2001).

6. CONCLUSION AND PERSPECTIVE

Increasing number of nanomaterials were introduced in delivering drugs, peptides, proteins, genes, and imaging probes to brain for diagnosis and treatment of brain disorders. Owing to the restriction of the BBB, surface modification with specific ligand is a most widely used strategy to enable the nanomaterials target brain, which is based on the specific internalization by receptors and transporters on the surface of cells. This chapter discussed the application of many kinds of nanomaterials in delivering small chemical drugs, peptides and proteins, and genes to brain for CNS disorder treatment. The application of imaging probes was also briefly discussed.

Although promising results were obtained from current studies, the brain targeting drug delivery efficiency is not high enough and the treatment outcomes still need to improve. As the most commonly used strategy, ligand modification displayed only modest benefit in improving targeting efficiency. More efficient ligands should be developed with high affinity and quick disassociation with target receptors or transporters. In addition, there is certainly room for improving specificity of the ligands, which is critical for enhancing selectivity of the targeting delivery systems.

In addition to the delivery efficiency, toxicity of these nanomaterials is a critical concern because of the important function of the brain. The mechanism of the toxicity should be well explored, which is useful for attenuating toxicity of the current nanomaterials, developing new materials with low toxicity and expanding the medical application of nanomaterials in brain.

REFERENCES

Abbott, N.J., 2005. Dynamics of CNS barriers: evolution, differentiation, and modulation. Cell. Mol. Neurobiol. 25, 5–23.

Aktas, Y., Yemisci, M., Andrieux, K., Gursoy, R.N., Alonso, M., Fernandez-Megia, E., Novoa-Carballal, R., Quinoa, E., Riguera, R., Sargon, M.F., Celik, H.H., Demir, A.S., Hincal, A.A., Alkara, T., Capan, Y., Couvreur, P., 2005. Development and brain delivery of chitosan-PEG nanoparticles functionalized with the monoclonal antibody OX26. Bioconjug. Chem. 16, 1503–1511.

Arndt-Jovin, D.J., Kantelhardt, S.R., Caarls, W., de Vries, A.H., Giese, A., Jovin Ast, T.M., 2009. Tumor-targeted quantum dots can help surgeons find tumor boundaries. IEEE Trans. Nanobioscience 8, 65–71.

Ballatori, N., Krance, S.M., Notenboom, S., Shi, S., Tieu, K., Hammond, C.L., 2009. Glutathione dysregulation and the etiology and progression of human diseases. Biol. Chem. 390, 191–214.

Barrett, G.L., Trieu, J., Naim, T., 2009. The identification of leptin-derived peptides that are taken up by the brain. Regul. Pept. 155, 55–61.

Begley, D.J., 2004. Delivery of therapeutic agents to the central nervous system: the problems and the possibilities. Pharmacol. Ther. 104, 29–45.

Birngruber, T., Raml, R., Gladdines, W., Gatschelhofer, C., Gander, E., Ghosh, A., Kroath, T., Gaillard, P.J., Pieber, T.R., Sinner, F., 2014. Enhanced doxorubicin delivery to the brain administered through glutathione PEGylated liposomal doxorubicin (2B3-101) as compared with generic caelyx®/Doxil® – a cerebral open flow microperfusion pilot study. J. Pharm. Sci. 103, 1945–1948.

Brasnjevic, I., Steinbusch, H.W., Schmitz, C., Martinez-Martinez, P., 2009. Delivery of peptide and protein drugs over the blood–brain barrier. Prog. Neurobiol. 87, 212–251.

Cabral, H., Matsumoto, Y., Mizuno, K., Chen, Q., Murakami, M., Kimura, M., Terada, Y., Kano, M.R., Miyazono, K., Uesaka, M., Nishiyama, N., Kataoka, K., 2011. Accumulation of sub-100 nm polymeric micelles in poorly permeable tumours depends on size. Nat. Nanotechnol. 6, 815–823.

Chen, Y.C., Chiang, C.F., Chen, L.F., Liao, S.C., Hsieh, W.Y., Lin, W.L., 2014. Polymersomes conjugated with des-octanoyl ghrelin for the delivery of therapeutic and imaging agents into brain tissues. Biomaterials 35, 2051–2065.

Chen, B., He, X.Y., Yi, X.Q., Zhuo, R.X., Cheng, S.X., 2015a. Dual-peptide-functionalized albumin-based nanoparticles with pH dependent self-assembly behavior for drug delivery. ACS Appl. Mater. Interfaces 7, 15148–15153.

Chen, J., Cun, X., Ruan, S., Wang, Y., Zhang, Y., He, Q., Gao, H., 2015b. Glioma cell-targeting doxorubicin delivery and redox-responsive release using angiopep-2 decorated carbonaceous nanodots. RSC Adv. 5 (48), 57045–57049.

Chowdhury, S.M., Surhland, C., Sanchez, Z., Chaudhary, P., Suresh Kumar, M.A., Lee, S., Peña, L.A., Waring, M., Sitharaman, B., Naidu, M., 2015. Graphene nanoribbons as a drug delivery agent for lucanthone mediated therapy of glioblastoma multiforme. Nanomedicine 11, 109–118.

Davson, H., 1976. The blood–brain barrier. J. Physiol. 255, 1–28.

Demeule, M., Currie, J.C., Bertrand, Y., Che, C., Nguyen, T., Regina, A., Gabathuler, R., Castaigne, J.P., Beliveau, R., 2008. Involvement of the low-density lipoprotein receptor-related protein in the transcytosis of the brain delivery vector angiopep-2. J. Neurochem. 106 (4), 1534–1544.

Dixit, S., Novak, T., Miller, K., Zhu, Y., Kenney, M.E., Broome, A.M., 2015. Transferrin receptor-targeted theranostic gold nanoparticles for photosensitizer delivery in brain tumors. Nanoscale 7, 1782–1790.

Elfinger, M., Maucksch, C., Rudolph, C., 2007. Characterization of lactoferrin as a targeting ligand for nonviral gene delivery to airway epithelial cells. Biomaterials 28, 3448–3455.

Feng, L., Long, H.Y., Liu, R.K., Sun, D.N., Liu, C., Long, L.L., Li, Y., Chen, S., Xiao, B., 2013. A quantum dot probe conjugated with Aβ antibody for molecular imaging of Alzheimer's disease in a mouse model. Cell. Mol. Neurobiol. 33, 759–765.

Fischer, D., Osburg, B., Petersen, H., Kissel, T., Bickel, U., 2004. Effect of poly(ethylene imine) molecular weight and pegylation on organ distribution and pharmacokinetics of polyplexes with oligodeoxynucleotides in mice. Drug Metab. Dispos. 32, 983–992.

Fornaguera, C., Feiner-Gracia, N., Calderó, G., García-Celma, M.J., Solans, C., 2015. Galantamine-loaded PLGA nanoparticles, from nano-emulsion templating, as novel advanced drug delivery systems to treat neurodegenerative diseases. Nanoscale 7, 12076–12084.

Fu, Q., Sun, J., Zhang, W., Sui, X., Yan, Z., He, Z., 2009. Nanoparticle albumin-bound (NAB) technology is a promising method for anti-cancer drug delivery. Recent Pat. Anticancer Drug Discov. 4 (3), 262–272.

Gao, H., Jiang, X., 2015. Brain delivery using nanotechnology. In: Di, L., Kerns, E.J. (Eds.), Blood–Brain Barrier in Drug Discovery. John Wiley & Sons Inc., New Jersey, pp. 521–534.

Gao, X., Tao, W., Lu, W., Zhang, Q., Zhang, Y., Jiang, X., Fu, S., 2006. Lectin-conjugated PEG-PLA nanoparticles: preparation and brain delivery after intranasal administration. Biomaterials 27, 3482–3490.

Gao, X., Wu, B., Zhang, Q., Chen, J., Zhu, J., Zhang, W., Rong, Z., Chen, H., Jiang, X., 2007b. Brain delivery of vasoactive intestinal peptide enhanced with the nanoparticles conjugated with wheat germ agglutinin following intranasal administration. J. Control. Release 121, 156–167.

Gao, X., Chen, J., Tao, W., Zhu, J., Zhang, Q., Chen, H., Jiang, X., 2007a. UEA I-bearing nanoparticles for brain delivery following intranasal administration. Int. J. Pharm. 340, 207–215.

Gao, H., Qian, J., Cao, S., Yang, Z., Pang, Z., Pan, S., Fan, L., Xi, Z., Jiang, X., Zhang, Q., 2012. Precise glioma targeting of and penetration by aptamer and peptide dual-functioned nanoparticles. Biomaterials 33 (20), 5115–5123.

Gao, H., Pang, Z., Jiang, X., 2013. Targeted delivery of nano-therapeutics for major disorders of the central nervous system. Pharm. Res. 30 (10), 2485–2498.

Gao, H., Wang, Y., Chen, C., Chen, J., Wei, Y., Cao, S., Jiang, X., 2014a. Incorporation of lapatinib into core-shell nanoparticles improves both the solubility and anti-glioma effects of the drug. Int. J. Pharm. 461 (1–2), 478–488.

Gao, H., Zhang, Q., Yu, Z., He, Q., 2014b. Cell-penetrating peptide-based intelligent liposomal systems for enhanced drug delivery. Curr. Pharm. Biotechnol. 15, 210–219.

Gao, H., Cao, S., Yang, Z., Zhang, S., Zhang, Q., Jiang, X., 2015. Preparation, characterization and anti-glioma effects of docetaxel-incorporated albumin-lipid nanoparticles. J. Biomed. Nanotechnol. 11 (12), 2137–2147.

Gozes, I., Morimoto, B.H., Tiong, J., Fox, A., Sutherland, K., Dangoor, D., Holser-Cochav, M., Vered, K., Newton, P., Aisen, P.S., Matsuoka, Y., van Dyck, C.H., Thal, L., 2005. NAP: research and development of a peptide derived from activity-dependent neuroprotective protein (ADNP). CNS Drug Rev. 11, 353–368.

Gu, G., Xia, H., Hu, Q., Liu, Z., Jiang, M., Kang, T., Miao, D., Tu, Y., Pang, Z., Song, Q., Yao, L., Chen, H., Gao, X., Chen, J., 2013. PEG-co-PCL nanoparticles modified with MMP-2/9 activatable low molecular weight protamine for enhanced targeted glioblastoma therapy. Biomaterials 34 (1), 196–208.

Gu, G., Hu, Q., Feng, X., Gao, X., Menglin, J., Kang, T., Jiang, D., Song, Q., Chen, H., Chen, J., 2014. PEG-PLA nanoparticles modified with APTEDB peptide for enhanced anti-angiogenic and anti-glioma therapy. Biomaterials 35 (28), 8215–8226.

Guo, X., Huang, L., 2012. Recent advances in nonviral vectors for gene delivery. Acc. Chem. Res. 45, 971–979.

He, H., Li, Y., Jia, C.R., Du, J., Ying, X., Lu, W.L., Lou, J.N., Wei, Y., 2010. PEGylated poly(amidoamine) dendrimer-based dual-targeting carrier for treating brain tumors. Biomaterials 32, 478–487.

Hu, Q., Gao, X., Gu, G., Kang, T., Tu, Y., Liu, Z., Song, Q., Yao, L., Pang, Z., Jiang, X., Chen, H., Chen, J., 2013. Glioma therapy using tumor homing and penetrating peptide-functionalized PEG-PLA nanoparticles loaded with paclitaxel. Biomaterials 34 (22), 5640–5650.

Hu, G., Cun, X., Wang, Y., He, Q., Gao, H., 2015a. Peptide mediated active targeting and intelligent particle size reduction-mediated enhanced penetrating of fabricated nanoparticles for triple-negative breast cancer treatment. Oncotarget 6, 41258–41274.

Hu, G., Wang, Y., He, Q., Gao, H., 2015b. Multistage drug delivery system based on microenvironment-responsive dendrimer-gelatin nanoparticles for deep tumor penetration. RSC Adv. 5, 85933–85937.

Huang, R., Ke, W., Liu, Y., Jiang, C., Pei, Y., 2008. The use of lactoferrin as a ligand for targeting the polyamidoamine-based gene delivery system to the brain. Biomaterials 29 (2), 238–246.

Huang, R., Ke, W., Liu, Y., Wu, D., Feng, L., Jiang, C., Pei, Y., 2010. Gene therapy using lactoferrin-modified nanoparticles in a rotenone-induced chronic Parkinson model. J. Neurol. Sci. 290 (1–2), 123–130.

Huang, R., Ma, H., Guo, Y., Liu, S., Kuang, Y., Shao, K., Li, J., Liu, Y., Han, L., Huang, S., An, S., Ye, L., Lou, J., Jiang, C., 2013. Angiopep-conjugated nanoparticles for targeted long-term gene therapy of Parkinson's disease. Pharm. Res. 30 (10), 2549–2559.

Hwang, D.W., Son, S., Jang, J., Youn, H., Lee, S., Lee, D., Lee, Y.S., Jeong, J.M., Kim, W.J., Lee, D.S., 2011. A brain-targeted rabies virus glycoprotein-disulfide linked PEI nanocarrier for delivery of neurogenic microRNA. Biomaterials 32, 4968–4975.

Hyun, H., Won, Y.W., Kim, K.M., Lee, J., Lee, M., Kim, Y.H., 2010. Therapeutic effects of a reducible poly (oligo-D-arginine) carrier with the heme oxygenase-1 gene in the treatment of hypoxic-ischemic brain injury. Biomaterials 31, 9128–9134.

Jaiswal, J.K., Mattoussi, H., Mauro, J.M., Simon, S.M., 2003. Long-term multiple color imaging of live cells using quantum dot bioconjugates. Nat. Biotechnol. 21, 47–51.

Kabanov, V.A., Zezin, A.B., Rogacheva, V.B., Gulyaeva, Z.G., Zansochova, M.F., Joosten, J.G.H., Brackman, J., 1998. Polyelectrolyte behavior of astramol poly(propyleneimine) dendrimers. Macromolecules 31, 5142–5144.

Kang, T., Gao, X., Hu, Q., Jiang, D., Feng, X., Zhang, X., Song, Q., Yao, L., Huang, M., Jiang, X., Pang, Z., Chen, H., Chen, J., 2014. iNGR-modified PEG-PLGA nanoparticles that recognize tumor vasculature and penetrate gliomas. Biomaterials 35 (14), 4319–4332.

Kibria, G., Hatakeyama, H., Ohga, N., Hida, K., Harashima, H., 2013. The effect of liposomal size on the targeted delivery of doxorubicin to Integrin alphavbeta3-expressing tumor endothelial cells. Biomaterials 34, 5617–5627.

Kievit, F.M., Veiseh, O., Fang, C., Bhattarai, N., Lee, D., Ellenbogen, R.G., Zhang, M., 2010. Chlorotoxin labeled magnetic nanovectors for targeted gene delivery to glioma. ACS Nano 4, 4587–4594.

Kim, S.S., Rait, A., Kim, E., Pirollo, K.F., Nishida, M., Farkas, N., Dagata, J.A., Chang, E.H., 2014. A nanoparticle carrying the p53 gene targets tumors including cancer stem cells, sensitizes glioblastoma to chemotherapy and improves survival. ACS Nano 8, 5495–5514.

Ko, Y.T., Bhattacharya, R., Bickel, U., 2009. Liposome encapsulated polyethylenimine/ODN polyplexes for brain targeting. J. Control. Release 133, 230–237.

Kobayashi, D., Zeller, M., Cole, T., Buttini, M., McConlogue, L., Sinha, S., Freedman, S., Morris, R.G., Chen, K.S., 2008. BACE1 gene deletion: impact on behavioral function in a model of Alzheimer's disease. Neurobiol. Aging 29, 861–873.

Kozielski, K.L., Tzeng, S.Y., Hurtado De Mendoza, B.A., Green, J.J., 2014. Bioreducible cationic polymer-based nanoparticles for efficient and environmentally triggered cytoplasmic siRNA delivery to primary human brain cancer cells. ACS Nano 8, 3232–3241.

Kreuter, J., 2014. Drug delivery to the central nervous system by polymeric nanoparticles: what do we know? Adv. Drug Deliv. Rev. 71, 2–14.

Kumar, P., Wu, H., McBride, J.L., Jung, K.E., Kim, M.H., Davidson, B.L., Lee, S.K., Shankar, P., Manjunath, N., 2007. Transvascular delivery of small interfering RNA to the central nervous system. Nature 448, 39–43.

Lai, F., Fadda, A.M., Sinico, C., 2013. Liposomes for brain delivery. Expert Opin. Drug Deliv. 10, 1003–1022.

Lalatsa, A., Schatzlein, A.G., Uchegbu, I.F., 2014. Strategies to deliver peptide drugs to the brain. Mol. Pharm. 11, 1081–1093.

Lee, T.J., Haque, F., Shu, D., Yoo, J.Y., Li, H., Yokel, R.A., Horbinski, C., Kim, T.H., Kim, S.H., Kwon, C.H., Nakano, I., Kaur, B., Guo, P., Croce, C.M., 2015. RNA nanoparticle as a vector for targeted siRNA delivery into glioblastoma mouse model. Oncotarget 6, 14766–14776.

Lei, Y., Wang, J., Xie, C., Wagner, E., Lu, W., Li, Y., Wei, X., Dong, J., Liu, M., 2013. Glutathione-sensitive RGD-poly(ethylene glycol)-SS-polyethylenimine for intracranial glioblastoma targeted gene delivery. J. Gene Med. 15, 291–305.

Li, J., Meng, Q., Lei, Y., Gu, B., Liu, Y., Lu, W., 2011. Benzamide analogue-conjugated polyethylenimine for brain-targeting and gene delivery. J. Drug Target. 19, 814–820.

Li, Y., He, H., Jia, X., Lu, W.L., Lou, J., Wei, Y., 2012. A dual-targeting nanocarrier based on poly(amidoamine) dendrimers conjugated with transferrin and tamoxifen for treating brain gliomas. Biomaterials 33, 3899–3908.

Li, J., Guo, Y., Kuang, Y., An, S., Ma, H., Jiang, C., 2013a. Choline transporter-targeting and co-delivery system for glioma therapy. Biomaterials 34 (36), 9142–9148.

Li, J., Zhang, C., Li, J., Fan, L., Jiang, X., Chen, J., Pang, Z., Zhang, Q., 2013b. Brain delivery of NAP with PEG-PLGA nanoparticles modified with phage display peptides. Pharm. Res. 30 (7), 1813–1823.

Li, J., Zhou, L., Ye, D., Huang, S., Shao, K., Huang, R., Han, L., Liu, Y., Liu, S., Ye, L., Lou, J., Jiang, C., 2013c. Choline-derivate-modified nanoparticles for brain-targeting gene delivery. Adv. Mater. 23 (39), 4516–4520.

Lin, L.X., Bo, X.Y., Tan, Y.Z., Sun, F.X., Song, M., Zhao, J., Ma, Z.H., Li, M., Zheng, K.J., Xu, S.M., 2014. Feasibility of β-sheet breaker peptide-H102 treatment for Alzheimer's disease based on β-amyloid hypothesis. PLoS One 9, e112052.

Lindqvist, A., Rip, J., Gaillard, P.J., Bjorkman, S., Hammarlund-Udenaes, M., 2013. Enhanced brain delivery of the opioid peptide DAMGO in glutathione pegylated liposomes: a microdialysis study. Mol. Pharm. 10, 1533–1541.

Liu, Y., Li, J., Shao, K., Huang, R., Ye, L., Lou, J., Jiang, C., 2010. A leptin derived 30-amino-acid peptide modified pegylated poly-L-lysine dendrigraft for brain targeted gene delivery. Biomaterials 31 (19), 5246–5257.

Liu, Z., Gao, X., Kang, T., Jiang, M., Miao, D., Gu, G., Hu, Q., Song, Q., Yao, L., Tu, Y., Chen, H., Jiang, X., Chen, J., 2013a. B6 Peptide-modified PEG-PLA nanoparticles for enhanced brain delivery of neuroprotective peptide. Bioconjug. Chem. 24, 997–1007.

Liu, Z., Jiang, M., Kang, T., Miao, D., Gu, G., Song, Q., Yao, L., Hu, Q., Tu, Y., Pang, Z., Chen, H., Jiang, X., Gao, X., Chen, J., 2013b. Lactoferrin-modified PEG-co-PCL nanoparticles for enhanced brain delivery of NAP peptide following intranasal administration. Biomaterials 34, 3870–3881.

Liu, Y., Ran, R., Chen, J., Kuang, Q., Tang, J., Mei, L., Zhang, Q., Gao, H., Zhang, Z., He, Q., 2014. Paclitaxel loaded liposomes decorated with a multifunctional tandem peptide for glioma targeting. Biomaterials 35 (17), 4835–4847.

Liu, Y., Xu, L.P., Dai, W., Dong, H., Wen, Y., Zhang, X., 2015. Graphene quantum dots for the inhibition of β amyloid aggregation. Nanoscale 7, 19060–19065.

Lu, Y.J., Yang, H.W., Hung, S.C., Huang, C.Y., Li, S.M., Ma, C.C., Chen, P.Y., Tsai, H.C., Wei, K.C., Chen, J.P., 2012. Improving thermal stability and efficacy of BCNU in treating glioma cells using PAA-functionalized graphene oxide. Int. J. Nanomedicine 7, 1737–1747.

Luciani, A., Olivier, J.C., Clement, O., Siauve, N., Brillet, P.Y., Bessoud, B., Gazeau, F., Uchegbu, I.F., Kahn, E., Frija, G., Cuenod, C.A., 2004. Glucose-receptor MR imaging of tumors: study in mice with PEGylated paramagnetic niosomes. Radiology 231, 135–142.

McAllister, M.S., Krizanac-Bengez, L., Macchia, F., Naftalin, R.J., Pedley, K.C., Mayberg, M.R., Marroni, M., Leaman, S., Stanness, K.A., Janigro, D., 2001. Mechanisms of glucose transport at the blood–brain barrier: an in vitro study. Brain Res. 904, 20–30.

van der Meel, R., Vehmeijer, L.J., Kok, R.J., Storm, G., van Gaal, E.V., 2013. Ligand-targeted particulate nanomedicines undergoing clinical evaluation: current status. Adv. Drug Deliv. Rev. 65 (10), 1284–1298.

Meng, Q., Yu, M., Gu, B., Li, J., Liu, Y., Zhan, C., Xie, C., Zhou, J., Lu, W., 2010. Myristic acid-conjugated polyethylenimine for brain-targeting delivery: in vivo and ex vivo imaging evaluation. J. Drug Target. 18, 438–446.

Menjoge, A.R., Kannan, R.M., Tomalia, D.A., 2010. Dendrimer-based drug and imaging conjugates: design considerations for nanomedical applications. Drug Discov. Today 15, 171–185.

Miura, Y., Takenaka, T., Toh, K., Wu, S., Nishihara, H., Kano, M.R., Ino, Y., Nomoto, T., Matsumoto, Y., Koyama, H., Cabral, H., Nishiyama, N., Kataoka, K., 2013. Cyclic RGD-linked polymeric micelles for targeted delivery of platinum anticancer drugs to glioblastoma through the blood–brain tumor barrier. ACS Nano 7, 8583–8592.

Morris, V.B., Labhasetwar, V., 2015. Arginine-rich polyplexes for gene delivery to neuronal cells. Biomaterials 60, 151–160.

Mu, K., Zhang, S., Ai, T., Jiang, J., Yao, Y., Jiang, L., Zhou, Q., Xiang, H., Zhu, Y., Yang, X., Zhu, W., 2015. Monoclonal antibody-conjugated superparamagnetic iron oxide nanoparticles for imaging of epidermal growth factor receptor-targeted cells and gliomas. Mol. Imaging 14.

Neuwelt, E., Abbott, N.J., Abrey, L., Banks, W.A., Blakley, B., Davis, T., Engelhardt, B., Grammas, P., Nedergaard, M., Nutt, J., 2008. Strategies to advance translational research into brain barriers. Lancet Neurol. 7, 84–96.

Nie, Y., Schaffert, D., Rodl, W., Ogris, M., Wagner, E., Gunther, M., 2011. Dual-targeted polyplexes: one step towards a synthetic virus for cancer gene therapy. J. Control. Release 152, 127–134.

Oh, B., Han, J., Choi, E., Tan, X., Lee, M., 2015. Peptide micelle-mediated delivery of tissue-specific suicide gene and combined therapy with avastin in a glioblastoma model. J. Pharm. Sci. 104, 1461–1469.

Onoshima, D., Yukawa, H., Baba, Y., 2015. Multifunctional quantum dots-based cancer diagnostics and stem cell therapeutics for regenerative medicine. Adv. Drug Deliv. Rev. 95, 2–14.

Ozpolat, B., Sood, A.K., Lopez-Berestein, G., 2014. Liposomal siRNA nanocarriers for cancer therapy. Adv. Drug Deliv. Rev. 66, 110–116.

Pang, Z., Lu, W., Gao, H., Hu, K., Chen, J., Zhang, C., Gao, X., Jiang, X., Zhu, C., 2008. Preparation and brain delivery property of biodegradable polymersomes conjugated with OX26. J. Control. Release 128, 120–127.

Pardridge, W.M., 2007. Drug targeting to the brain. Pharm. Res. 24 (9), 1733–1744.

Park, T.E., Singh, B., Li, H., Lee, J.Y., Kang, S.K., Choi, Y.J., Cho, C.S., 2015. Enhanced BBB permeability of osmotically active poly(mannitol-co-PEI) modified with rabies virus glycoprotein via selective stimulation of caveolar endocytosis for RNAi therapeutics in Alzheimer's disease. Biomaterials 38, 61–71.

Qian, Y., Zha, Y., Feng, B., Pang, Z., Zhang, B., Sun, X., Ren, J., Zhang, C., Shao, X., Zhang, Q., Jiang, X., 2013. PEGylated poly(2-(dimethylamino) ethyl methacrylate)/DNA polyplex micelles decorated with phage-displayed TGN peptide for brain-targeted gene delivery. Biomaterials 34, 2117–2129.

Qian, J., Ruan, S., Cao, X., Cun, X., Chen, J., Shen, S., Jiang, X., He, Q., Zhu, J., Gao, H., 2014a. Fluorescent carbonaceous nanospheres as biological probe for noninvasive brain imaging. J. Colloid Interface Sci. 436, 227–233.

Qian, X., Long, L., Shi, Z., Liu, C., Qiu, M., Sheng, J., Pu, P., Yuan, X., Ren, Y., Kang, C., 2014b. Star-branched amphiphilic PLA-b-PDMAEMA copolymers for co-delivery of miR-21 inhibitor and doxorubicin to treat glioma. Biomaterials 35, 2322–2335.

Ren, J., Shen, S., Wang, D., Xi, Z., Guo, L., Pang, Z., Qian, Y., Sun, X., Jiang, X., 2012. The targeted delivery of anticancer drugs to brain glioma by PEGylated oxidized multi-walled carbon nanotubes modified with angiopep-2. Biomaterials 33 (11), 3324–3333.

Rotman, M., Welling, M.M., Bunschoten, A., de Backer, M.E., Rip, J., Nabuurs, R.J.A., Gaillard, P.J., van Bunchem, M.A., van der Maarel, S.M., van der Weerd, L., 2015. Enhanced glutathione PEGylated liposomal brain delivery of an anti-amyloid single domain antibody fragment in a mouse model for Alzheimer's disease. J. Control. Release 203, 40–50.

Ruan, S., Qian, J., Shen, S., Chen, J., Zhu, J., Jiang, X., He, Q., Yang, W., Gao, H., 2014a. Fluorescent carbonaceous nanodots for noninvasive glioma imaging after angiopep-2 decoration. Bioconjug. Chem. 25 (12), 2252–2259.

Ruan, S., Qian, J., Shen, S., Zhu, J., Jiang, X., He, Q., Gao, H., 2014b. A simple one-step method to prepare fluorescent carbon dots and their potential application in non-invasive glioma imaging. Nanoscale 6 (17), 10040–10047.

Ruan, S., He, Q., Gao, H., 2015a. Matrix metalloproteinase triggered size-shrinkable gelatin-gold fabricated nanoparticles for tumor microenvironment sensitive penetration and diagnosis of glioma. Nanoscale 7 (21), 9487–9496.

Ruan, S., Yuan, M., Zhang, L., Hu, G., Chen, J., Cun, X., Zhang, Q., Yang, Y., He, Q., Gao, H., 2015b. Tumor microenvironment sensitive doxorubicin delivery and release to glioma using angiopep-2 decorated gold nanoparticles. Biomaterials 37, 425–435.

Semete, B., Booysen, L., Lemmer, Y., Kalombo, L., Katata, L., Verschoor, J., Swai, H.S., 2010. In vivo evaluation of the biodistribution and safety of PLGA nanoparticles as drug delivery systems. Nanomedicine 6, 662–671.

Shen, J., Yu, M., Meng, Q., Li, J., Lv, Y., Lu, W., 2013. Fatty acid-based strategy for efficient brain targeted gene delivery. Pharm. Res. 30, 2573–2583.

Shevtsov, M.A., Nikolaev, B.P., Yakovleva, L.Y., Marchenko, Y.Y., Dobrodumov, A.V., Mikhrina, A.L., Martynova, M.G., Bystrova, O.A., Yakovenko, I.V., Ischenko, A.M., 2014. Superparamagnetic iron oxide nanoparticles conjugated with epidermal growth factor (SPION-EGF) for targeting brain tumors. Int. J. Nanomedicine 9, 273–287.

Shevtsov, M.A., Nikolaev, B.P., Ryzhov, V.A., Yakovleva, L.Y., Marchenko, Y.Y., Parr, M.A., Rolich, V.I., Mikhrina, A.L., Dobrodumov, A.V., Pitkin, E., Multhoff, G., 2015a. Ionizing radiation improves glioma-specific targeting of superparamagnetic iron oxide nanoparticles conjugated with cmHsp70.1 monoclonal antibodies (SPION-cmHsp70.1). Nanoscale 7 (48), 20652–20664.

Shevtsov, M.A., Nikolaev, B.P., Yakovleva, L.Y., Dobrodumov, A.V., Zhakhov, A.V., Mikhrina, A.L., Pitkin, E., Parr, M.A., Rolich, V.I., Simbircev, A.S., Ischenko, A.M., 2015b. Recombinant interleukin-1 receptor antagonist conjugated to superparamagnetic iron oxide nanoparticles for theranostic targeting of experimental glioblastoma. Neoplasia 17 (1), 32–42.

Shi, N., Pardridge, W.M., 2000. Noninvasive gene targeting to the brain. Proc. Natl. Acad. Sci. U.S.A. 97, 7567–7572.

Shi, K., Li, J., Cao, Z., Yang, P., Qiu, Y., Yang, B., Wang, Y., Long, Y., Liu, Y., Zhang, Q., Qian, J., Zhang, Z., Gao, H., He, Q., 2015. A pH-responsive cell-penetrating peptide-modified liposomes with active recognizing of integrin alphabeta for the treatment of melanoma. J. Control. Release 217, 138–150.

Shityakov, S., Salvador, E., Pastorin, G., Förster, C., 2015. Blood–brain barrier transport studies, aggregation, and molecular dynamics simulation of multiwalled carbon nanotube functionalized with fluorescein isothiocyanate. Int. J. Nanomedicine 10, 1703–1713.

Sillerud, L.O., Solberg, N.O., Chamberlain, R., Orlando, R.A., Heidrich, J.E., Brown, D.C., Brady, C.I., Vander Jaqt, T.A., Garwood, M., Vander Jaqt, D.L., 2013. SPION-enhanced magnetic resonance imaging of Alzheimer's disease plaques in AβPP/PS-1 transgenic mouse brain. J. Alzheimers Dis. 34 (2), 349–365.

Smolny, M., Rogers, M.L., Shafton, A., Rush, R.A., Stebbing, M.J., 2014. Development of non-viral vehicles for targeted gene transfer into microglia via the integrin receptor CD11b. Front. Mol. Neurosci. 7, 79.

Somani, S., Dufes, C., 2014. Applications of dendrimers for brain delivery and cancer therapy. Nanomedicine (Lond.) 9, 2403–2414.

Somani, S., Blatchford, D.R., Millington, O., Stevenson, M.L., Dufes, C., 2014. Transferrin-bearing polypropylenimine dendrimer for targeted gene delivery to the brain. J. Control. Release 188, 78–86.

Song, L.Y., Ahkong, Q.F., Rong, Q., Wang, Z., Ansell, S., Hope, M.J., Mui, B., 2002. Characterization of the inhibitory effect of PEG-lipid conjugates on the intracellular delivery of plasmid and antisense DNA mediated by cationic lipid liposomes. Biochim. Biophys. Acta 1558, 1–13.

Soni, V., Kohli, D.V., Jain, S.K., 2008. Transferrin-conjugated liposomal system for improved delivery of 5-fluorouracil to brain. J. Drug Target. 16, 73–78.

Su, Z., Xing, L., Chen, Y., Xu, Y., Yang, F., Zhang, C., Ping, Q., Xiao, Y., 2014. Lactoferrin-modified poly(ethylene glycol)-grafted BSA nanoparticles as a duel-targeting carrier for treating brain gliomas. Mol. Pharm. 11, 1823–1834.

Sun, X., Pang, Z., Ye, H., Qiu, B., Guo, L., Li, J., Ren, J., Qian, Y., Zhang, Q., Chen, J., Jiang, X., 2012. Co-delivery of pEGFP-hTRAIL and paclitaxel to brain glioma mediated by an angiopep-conjugated liposome. Biomaterials 33, 916–924.

Suzuki, Y.A., Lopez, V., Lonnerdal, B., 2005. Mammalian lactoferrin receptors: structure and function. Cell. Mol. Life Sci. 62, 2560–2575.

Taskar, K.S., Rudraraju, V., Mittapalli, R.K., Samala, R., Thorsheim, H.R., Lockman, J., Gril, B., Hua, E., Palmieri, D., Polli, J.W., Castellino, S., Rubin, S.D., Lockman, P.R., Steeg, P.S., Smith, Q.R., 2012. Lapatinib distribution in HER2 overexpressing experimental brain metastases of breast cancer. Pharm. Res. 29, 770–781.

Toman, P., Lien, C.F., Ahmad, Z., Dietrich, S., Smith, J.R., An, Q., Molnár, É., Pilkington, G.J., Górecki, D.C., Tsibouklis, J., Barbu, E., 2015. Nanoparticles of alkylglyceryl-dextran-graft-poly(lactic acid) for drug delivery to the brain: preparation and in vitro investigation. Acta Biomater. 23, 250–262.

Tuladhar, A., Morshead, C.M., Shoichet, M.S., 2015. Circumventing the blood–brain barrier: local delivery of cyclosporin A stimulates stem cells in stroke-injured rat brain. J. Control. Release 215, 1–11.
Wang, J.T., Al-Jamal, K.T., 2015. Functionalized carbon nanotubes: revolution in brain delivery. Nanomedicine (Lond.) 10, 2639–2642.
Wang, J., Lei, Y., Xie, C., Lu, W., Yan, Z., Gao, J., Xie, Z., Zhang, X., Liu, M., 2013. Targeted gene delivery to glioblastoma using a C-end rule RGERPPR peptide-functionalised polyethylenimine complex. Int. J. Pharm. 458, 48–56.
Wang, J., Lei, Y., Xie, C., Lu, W., Wagner, E., Xie, Z., Gao, J., Zhang, X., Yan, Z., Liu, M., 2014. Retro-inverso CendR peptide-mediated polyethyleneimine for intracranial glioblastoma-targeting gene therapy. Bioconjug. Chem. 25, 414–423.
Wang, F., Zhang, W., Shen, Y., Huang, Q., Zhou, D., Guo, S., 2015. Efficient RNA delivery by integrin-targeted glutathione responsive polyethyleneimine capped gold nanorods. Acta Biomater. 23, 136–146.
Weissig, V., Pettinger, T.K., Murdock, N., 2014. Nanopharmaceuticals (part 1): products on the market. Int. J. Nanomedicine 9, 4357–4373.
van de Wetering, P., Cherng, J.Y., Talsma, H., Crommelin, D.J.A., Hennink, W.E., 1998. 2-(Dimethylamino)ethyl methacrylate based (co)polymer as gene transfer agents. J. Control. Release 53, 145–153.
Wohlfart, S., Gelperina, S., Kreuter, J., 2012. Transport of drugs across the blood–brain barrier by nanoparticles. J. Control. Release 161, 264–273.
Wong, C., Stylianopoulos, T., Cui, J., Martin, J., Chauhan, V.P., Jiang, W., Popovic, Z., Jain, R.K., Bawendi, M.G., Fukumura, D., 2011. Multistage nanoparticles delivery system for deep penetration into tumor tissue. Proc. Natl. Acad. Sci. U.S.A. 108, 2426–2431.
Wong, B.S., Yoong, S.L., Jagusiak, A., Panczyk, T., Ho, H.K., Ang, W.H., Pastorin, G., 2013. Carbon nanotubes for delivery of small molecule drugs. Adv. Drug Deliv. Rev. 65, 1964–2015.
Xie, H., Zhu, Y., Jiang, W., Zhou, Q., Yang, H., Gu, N., Zhang, Y., Xu, H., Xu, H., Yang, X., 2011. Lactoferrin-conjugated superparamagnetic iron oxide nanoparticles as a specific MRI contrast agent for detection of brain glioma in vivo. Biomaterials 32 (2), 495–502.
Yang, Z., Zhang, Y., Yang, Y., Sun, L., Han, D., Li, H., Wang, C., 2010. Pharmacological and toxicological target organelles and safe use of single-walled carbon nanotubes as drug carriers in treating Alzheimer disease. Nanomedicine 6 (3), 427–441.
Yang, H.W., Huang, C.Y., Lin, C.W., Liu, H.L., Huang, C.W., Liao, S.S., Chen, P.Y., Lu, Y.J., Wei, K.C., Ma, C.C., 2014. Gadolinium-functionalized nanographene oxide for combined drug and microRNA delivery and magnetic resonance imaging. Biomaterials 35, 6534–6542.
Yi, X., Manickam, D.S., Brynskikh, A., Kabanov, A.V., 2014. Agile delivery of protein therapeutics to CNS. J. Control. Release 190, 637–663.
Ying, X., Wen, H., Lu, W., Du, J., Guo, J., Tian, W., Men, Y., Zhang, Y., Li, R., Yang, T., Shang, D., Lou, J., Zhang, L., Zhang, Q., 2010. Dual-targeting daunorubicin liposomes improve the therapeutic efficacy of brain glioma in animals. J. Control. Release 141, 183–192.
Zhan, C., Qian, J., Feng, L., Zhong, G., Zhu, J., Lu, W., 2011. Cyclic RGD-poly(ethylene glycol)-polyethyleneimine is more suitable for glioblastoma targeting gene transfer in vivo. J. Drug Target. 19, 573–581.
Zhang, Y., Calon, F., Zhu, C., Boado, R.J., Pardridge, W.M., 2003. Intravenous nonviral gene therapy causes normalization of striatal tyrosine hydroxylase and reversal of motor impairment in experimental parkinsonism. Hum. Gene Ther. 14, 1–12.
Zhang, L., Zhu, S., Qian, L., Pei, Y., Qiu, Y., Jiang, Y., 2011. RGD-modified PEG-PAMAM-DOX conjugates: in vitro and in vivo studies for glioma. Eur. J. Pharm. Biopharm. 79 (2), 232–240.

Zhang, Y., Nayak, T.R., Hong, H., Cai, W., 2012. Graphene: a versatile nanoplatform for biomedical applications. Nanoscale 4, 3833–3842.

Zhang, B., Shen, S., Liao, Z., Shi, W., Wang, Y., Zhao, J., Hu, Y., Yang, J., Chen, J., Mei, H., Hu, Y., Pang, Z., Jiang, X., 2014a. Targeting fibronectins of glioma extracellular matrix by CLT1 peptide-conjugated nanoparticles. Biomaterials 35 (13), 4088–4098.

Zhang, B., Wang, H., Liao, Z., Wang, Y., Hu, Y., Yang, J., Shen, S., Chen, J., Mei, H., Shi, W., Hu, Y., Pang, Z., Jiang, X., 2014b. EGFP-EGF1-conjugated nanoparticles for targeting both neovascular and glioma cells in therapy of brain glioma. Biomaterials 35 (13), 4133–4145.

Zhang, C., Chen, J., Feng, C., Shao, X., Liu, Q., Zhang, Q., Pang, Z., Jiang, X., 2014c. Intranasal nanoparticles of basic fibroblast growth factor for brain delivery to treat Alzheimer's disease. Int. J. Pharm. 461 (1–2), 192–202.

Zhang, C., Zheng, X., Wan, X., Shao, X., Liu, Q., Zhang, Z., Zhang, Q., 2014d. The potential use of H102 peptide-loaded dual-functional nanoparticles in the treatment of Alzheimer's disease. J. Control. Release 192, 317–324.

Zheng, M., Ruan, S., Liu, S., Sun, T., Qu, D., Zhao, H., Xie, Z., Gao, H., Jing, X., Sun, Z., 2015a. Self-targeting fluorescent carbon dots for diagnosis of brain cancer cells. ACS Nano 9 (11), 11455–11461.

Zheng, X., Shao, X., Zhang, C., Tan, Y., Liu, Q., Wan, X., Zhang, Q., Xu, S., Jiang, X., 2015b. Intranasal H102 peptide-loaded liposomes for brain delivery to treat Alzheimer's disease. Pharm. Res. 32, 3837–3849.

Zhou, J., Zhang, X., Li, M., Wu, W., Sun, X., Zhang, L., Gong, T., 2013. Novel lipid hybrid albumin nanoparticle greatly lowered toxicity of pirarubicin. Mol. Pharm. 10, 3832–3841.

Zhu, S., Hong, M., Zhang, L., Tang, G., Jiang, Y., Pei, Y., 2010. PEGylated PAMAM dendrimer-doxorubicin conjugates: in vitro evaluation and in vivo tumor accumulation. Pharm. Res. 27 (1), 161–174.

Zloković, B.V., Lipovac, M.N., Begley, D.J., Davson, H., Rakić, L., 1987. Transport of leucine-enkephalin across the blood–brain barrier in the perfused guinea pig brain. J. Neurochem. 49, 310–315.

Zlokovic, B.V., Hyman, S., McComb, J.G., Lipovac, M.N., Tang, G., Davson, H., 1990. Kinetics of arginine-vasopressin uptake at the blood–brain barrier. Biochim. Biophys. Acta 1025, 191–198.

Chapter 2

The Route of Nanomaterials Entering Brain

Y. Liu, Q. He
Sichuan University, Chengdu, China

Chapter Outline

1. INTRODUCTION

The central nervous system (CNS) is the most important center for regulating and controlling the physiological functions of mammals. To protect and keep a stable microenvironment for the normal activities of neurons, several barriers play a key role in guarding the brain parenchyma against invading substances or organisms (Neuwelt et al., 2008; Tuladhar et al., 2015). These barrier systems called the blood–brain barrier (BBB) are a dynamic structure that adjusts and regulates the balance between blood and brain to maintain the normal function of the CNS (Abbott, 2005).

Neurotoxicity of Nanomaterials and Nanomedicine. http://dx.doi.org/10.1016/B978-0-12-804598-5.00002-7

The BBB consists of brain capillary endothelial cells (BCECs), astrocytes, microglia, pericytes, and extracellular base membranes (Hawkins and Egleton, 2007; Pardridge, 2007a,b). Apart from these anatomical structures, several unique histological characters also contribute to the low permeability of the BBB (Chen and Liu, 2012). Unlike other endothelial walls, the pore size of the BCECs is tiny and with few pinocytotic vesicles (Stewart, 2000). The endothelial cells overlap with each other and form a continuous membranous structure with tight junctions (TJ) (Persidsky et al., 2006), adherent junctions, and junctional adhesion molecules (Hawkins and Davis, 2005). These adhesion proteins, including occludin, claudins, zonula occludens, and cingulin, link to the actin cytoskeleton and notably reduce the permeability of brain endothelial cells (Hawkins and Davis, 2005; Wolburg et al., 2003, 2009). As a result, the transendothelial electrical resistance of BCECs can be as high as $1000\,\Omega\,cm^2$ instead of $2–20\,\Omega\,cm^2$ in peripheral vascular endothelial cells (Butt et al., 1990). Besides, the effective efflux transporters like P-glycoprotein (P-gp) and specialized enzyme systems also prevent the entry of many substances (Demeule et al., 2002; Qosa et al., 2015). Based on all these characters, the BBB can selectively take up essential nutrients of the brain and discharge harmful and surplus substances.

The existence of the BBB provides a safe and stable environment for the CNS under normal conditions, but on the other hand, the BBB becomes a huge obstacle of the transporting of therapeutic drugs, and leads to a bottleneck of the treatment of CNS diseases as well (Azad et al., 2015). Generally, drugs that can get through the BBB must be highly liposoluble and with a small molecular weight below 400–500 kDa (Pardridge, 2007b). According to the statistics, nearly all large-molecule and 98% of the small-molecule drugs are prohibited from entering the brain. Only a very few drugs used for treating depression, schizophrenia, and insomnia possess a molecular weight below 400–500 kDa among the over 7000 kinds of drugs in the comprehensive medicinal chemistry. That means most of the drugs for CNS diseases such as Alzheimer disease, Parkinson disease, and brain tumor can hardly penetrate across the BBB, especially peptides, proteins, or genes (Banks, 2015; Pardridge, 2015b; Yao et al., 2015). Therefore it is necessary to investigate the routes for drugs to enter into brain based on biochemical and cytobiological properties of the BBB, and achieve an improved application of therapeutic agents for CNS diseases (Pardridge, 2003b).

Nano-based brain targeting drug delivery systems have been considered as superior carriers for enhancing the delivery of a broad range of drugs in brain. Based on reasonable construction or modification, nanocarriers or nanomaterials can increase the drug concentration at the BBB by passive or active targeting and penetrate across the BBB through different transcytosis pathways (Blasi et al., 2007; Cheng et al., 2014; Kreuter, 2014). Penetration of the BBB is an energy-dependent and highly regulated process that allows the whole nanocarrier together with the therapeutic agents loaded in to enter brain parenchyma

(Mehmood et al., 2015). Besides, nanocarriers can also bypass or even inhibit the drug efflux expressed on the BBB, thus further promoting drug accumulation in the brain (Choo et al., 2000). The most frequently used pathways for nanomaterials to transport across the BBB include transporter-mediated transcytosis, receptor-mediated transcytosis (RMT), and adsorptive-mediated transcytosis (AMT) (Chen and Liu, 2012). There are also other routes to enter the brain such as intranasal drug delivery (Illum, 2015), or temporarily "open" the TJs by disturbing the function of the BBB (Salahuddin et al., 1988), which can both directly lead to drug accumulation in the brain (Fig. 2.1). In this chapter, we mainly describe transport routes of nanomaterials entering brain separately on the basis of different mechanisms.

2. TRANSPORTER-MEDIATED TRANSCYTOSIS

The brain tissue needs abundant nutrients, which can be transported into the brain depending on the nutrient transporter systems on BCECs (Gynther et al., 2009). The transporters can carry essential nutrients and metabolic substrates including saccharides, amino acids, monocarboxylic acids, fatty acids, vitamins, and nucleosides to maintain the normal physiological functions of CNS (Abbott et al., 2010; Pardridge, 2015a). Generally these transporters do not require adenosine triphosphate for their functions (Pardridge, 2003a). There are nearly 30 kinds of specific transporters expressed on both luminal and abluminal membranes of BCECs, such as monocarboxylate transporters, amino acid transporters, hexose transporters, choline transporter (ChT), and organic anion and cation transporters (Campos-Bedolla et al., 2014), and some of them had already been identified at the gene or protein level (Dahlin et al., 2009; Redzic, 2011). Nanomaterials modified with endogenous analogs thus can enter the brain through these transport systems and this pathway is called transporter-mediated transcytosis or carrier-mediated transcytosis.

2.1 Hexose Transporters

As basic nutrients for supporting brain physiological functions, saccharides can transport across the BBB through hexose transporters. There are two main groups of hexose transporters including sodion-dependent transporters and sodion-independent glucose transporters (GLUTs), and GLUTs are considered to be the main transporter in the brain. Among all the GLUT isomers, GLUT-1 is overexpressed on the luminal surface of the brain capillaries and choroid plexus, which takes main responsibility for selectively transporting D-glucose into the brain (Nualart et al., 1999). GLUT-3 is the predominant transporter on the cytomembrane of neurons, whereas GLUT-2 is expressed on astrocytes in the brain. The capacity of GLUT-1 to transport its substrates such as glucose, 2-deoxyglucose, mannose, galactose, and their analogs through the BBB is nearly 1420 nmol/min/g tissue, which is much higher than that of other nutrient transporters (Béduneau et al., 2007);

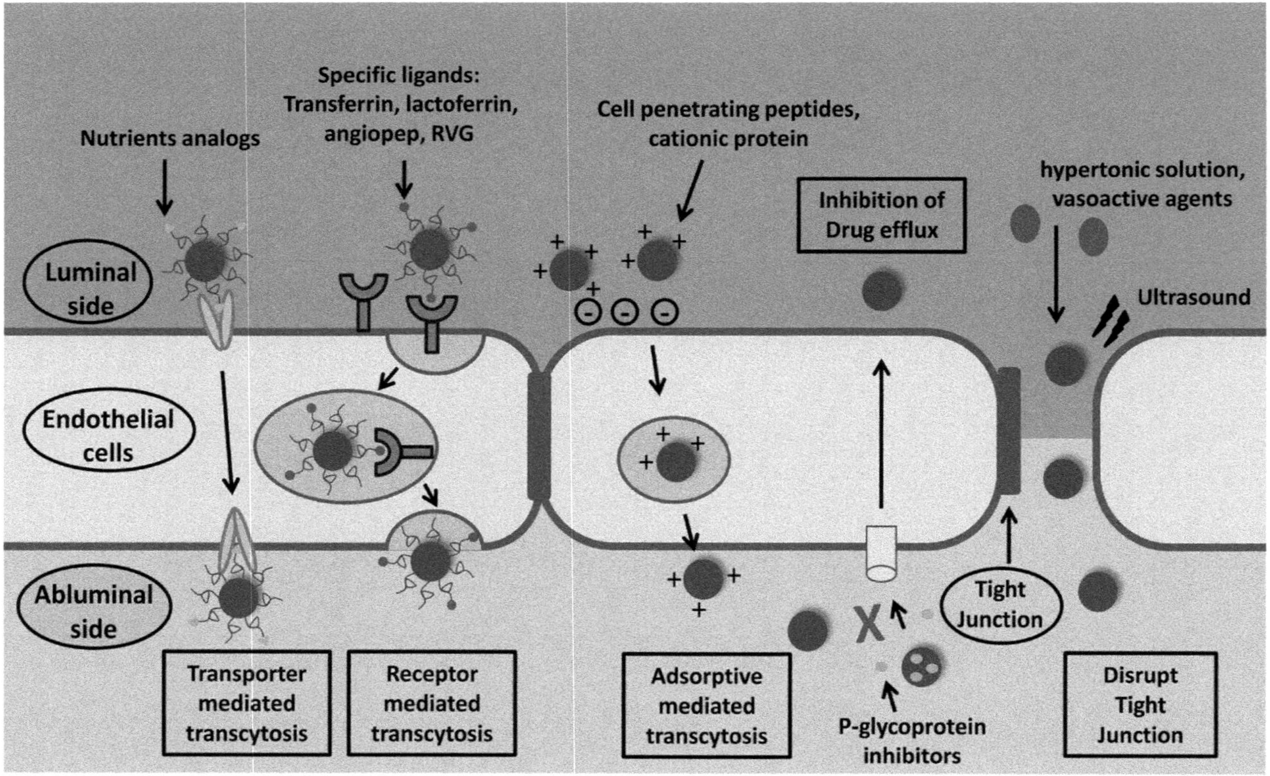

FIGURE 2.1 Different routes for nanomaterials to enter the brain including transporter-mediated transcytosis, receptor-mediated transcytosis, adsorptive-mediated transcytosis, inhibition of drug efflux pumps, and disrupting the tight junctions.

thus GLUT-1 is a compromising target for nanomaterials to penetrate across the BBB. Ying et al. (2010) used *p*-aminophenyl-α-D-manno-pyranoside (MAN), a mannose analog to modify daunorubicin-loaded liposomes. MAN-modified liposomes showed enhanced transporting ratios across the BBB model in vitro and improved the tumor inhibition effect in vivo when used in glioma treatment. Xie et al. synthesized glucose-conjugated cholesterol-anchored polyethylene glycol (PEG) materials with different chain lengths at the C-6 position of glucose for liposomes preparation. The results showed that glucose-modified liposomes achieved enhanced accumulation in the brain among which PEG_{1000}-linked modified liposomes were the most efficient (Xie et al., 2012).

2.2 Choline Transporters

As one of the components of cell membranes and the precursor of the neurotransmitter acetylcholine, choline is also an essential nutrient in the CNS. The brain demands a high concentration of choline to exert physiological functions under normal condition, whereas the synthesized choline in situ cannot meet the requirement. Thus the cholinergic function of the brain greatly depends on choline transport from the blood circulation to the brain tissue (Allen and Smith, 2001). Meanwhile, choline concentration in the plasma is low enough that it cannot occupy all the binding sites on the BBB (Allen and Lockman, 2003). Therefore the ChT on the BBB is considered as another superior target for transporting choline and its derivatives into the brain tissue without competitively inhibiting the normal endogenous cholinergic function. The structure–function relationship of choline derivatives showed that a quaternary ammonium group and a free hydroxyl group were required as essential factors of BBB ChT affinity, and the binding efficiency was even higher when a bis-quaternary ammonium group was joined (Ayers et al., 2002). Besides, some more factors such as the lipophilicity of quaternary ammonium moieties and the length and rigidity of the linkers between the two quaternary ammonium groups would influence the ChT affinity as well (Zheng et al., 2010). Based on these, Li et al. synthesized a choline-derivate ligand for enhancing the penetration behavior of PEGylated dendrigraft poly-L-lysines gene delivery dendrimers across the BBB. All the data suggested that the modified dendrimers could penetrate into the brain through the ChT-mediated pathway and lead to increased gene delivery and expression in the whole brain including cortical layer, caudate putamen, hippocampus, substantia nigra, and fourth ventricle (Li et al., 2011). Although the understanding of ChT-based BBB transporting has developed rapidly, more studies focused on the mechanism are still required, and the potential toxicity of positively charged quaternary ammonium nanomaterials should be taken into account.

2.3 Amino Acid Transporters

Amino acid transporters are usually used in the design of brain targeting prodrugs, whereas the utilization in transporting nanomaterials is infrequent; thus we only give a brief introduction here. Essential amino acids are commonly transported across the BBB through specific transporters to participate in brain amino acid metabolism, such as the synthesis of neurotransmitter. Based on the difference of the substrates, amino acid transporters are divided into cationic, anionic, and neutral amino acid transporters. Large neutral amino acid transporter (LAT1) is the most abundant carrier for amino acids, which is expressed on both luminal and abluminal membranes of BCECs. LAT1 carries large neutral amino acids such as leucine, tryptophan, tyrosine, and phenylalanine across the BBB in the sodion-independent pathway, and is involved in the brain delivery of several central active substances including L-dopa, melphalan, gabapentin, and baclofen (Campos-Bedolla et al., 2014).

3. RECEPTOR-MEDIATED TRANSCYTOSIS

The utilization of specific receptors is the most commonly used strategy to enhance brain drug delivery. Unlike transporter-mediated transcytosis, receptors expressed on BCECs mainly lead to the BBB penetration of large-molecule peptides or proteins instead of small-molecule nutrients (Chen et al., 2004; Khalil et al., 2006). RMT is highly selective that provides an opportunity for drugs or nanomaterials to actively target the BBB, especially when the receptors are upregulated under pathological conditions such as brain tumor or inflammation (Chen and Liu, 2012). Most of the RMT pathways involve the formation of clathrin-coated pits (Pelkmans et al., 2001). Once the ligand-modified nanomaterials in the blood circulation recognize and bind to corresponding receptors on the luminal side of BCECs, the whole nanocarrier would undergo an endocytosis process, invaginate into the cytoplasm, and then shed from the cell membrane to form transport vesicles (Pardridge, 2002). Then the ligands will detach the receptors after acidification in the endosomes several minutes later, and release from the abluminal side (Pardridge, 2005). This receptor-mediated "drug trafficking" process is highly regulated and energy dependent, which allows the intact nanomaterials to penetrate across the BBB, even bypass the efflux pumps (Chen and Liu, 2012). The whole process through the BBB is rapid as the distance of endothelial cells is only 200–300 nm (Abbott et al., 2010).

During the RMT process, the particle sizes of nanomaterials can influence the efficiency of transcytosis to some extent. It is reported that particles smaller than 100 nm can achieve a better mimicry of membrane receptors and thus can be transported via the RMT more efficiently (Wong et al., 2010). Dhanikula et al. (2008) prepared methotrexate-loaded nanoparticles with a series of particle sizes from 70 to 350 nm. The targeting efficiency study showed that nanoparticle of 70 nm size could significantly enhance drug accumulation in both cerebrum and cerebellum compared with the larger ones. Another study by MacKay et al. (2005) proved smaller nanoparticles could not only improve the brain delivery property, but also penetrate deeper into the brain tissues. As molecular biology and proteomics and genomics of the BBB were rapidly discovered in recent years, a large number of specific ligands were used for nanomaterial modification to deliver into the brain successfully.

3.1 Transferrin Receptor

Transferrin receptor (TfR) is a transmembrane glycoprotein that is overexpressed on tumor cells and BCECs. The natural endogenous ligand of TfR is transferrin (Tf), which contributes to the uptake of iron in the brain tissue (Pardridge et al., 1987). Commonly TfR is combined with two 90-kDa subunits through disulfide bond, and each subunit can bind to Tf selectively (Moos and Morgan, 2000). In 1987, the BBB penetration capacity of Tf through RMT was shown. With the rapid development of nano-based drug delivery system, Tf was also widely used for transporting nanomaterials across the BBB. Tf-coated poly lactic-co-glycolic acid nanoparticles exhibited significantly enhanced endocytosis on BCECs, and this enhancement could be competitively inhibited by the excess free Tf (Chang et al., 2009). The brain accumulation of polyamidoamine (PAMAM)-based gene delivery system was increased by 2.3-fold after the modification of Tf, and the transfection efficiency in the brain was about twofold compared with unmodified PAMAM after systemic administration through tail vein into male BALB/c mice (Huang et al., 2008). Tf could also benefit the BBB penetration of daunorubicin-loaded liposomes and achieved prolonged survival time when used in antiglioma therapeutics of brain tumor–bearing rats (Ying et al., 2010).

However, when Tf-modified nanomaterials enter the blood circulation, endogenous Tf would inevitably compete with the drug delivery system and thus to some degree leads to decreased targeting efficiency (Ji et al., 2006). Injection of exogenous Tf would influence the normal physiological function of Tf at the same time. Therefore, a better selection is using antibodies directly against TfR or peptides that recognize TfR. The monoclonal antibody could specifically bind to TfR but with different binding sites to prevent competitive inhibition. Among the antibodies that recognize TfRs, the most representative one is an antirat TfR monoclonal antibody, OX26. The immunoliposomes modified with OX26 could deliver plasmid into the CNS, and the gene expressed protein was highly detected in the brain tissue (Shi et al., 2001). Pang et al. prepared biodegradable polyethylene glycol-poly ε-caprolactone (PEG-PCL) polymersomes conjugated with OX26 to deliver NC-1900 to the brain. In this study, the learning and memory deficits induced by scopolamine was obviously improved by OX26-modified NC-1900-loaded nanoparticles and the pharmacokinetics property was better than unmodified group or free drug (Pang et al., 2008). Other anti-TfR antibodies such as 8D3, R17-217 had all been investigated as potential brain-targeting ligands (Lee et al., 2000; van Rooy et al., 2011). But the clinical usage of anti-TfR-based drug delivery systems is infeasible so far because all these antibodies are antimouse or antirat antibodies that cannot recognize human TfRs. A peptide obtained from phage display named 7 peptide or T7 peptide (HAIYPRH) had exhibited high affinity to TfR (Lee et al., 2001). PEG-PCL nanoparticles modified with T7 peptide showed enhanced cellular uptake on TfR overexpressed cells and the capacity of nanoparticles to transport across physiological barriers was significantly improved in the presence of T7 modification (Du et al., 2013). Zong et al. (2014) conjugated T7 peptide to liposomes through distearoyl-*sn*-glycero-3-phosphoethanolamine-polyethylene glycol-2000 for brain tumor targeting. The cellular uptake of T7-Lip on brain tumor cells and BCECs were both increased. In vivo images displayed the enhanced brain entry and brain tumor accumulation of T7-Lip as well. To further enhance the penetrating efficiency, they used a cell-penetrating peptide transactivator of transcription (CPP TAT) to comodify the liposomes, and the dual-ligand doxorubicin-loaded liposomes exhibited a satisfactory antiglioma therapeutic effect. Similarly, Sharma et al. (2013) designed liposomes comodified with Tf and poly-arginine, and finally achieved enhanced brain accumulation of liposomes as well.

3.2 Insulin Receptor

The insulin receptor is a glycoprotein with molecular weight of about 300kDa. Consisting of two α subunits and two β subunits, insulin receptor is a heterotetramer with subunits linked through disulfide bonds to form columnar structures (De Boer and Gaillard, 2007). When insulin binds to the receptor, the conformation of insulin receptor will change and tyrosine kinase is activated to induce the internalization of insulin. The insulin receptor is highly expressed on the luminal side of BCECs and many other regions of the CNS such as olfactory bulb, choroid plexus, hypothalamus, hippocampus, cerebellum, and cerebral cortex (Pardridge et al., 1985), and the insulin-binding affinity of these regions differs from each other corresponding to the physiological requirement of the brain. Among all these regions, the insulin receptor on BCECs can bind to insulin or its analogs with a high-affinity dissociation constant (K_D) of 1.2 ± 0.5 nM (Pardridge et al., 1985); thus insulin receptor can also be utilized to mediate brain entry.

Unlike Tf, endogenous insulin is not suitable as targeting functional ligands because insulin can be degraded rapidly in blood circulation and may cause hypoglycemia in vivo (Bickel et al., 2001). Therefore, some researchers carried out the investigation of the human insulin receptor monoclonal antibody (HIRMAb); 83-14 is a mouse anti-HIRMAb. Wu et al. (1997) developed a ^{125}I radiolabeled $A\beta_{1\text{-}40}$ that was modified with 83-14; the results showed 83-14 could enhance the brain accumulation of $A\beta_{1\text{-}40}$ and could be used as a diagnostic probe for Alzheimer disease. As for carrying nanomaterials into the brain, Zhang et al. (2002) had reported an immunoliposome modified with 83-14 on the surface for gene therapy against brain tumor. In vivo data showed that the antiepidermal growth factor receptor expression efficiency of 83-14-modified liposomes was significantly stronger and the survival time of 83-14-modified formulation treated mice increased by 88% compared with other groups (Zhang et al., 2004). However, the insulin receptor-based approach is still of great risk as insulin receptor is one of the most important mechanisms of human beings involved in physiological glucose homeostasis (Chen and Liu, 2012).

3.3 Low-Density Lipoprotein Receptor-Related Proteins

Low-density lipoprotein receptor-related proteins (LRP) including LRP-1 and LRP-2 are similar structures to low-density lipoprotein receptors, which can be utilized in brain targeting drug delivery. LRP can bind to a diverse range of ligands including proteins, proteases, and apolipoproteins, and lead to receptor-mediated BBB transporting (Bell et al., 2007). LRP is highly expressed in the CNS, including brain capillary endothelial walls, and more importantly, overexpressed in diseased site in the brain, such as the nidus of Alzheimer disease and brain tumor. Therefore, drug delivery systems based on the LRP pathway can not only penetrate across the BBB, but also further accumulate in the diseased regions (Herz and Strickland, 2001).

Lactoferrin (Lf) is an ion-binding glycoprotein that participates in several physiological functions including the transporting of iron, antioxidation, and immunoregulation (Ward et al., 2002). Lf belongs to the Tf family and functions through specific binding to the receptors on the cell surface. Interestingly, there are several kinds of receptors involved in the transporting process of Lf and different receptors exert their corresponding functions in different tissues or cells (Suzuki et al., 2005). Fillebeen showed Lf could carry cargoes to cross the BBB, whereas this process was specifically inhibited by nearly 70% by LRP, thus confirming that LRP was highly associated with the transportation of Lf (Fillebeen et al., 1999). It needs to be stressed that several unique advantages exist in Lf-mediated brain targeting delivery. The blood concentration of Lf is lower than the dissociation constant of Lf binding to BCECs (Huang et al., 2007), thus the Lf receptor would not be saturated by endogenous Lfs under normal conditions. As a result, the exogenous injected Lf can bind to the receptors on brain capillary vessels efficiently without competitive inhibition. Besides, the BBB transporting induced by Lf is a unidirectional transmission process from the luminal side to brain tissue to achieve drug accumulation in brain parenchyma (Fillebeen et al., 1999). Meanwhile, the expression of Lf receptors would progressively increase in neurons and capillaries of patients with neurodegenerative diseases such as Alzheimer disease and Parkinson disease. Chen et al. (2010) designed an Lf-modified procationic liposomeal drug delivery system for brain delivery. In vivo data showed the content of Lf-modified liposomes detected in the brain was nearly twofold higher than that of conventional liposomes. This Lf-modified system also displayed an enhanced therapeutic effect when doxorubicin was loaded in the liposomes for glioma treatment (Chen et al., 2011). Another Lf-based nanocarrier for gene delivery was Lf-modified DNA/PAMAM complex (Huang et al., 2008). The brain gene expression of Lf-modified DNA/PAMAM complex was even 2.3-fold higher than Tf-modified ones. Their further study used the same carrier for treating Parkinson disease. The therapeutic gene hGDNF loaded in Lf-modified complex showed a satisfying neuroprotective effect including improved locomotor activity, reduced dopaminergic neuronal loss, and enhanced monoamine neurotransmitter levels in rats with Parkinson disease (Huang et al., 2009).

Angiopeps are also known for their effective BBB targeting properties. The most investigated one is called angiopep-2 (TFFYGGSRGKRNNFKTEEY) derived from Kunitz domain of aprotinin. It has been confirmed that the brain entry mediated by angiopep-2 is closely related to LRP receptors and the transporting efficiency is even stronger than traditional protein ligands such as Tf and Lf (Demeule et al., 2008). Yang et al. prepared glioma targeting liposomes comodified with angiopep-2 and another tumor homing peptide tLyP-1. As LRP receptor was overexpressed on both the BBB and glioblastoma cells, the liposomes modified with angiopep-2 displayed enhanced cellular uptake on BCECs and three glioma cell lines (Yang et al., 2014). Ruan et al. evaluated the glioma targeting efficiency of fluorescence carbonaceous nanodots (CDs). The distribution of angiopep-2-modified CDs in the brain was significantly enhanced and could be colocalized with LRP-1 receptor (Ruan et al., 2014). Another study of the same group designed a gold nanoparticle-based tumor microenvironment–responsive doxorubicin delivery system, which was also decorated with angiopep-2. Similarly, the modification of angiopep-2 greatly improved the brain distribution of nanoparticles and achieved better therapeutic effects (Ruan et al., 2015).

Some researchers reported that polysorbate 80 could induce LRP-associated BBB penetration as well. The modification of polysorbate 80 on the surface of nanoparticles could promote the brain entry of both small-molecule and macromolecule drugs (Kurakhmaeva et al., 2009). The BBB targeting mechanism study showed that polysorbate 80 could adsorb apolipoproteins E in the circulation system, followed by transcytosis through LRP (Kreuter et al., 2002). Polybutyl cyanoacrylate (PBCA) nanoparticles coated with polysorbate 80 were able to cross the BBB after intravenous administration and could be detected in both brain vessels and brain parenchyma 1 h post injection. The accumulation of nanoparticles in the brain reached nearly 40% of the injection dose, whereas conventional nanoparticles without modification were mainly found in organs of the reticuloendothelial system (Ambruosi et al., 2005).

3.4 Nicotine Acetylcholine Receptor

The nicotine acetylcholine receptor (nAChR) is widely expressed in the brain and leads to the brain entry of several virus glycoprotein and neurotoxins. The rabies virus glycoprotein was reported to specifically bind to nAChR and enable virus entering neuronal cells. Studies showed that a 29-amino acid peptide named RVG (YTIWMPENPRPGTPCDIFTNSRGKRASNGC), which was derived from rabies virus glycoprotein, inherited the ability to recognize nAChR and could induce BBB penetration of small interfering RNA (Kumar et al., 2007). Based on this, Liu et al. established RVG-modified PAMAM as vectors for brain targeting gene delivery. RVG-PAMAM exhibited superior brain targeting and gene transfection efficiency both in vitro and in vivo, and the enhanced cellular internalization could be competitively inhibited by the addition of free RVG peptide, confirming the selective binding property of RVG (Liu et al., 2009). RVG-modified poly-arginine/DNA complex also showed improved targeting capacity to neuronal cells in vitro and brain distribution in vivo (Gong et al., 2012). nAChR could also bind to neurotoxins. Neurotoxins such as candoxin from the Malayan krait (*Bungarus candidus*) would bind to neuronal nAChR in nanomolar concentrations with very poor reversibility (Nirthanan et al., 2002). CDX, a 16-residue peptide derived from the loop II region of candoxin, exhibited high affinity to nAChR. PEG-PLA nanoparticles modified with CDX could achieve specific endocytosis on BCECs, and finally efficiently deliver paclitaxel into the brain to treat U87 glioblastoma (Zhan et al., 2011).

4. ADSORPTIVE-MEDIATED TRANSCYTOSIS

The BBB is negatively charged due to anionic barriers such as heparin sulfate and chondroitin sulfate on the base membrane and sialic acid on the luminal side of brain capillary endothelial walls (Hardebo and Kåhrström, 1985; Hervé et al., 2008). Positively charged substances can interact with the BCECs through electrostatic adsorption and penetrate across the BBB (Xu et al., 2009). The continuous distribution of these negatively charged substances form the most important factor for cationic nanomaterials to enter the brain. The carboxy of sialic acids on the luminal side of vascular can contribute to adsorption and internalization of cationic materials, while the sulfate groups on the base membrane benefit the exocytosis and transportation into brain parenchyma (Clark, 2003; Vorbrodt, 1989).

The mechanism of AMT is not yet clear. A few researches showed that caveolin was involved in the endocytosis of cationic nanomaterials (Girod et al., 1999). The AMT process is both energy and concentration dependent with a transport speed much slower than transporter- or receptor-mediated transcytosis (Bickel et al., 2001). Compared with the other two pathways described earlier, AMT is nonspecific without competitive inhibition, with low binding affinity but high cellular entry capacity (Chen and Liu, 2012).

There are plenty of evidences that suggested AMT enabled poor brain penetrating drugs or nanomaterials to accumulate in the brain. However, the promoted penetration process could occur not only in the brain but also in other normal vessels and organs due to the nonselectivity of cationic substances (Lu et al., 2007; Saar et al., 2005). Cationic nanomaterials usually formed aggregates in the presence of negatively charged serum proteins and accumulated in the lung or the liver passively (Li and Huang, 2008). The off-target might cause unwanted side effects or toxicity in normal organs. Although the potential risk indeed existed, AMT gained remarkable importance as efficient route of brain targeting delivery. The balance of maintaining penetrating ability at the BBB while reducing distribution to other organs is a great challenge for researchers.

Here we describe CPPs and cationic proteins, which are the most typical AMT-based brain targeting drug delivery methods, and how researchers overcame the bottleneck of AMT.

4.1 Cell-Penetrating Peptides

CPPs are short peptides that can enter living cells efficiently in an energy-dependent way, always less than 30 amino acid residues (Milletti, 2012). Most CPPs are positively charged with amphipathic characters, and can interact with lipid membranes with helical conformation (Deshayes et al., 2005; Zorko and Langel, 2005). Since the first CPP TAT discovered by Frankel et al. in the 1980s, more and more CPPs were identified and used for targeting drug delivery, especially for those macromolecules with poor biological barrier penetration capacity (Lee et al., 2013). The development of CPP-mediated brain targeting drug delivery systems kept increasing and it is proved that several CPPs including TAT, Syn-B, poly-arginine, and penetratin could guide drugs or nanomaterials to penetrate across the BBB and enter the brain (Rao et al., 2008; Rousselle et al., 2001; Schwarze et al., 1999). Early researches considered CPP could directly enter the cells without energy dissipation. However, more and more studies have confirmed the penetrating process of CPPs was both energy and temperature dependent (Hervé et al., 2008; Huang et al., 2013b). One of the hypotheses was that the endocytosis of CPP was highly related to heparin sulfate on the cell membrane and involved the formation of caveola (Wang et al., 2014).

TAT was derived from the trans-activating transcriptor protein of human immunodeficiency virus type 1 with a basic region of two lysine and six arginine residues (Banks et al., 2005; Loret et al., 1991). It was proved that the cationic charges of TAT facilitated the interaction with cellular membranes, and triggered penetration of several biological barriers including the BBB (Liu et al., 2008a; Santra et al., 2005b). Some researches claimed TAT and its cargo entered the cells primarily by lipid raft-mediated macropinocytosis (Wadia et al., 2004). Qin et al. prepared TAT-modified doxorubicin-loaded liposomes for brain drug delivery. TAT-LIP achieved a rapid accumulation in the brain within 1 h after systemic administration via tail vein into Kunming mice and the distribution of TAT-LIP was significantly enhanced in every region of the brain including cortex, striatum, hippocampus, cerebellum, and brainstem (Qin et al., 2011a,b). Another study of TAT-quantum dots conjugation exhibited that quantum dots modified with TAT could migrate across the endothelial cells and reached the brain parenchyma (Santra et al., 2005a). TAT-conjugated PEG-b-cholesterol micelles were also able to transport across the BBB and delivered antibiotics to the nucleus of neurons (Liu et al., 2008b). Another classic CPP for AMT of the BBB is Syn-B. It was derived from mammalian antimicrobial protein protegrin-1 which had exposed promoting intracellular delivery capacity both in vitro and in vivo (De Boer and Gaillard, 2007). Among the Syn-B peptides, Syn-B3 (RRLSYSRRRF) was used for brain targeting, which enhanced the brain distribution of doxorubicin by 30-fold after in situ perfusion from carotid artery (Rousselle et al., 2001). It is worth noting for CPPs-mediated brain targeting that as strongly positively charged penetrating peptides, CPPs especially arginine-rich ones could lead to endosome escape following the endocytosis (Wadia et al., 2004; Wender et al., 2008). Gong et al. designed nona-arginine (r9)-based nanoparticles to transfer DNA across the BBB. The positively charged oligoarginine could be both the carrier of DNA and the impetus for BBB penetration. In vitro cellular uptake study showed that the nanoparticles could successfully escape from endosomes into cytoplasm 60 min after the fluorescence-labeled R9/DNA complex was added, even faster than Lipofectamine (Gong et al., 2012). As gene therapy is attracting more and more attention in treating brain diseases such as Alzheimer disease and gliomas, CPPs have superior potential for brain targeting gene delivery systems as well.

4.2 Cationic Proteins

Cationic proteins can also dramatically increase the BBB penetration through adsorption-mediated pathway. One of the most classic cationic proteins is the cationic bovine serum albumin (CBSA). CBSA had already been used as BBB targeting ligand back in 1987 when Kumagai et al. (1987) conjugated β-endorphin to CBSA through disulfide bond and the central analgesia effect was significantly improved. For nanomaterials, CBSA also acted as a potential ligand that could guide the cargo entering the brain. Lu et al. reported CBSA-modified PEG-PLA nanoparticles for brain targeting drug delivery. Compared with neutral bovine serum albumin-conjugated nanoparticles (BSA-NP) and nanoparticles without modification (NP), CBSA-NP enhanced both the cellular uptake on BCECs in vitro and the biodistribution of fluorescence-labeled nanoparticles in the brain in vivo by almost 2.3-fold (Lu et al., 2005, 2007). Meanwhile, when an antitumor agent hTRAIL (tumor necrosis factor-related apoptosis inducing ligand) was loaded in CBSA-NP, the modified nanoparticles displayed effective gene delivery in the brain. In vivo data showed that the transfection efficiency of CBSA-NP-hTRAIL in cortex, diacele, and brain tumor site were all obviously increased, whereas NP-hTRAIL could hardly get into the brain. As a result, CBSA-conjugated gene delivery system displayed the strongest antitumor effect, which significantly prolonged the median survival time of brain tumor–bearing mice (Lu et al., 2006). Apart from the antibrain tumor application, CBSA also functioned in treating Alzheimer disease. Using CBSA-NP to carry a vasopressin fragment analog NC-1900 could facilitate memory recovery of animals that had been treated with scopolamine to induce memory impairments (Xie et al., 2006).

4.3 Methods That Improve the Nonselectivity of AMT

As mentioned earlier, the application of AMT in brain targeting delivery is limited because of the nonselectivity. One successful method to overcome this problem is to provide stimulation-responsive property to cationic structures. Huang et al. (2013a) designed an activated CPP for brain tumor targeting and therapy. The so-called dtACPP was sensitive to both acidic pH microenvironment and matrix metalloproteinase-2 (MMP-2) through the structural modification of r9. The positive charge of arginine was shielded under normal condition to prevent the uptake of the reticuloendothelial system. However, when the dtACPP-modified nanoparticles were accumulated at the brain tumor foci through the enhanced permeability and retention effect, the linkage of r9 and the shielding sequence was cleaved under the effect of overexpressed MMP-2 in the tumor, and the shielding sequence would also transfer to positive electricity in acidic microenvironment. As a result, the dtACPP would exert the CPP-mediated penetration and deliver therapeutic agents into the brain tumor efficiently. This modification by shielding the positively charged cationic motifs is of great potential in tumor targeting drug delivery, but the penetrating capacity would not be activated until the drug delivery system is transported to the tumor tissue. Therefore, when this system is used in brain tumor targeting, it requires the development of brain tumor till the integrity of the BBB is already disrupted. Another alternative solution is to endue active targeting ability to CPP directly. In the research of Liu et al. (2014), a specific ligand cyclic RGDfK was conjugated to octa-arginine to form a multifunctional tandem peptide R8-RGD to decorate paclitaxel loaded liposomes. As RGD could recognize integrin $\alpha_v\beta_3$ expressed on both BCECs and brain tumor cells, the R8-RGD could bind to integrin receptors on the BBB and penetrate into the brain parenchyma through synergistic effect of RMT and AMT. The in vitro BBB model study exhibited R8-RGD-modified liposomes could highly transport across the BCEC monolayer, and the in vivo images of brain tumor–bearing mice proved that R8-RGD-modified liposomes could not only enter the brain efficiently, but also accumulate in the brain tumor site selectively based on the active targeting of RGD motif. The medium survival time of brain tumor–bearing mice was significantly enhanced after the treatment of paclitaxel-loaded liposomes decorated with R8-RGD. Therefore, although AMT had been investigated for decades, it still plays a key role in brain targeting drug delivery systems.

5. INTRANASAL DRUG DELIVERY

As a unique administration route, intranasal drug delivery had been proved to be a noninvasive pathway to circumvent the BBB and transport biologics into the brain tissue. The nasal mucosal system is formed by respiratory mucosa and olfactory mucosa. The olfactory mucosa is a layer of olfactory epithelium together with the underlying lamina propria, which displays much weaker membrane barrier functions than the BBB (Lochhead and Thorne, 2012). It is commonly approved that most drugs get in the circulation system through the adsorption of respiratory mucosa after intranasal administration, whereas a small part of drugs will transport across the olfactory mucosa and enter the brain parenchyma directly (Illum, 2000). Furthermore, to overcome the rapid nasal mucociliary clearance within the lumen of the nasal cavity, especially for peptide or protein drugs, drug-loaded nanoparticles or nanomaterials were developed for brain delivery through the intranasal pathway. Nanoparticles with superior physicochemical characters can provide appropriate particle sizes for transporting across physiological barriers, benefit the solubility of lipophilic drugs, and prevent the unstable drugs from being degraded before entering the brain (Kubek et al., 2009). Shah et al. (2015) carried out an investigation based on a phenyl carbamate derivate RHT-loaded solid lipid nanoparticle system for intranasal administration, which exhibited higher in vitro and ex vivo diffusion properties during intranasal delivery. Meanwhile, modification with specific targeting or highly penetrating ligands described earlier can also be used in intranasal nanoparticles for brain targeting. For the treatment of Alzheimer disease, several researches had confirmed that decoration with wheat germ agglutinin was beneficial to the enhanced therapeutic effects of drug-loaded nanomaterials after intranasal administration (Liu et al., 2011; Zhang et al., 2014). Lf-modified PEG-PCL nanoparticles also displayed enhanced brain distribution by increasing the nanoparticles accumulation in the rat cerebrum by 1.36-fold following intranasal administration (Liu et al., 2013). Xia et al. (2011) prepared low-molecular-weight protamine (a kind of CPPs)-modified PEG-PLA nanoparticles and they turned out to be an effective and safe carrier for nose-to-brain drug delivery.

Intranasal drug delivery is an attractive pathway to overcome the obstacles imposed by the BBB and is of great potential because of the noninvasiveness. However, some problems also limit the application of this strategy. The main limitation is the relative low dosage of drugs, of which the dose volume through nasal cavity should always be less than 100 μL. Besides, some other requirements on drug properties such as appropriate molecular weight and oil–water partition coefficient, stability in nasal cavity and mucosa, and nontoxicity to nasal cilia system should also be taken into consideration.

6. INHIBITING THE FUNCTION OF THE BLOOD–BRAIN BARRIER

Based on the understanding of the physical structure of the BBB, several methods that inhibit or disturb the function of the BBB can also increase the brain entry of nanomaterials.

6.1 Inhibition of Efflux Pumps

Studies over the past decades explicitly showed that efflux pumps like P-gp are one of the most important structures for maintaining a stable microenvironment of brain tissues. P-gp was first identified to bind to drugs reversibly and pump the intracellular drugs out to the extracellular matrix against the concentration gradient in the research of multidrug resistance of tumor cells (Miller et al., 2008). Different kinds of substrates can competitively bind to P-gp, thus competitive inhibition of P-gp can be used to enhance drug delivery into the brain. It is commonly considered that ligand-modified nanocarriers were able to enter the BCECs by endocytosis and thus circumvent the efflux pumps on the BBB. However, most of the drugs released from the vectors might still be expelled by P-gp. Therefore, coencapsulation of P-gp inhibitors and therapeutic agents in the nanomaterials became a promising alternative for brain targeting drug delivery. Tang et al. entrapped paclitaxel together with borneol in lipid-albumin nanoparticles for enhancing drug delivery into glioma. Borneol is a simple bicyclic monoterpene that can inhibit the activity of P-gp on the cell membrane. Codelivery of borneol increased the uptake of PTX on glioma C6 cells, and in vivo images showed the fluorescence signal of borneol-loaded nanoparticles was stronger than that of conventional nanoparticles (Tang et al., 2015). Pluronic block copolymer, polymer of hydrophilic ethylene oxide and hydrophobic propylene oxide blocks, is another P-gp modulator that promoted the BBB penetration of drugs (Kabanov et al., 2003). More importantly, as an amphipathic material, Pluronic block copolymer is also a drug delivery nanocarrier itself by assembly forming micelles in water. The brain penetration of a model P-gp substrate digoxin was increased by fourfold after the treatment of Pluronic P85 (Batrakova et al., 2001). However, most competitive inhibitors of P-gp were of high pharmacological activities and might cause side effects. Besides, P-gp is widely expressed on many normal tissues as well, thus P-gp inhibitors might also strongly influence the pharmacokinetic properties of therapeutic agents. Therefore, the inhibition of efflux pumps was still limited in clinical applications.

6.2 Disturbing the Structure of the Blood–Brain Barrier

The integrity of the BBB is the key point that prevents drugs entering the brain. Thus the permeability into the brain will be increased when the structure of the BBB is disturbed, for example, disrupting TJs momentarily. Traditional disturbing methods include using hypertonic solution, vasoactive agents, cytokines, or surfactants. Hypertonic solutions such as mannitol, glucose, or dextran can induce a sudden increase of blood osmotic pressure near the brain capillaries; the endothelial cells will then suffer a dehydration process and finally lead to reversible loosening of TJs (Salahuddin et al., 1988). Vasoactive agents such as bradykinin and its analogs can function to shrink the BCECs by binding to corresponding receptors on the endothelial cells (Qin et al., 2009). Interleukin-1β and other cytokines may disrupt TJ-related adhesion proteins such as occludin and zonula occludens (Bolton et al., 1998). However, the application of these methods was limited as they were not able to selectively open the BBB around the diseased region, which would probably affect the function of normal brain tissue. A new technology based on ultrasound to enhance brain drug delivery was developed. This kind of BBB opening can be achieved at the precise place by focused ultrasound under the guidance of magnetic resonance imaging, and reduce side effects resulting from the disruption of the BBB. Ultrasound-induced oscillation of gas-filled microbubbles can cause BBB disruption temporarily (O'Brien et al., 2014), and it is reported that self-assembled microbubbles could be used as focused ultrasound contrast agents and for facilitating PBCA nanoparticles delivery into the brain (Åslund et al., 2015).

7. NANOMATERIALS ENTERING THE BRAIN UNDER PATHOLOGICAL CONDITIONS

Through all the methods described above, drug-loaded nanoparticles could be delivered into the CNS purposefully. However, nanomaterials can also enter the brain tissue unexpectedly when the BBB is suffering impact under pathological conditions. The integrity and functions of the BBB would significantly change in the presence of neurodegenerative disease, inflammation, or infection. It was reported that the BBB was disrupted in a subgroup of patients with Alzheimer disease, and the main reason for the disruption was amyloid angiopathy (Bowman et al., 2007). Another research showed the permeability across the BBB was significantly enhanced in Alzheimer disease, multiple sclerosis, and experimental autoimmune encephalomyelitis model mice (Cheng et al., 2010). In the case of these BBB destroyed conditions, the treatment plan should put greater emphasis on restoring the integrity of BBB than taking advantage of the impaired BBB for nanomaterials transporting into the brain.

8. SUMMARY AND PROSPECTS

There is an increasing number of investigations focused on brain targeting drug delivery, and the routes for nanomaterials to enter the brain never stopped improving. Several brain targeting constructions and designs were established based on rapid development of materials, molecular biology, and cytobiology. However, several limitations still remain in transporting drugs into the CNS. The first and the most important problem is the relatively low targeting efficiency. Although nanomaterials based on the three main BBB penetrating pathways including transporter-mediated transcytosis, RMT, and AMT exhibited significantly enhanced brain targeting efficiency in many researches, drug accumulation in the brain is still extremely limited especially compared with the liver or the spleen where the reticular epithelial system exists. Besides, there are drawbacks in all the existing routes for brain targeting. Transporter- or receptor-mediated transcytosis always displayed high affinity but low capacity. Adsorptive-mediated drug delivery systems can lead to the penetrating into all the organs they contact because of the nonselectivity of cationic molecules. Intranasal drug delivery is only suitable for drugs with low dose, and the disruption of BBB function can be risky and harmful in the clinical setting. Therefore, an overall consideration about how to optimize targeting functional ligands or combine the advantages of different BBB transporting mechanisms is necessary in the future researches. On the other hand, most current studies only focus on the improvement of drug delivery into the brain, but are not concerned about the precise distribution after the drug entered the brain parenchyma. In fact, as the brain controls basic physiological activities of human beings, drug delivery at disease foci would increase therapeutic effects, whereas drug accumulation in normal brain tissues might cause unwanted severer side effects. Thus it is important to further promote therapeutic agents to accumulate at the diseased region after brain entry. A successful example of this point is two-ordered targeting system for brain tumor treatment, which means using two kinds of ligands to modify drug delivery systems, one for BBB targeting and the other for brain tumor targeting. The therapeutic agents thus could be precisely delivered to the brain tumor foci. Last but not least, the expression level of receptors may vary a lot when diseases happen in the CNS, thus the BBB penetration ability of nanomaterials can be greatly influenced when the CNS is under pathological conditions, especially for those drug delivery systems that are based on transporter and receptor-mediated transcytosis. The progress of CNS diseases such as brain tumor and Alzheimer disease will also affect the penetration effect. In conclusion, all the aforementioned problems are demanded to be solved and a multidisciplinary cross-connection is required to achieve more successful brain targeting drug delivery in the future investigations.

REFERENCES

Abbott, N.J., Patabendige, A.A., Dolman, D.E., Yusof, S.R., Begley, D.J., 2010. Structure and function of the blood–brain barrier. Neurobiol. Dis. 37, 13–25.

Abbott, N.J., 2005. Dynamics of CNS barriers: evolution, differentiation, and modulation. Cell Mol. Neurobiol. 25, 5–23.

Allen, D., Lockman, P., 2003. The blood–brain barrier choline transporter as a brain drug delivery vector. Life Sci. 73, 1609–1615.

Allen, D.D., Smith, Q.R., 2001. Characterization of the blood–brain barrier choline transporter using the in situ rat brain perfusion technique. J. Neurochem. 76, 1032–1041.

Ambruosi, A., Yamamoto, H., Kreuter, J., 2005. Body distribution of polysorbate-80 and doxorubicin-loaded [14C] poly(butyl cyanoacrylate) nanoparticles after IV administration in rats. J. Drug Target. 13, 535–542.

Åslund, A.K., Berg, S., Hak, S., Mørch, Ý., Torp, S.H., Sandvig, A., Widerøe, M., Hansen, R., de Lange Davies, C., 2015. Nanoparticle delivery to the brain—by focused ultrasound and self-assembled nanoparticle-stabilized microbubbles. J. Control. Release 220, 287–294.

Ayers, J.T., Dwoskin, L.P., Deaciuc, A.G., Grinevich, V.P., Zhu, J., Crooks, P.A., 2002. bis-Azaaromatic quaternary ammonium analogues: ligands for $\alpha4\beta2^*$ and $\alpha7^*$ subtypes of neuronal nicotinic receptors. Bioorg. Med. Chem. Lett. 12, 3067–3071.

Azad, T.D., Pan, J., Connolly, I.D., Remington, A., Wilson, C.M., Grant, G.A., 2015. Therapeutic strategies to improve drug delivery across the blood–brain barrier. Neurosurg. Focus 38, E9.

Banks, W.A., Robinson, S.M., Nath, A., 2005. Permeability of the blood–brain barrier to HIV-1 Tat. Exp. Neurol. 193, 218–227.

Banks, W.A., 2015. Peptides and the blood–brain barrier. Peptides 72, 16–19.

Batrakova, E.V., Miller, D.W., Li, S., Alakhov, V.Y., Kabanov, A.V., Elmquist, W.F., 2001. Pluronic P85 enhances the delivery of digoxin to the brain: in vitro and in vivo studies. J. Pharmacol. Exp. Ther. 296, 551–557.

Béduneau, A., Saulnier, P., Benoit, J.-P., 2007. Active targeting of brain tumors using nanocarriers. Biomaterials 28, 4947–4967.

Bell, R.D., Sagare, A.P., Friedman, A.E., Bedi, G.S., Holtzman, D.M., Deane, R., Zlokovic, B.V., 2007. Transport pathways for clearance of human Alzheimer's amyloid β-peptide and apolipoproteins E and J in the mouse central nervous system. J. Cereb. Blood Flow Metab. 27, 909–918.

Bickel, U., Yoshikawa, T., Pardridge, W.M., 2001. Delivery of peptides and proteins through the blood–brain barrier. Adv. Drug Deliv. Rev. 46, 247–279.

Blasi, P., Giovagnoli, S., Schoubben, A., Ricci, M., Rossi, C., 2007. Solid lipid nanoparticles for targeted brain drug delivery. Adv. Drug Deliv. Rev. 59, 454–477.

Bolton, S., Anthony, D., Perry, V., 1998. Loss of the tight junction proteins occludin and zonula occludens-1 from cerebral vascular endothelium during neutrophil-induced blood–brain barrier breakdown in vivo. Neuroscience 86, 1245–1257.

Bowman, G., Kaye, J., Moore, M., Waichunas, D., Carlson, N., Quinn, J., 2007. Blood–brain barrier impairment in Alzheimer disease stability and functional significance. Neurology 68, 1809–1814.

Butt, A.M., Jones, H.C., Abbott, N.J., 1990. Electrical resistance across the blood–brain barrier in anaesthetized rats: a developmental study. J. Physiol. 429, 47–62.

Campos-Bedolla, P., Walter, F.R., Veszelka, S., Deli, M.A., 2014. Role of the blood–brain barrier in the nutrition of the central nervous system. Arch. Med. Res. 45, 610–638.

Chang, J., Jallouli, Y., Kroubi, M., Yuan, X.-B., Feng, W., Kang, C.-S., Pu, P.-Y., Betbeder, D., 2009. Characterization of endocytosis of transferrin-coated PLGA nanoparticles by the blood–brain barrier. Int. J. Pharm. 379, 285–292.

Chen, Y., Liu, L., 2012. Modern methods for delivery of drugs across the blood–brain barrier. Adv. Drug Deliv. Rev. 64, 640–665.

Chen, Y., Dalwadi, G., Benson, H., 2004. Drug delivery across the blood–brain barrier. Curr. Drug Deliv. 1, 361–376.

Chen, H., Tang, L., Qin, Y., Yin, Y., Tang, J., Tang, W., Sun, X., Zhang, Z., Liu, J., He, Q., 2010. Lactoferrin-modified procationic liposomes as a novel drug carrier for brain delivery. Eur. J. Pharm. Sci. 40, 94–102.

Chen, H., Qin, Y., Zhang, Q., Jiang, W., Tang, L., Liu, J., He, Q., 2011. Lactoferrin modified doxorubicin-loaded procationic liposomes for the treatment of gliomas. Eur. J. Pharm. Sci. 44, 164–173.

Cheng, Z., Zhang, J., Liu, H., Li, Y., Zhao, Y., Yang, E., 2010. Central nervous system penetration for small molecule therapeutic agents does not increase in multiple sclerosis- and Alzheimer's disease-related animal models despite reported blood–brain barrier disruption. Drug Metab. Dispos. 38, 1355–1361.

Cheng, Y., Morshed, R.A., Auffinger, B., Tobias, A.L., Lesniak, M.S., 2014. Multifunctional nanoparticles for brain tumor imaging and therapy. Adv. Drug Deliv. Rev. 66, 42–57.

Choo, E.F., Leake, B., Wandel, C., Imamura, H., Wood, A.J., Wilkinson, G.R., Kim, R.B., 2000. Pharmacological inhibition of P-glycoprotein transport enhances the distribution of HIV-1 protease inhibitors into brain and testes. Drug Metab. Dispos. 28, 655–660.

Clark, D.E., 2003. In silico prediction of blood–brain barrier permeation. Drug Discov. Today 8, 927–933.

Dahlin, A., Royall, J., Hohmann, J.G., Wang, J., 2009. Expression profiling of the solute carrier gene family in the mouse brain. J. Pharmacol. Exp. Ther. 329, 558–570.

De Boer, A., Gaillard, P., 2007. Drug targeting to the brain. Annu. Rev. Pharmacol. Toxicol. 47, 323–355.

Demeule, M., Régina, A., Jodoin, J., Laplante, A., Dagenais, C., Berthelet, F., Moghrabi, A., Béliveau, R., 2002. Drug transport to the brain: key roles for the efflux pump P-glycoprotein in the blood–brain barrier. Vascul. Pharmacol. 38, 339–348.

Demeule, M., Currie, J.C., Bertrand, Y., Ché, C., Nguyen, T., Régina, A., Gabathuler, R., Castaigne, J.P., Béliveau, R., 2008. Involvement of the low-density lipoprotein receptor-related protein in the transcytosis of the brain delivery vector Angiopep-2. J. Neurochem. 106, 1534–1544.

Deshayes, S., Morris, M., Divita, G., Heitz, F., 2005. Cell-penetrating peptides: tools for intracellular delivery of therapeutics. Cell Mol. Life Sci. 62, 1839–1849.

Dhanikula, R.S., Argaw, A., Bouchard, J.-F., Hildgen, P., 2008. Methotrexate loaded polyether-copolyester dendrimers for the treatment of gliomas: enhanced efficacy and intratumoral transport capability. Mol. Pharm. 5, 105–116.

Du, W., Fan, Y., Zheng, N., He, B., Yuan, L., Zhang, H., Wang, X., Wang, J., Zhang, X., Zhang, Q., 2013. Transferrin receptor specific nanocarriers conjugated with functional 7 peptide for oral drug delivery. Biomaterials 34, 794–806.

Fillebeen, C., Descamps, L., Dehouck, M.-P., Fenart, L., Benaïssa, M., Spik, G., Cecchelli, R., Pierce, A., 1999. Receptor-mediated transcytosis of lactoferrin through the blood–brain barrier. J. Biol. Chem. 274, 7011–7017.

Girod, J., Fenart, L., Régina, A., Dehouck, M.-P., Hong, G., Scherrmann, J.-M., Cecchelli, R., Roux, F., 1999. Transport of cationized anti-tetanus Fab'2 fragments across an in vitro blood–brain barrier model: involvement of the transcytosis pathway. J. Neurochem. 73, 2002–2008.

Gong, C., Li, X., Xu, L., Zhang, Y.-H., 2012. Target delivery of a gene into the brain using the RVG29-oligoarginine peptide. Biomaterials 33, 3456–3463.

Gynther, M., Ropponen, J., Laine, K., Leppanen, J., Haapakoski, P., Peura, L., Jarvinen, T., Rautio, J., 2009. Glucose promoiety enables glucose transporter mediated brain uptake of ketoprofen and indomethacin prodrugs in rats. J. Med. Chem. 52, 3348–3353.

Hardebo, J., Kåhrström, J., 1985. Endothelial negative surface charge areas and blood–brain barrier function. Acta Physiol. Scand. 125, 495–499.

Hawkins, B.T., Davis, T.P., 2005. The blood–brain barrier/neurovascular unit in health and disease. Pharmacol. Rev. 57, 173–185.

Hawkins, B.T., Egleton, R.D., 2007. Pathophysiology of the blood–brain barrier: animal models and methods. Curr. Top. Dev. Biol. 80, 277–309.

Hervé, F., Ghinea, N., Scherrmann, J.-M., 2008. CNS delivery via adsorptive transcytosis. AAPS J. 10, 455–472.

Herz, J., Strickland, D.K., 2001. LRP: a multifunctional scavenger and signaling receptor. J. Clin. Invest. 108, 779.

Huang, R.-Q., Ke, W.-L., Qu, Y.-H., Zhu, J.-H., Pei, Y.-Y., Jiang, C., 2007. Characterization of lactoferrin receptor in brain endothelial capillary cells and mouse brain. J. Biomed. Sci. 14, 121–128.

Huang, R., Ke, W., Liu, Y., Jiang, C., Pei, Y., 2008. The use of lactoferrin as a ligand for targeting the polyamidoamine-based gene delivery system to the brain. Biomaterials 29, 238–246.

Huang, R., Han, L., Li, J., Ren, F., Ke, W., Jiang, C., Pei, Y., 2009. Neuroprotection in a 6-hydroxydopamine-lesioned Parkinson model using lactoferrin-modified nanoparticles. J. Gene Med. 11, 754–763.

Huang, S., Shao, K., Liu, Y., Kuang, Y., Li, J., An, S., Guo, Y., Ma, H., Jiang, C., 2013a. Tumor-targeting and microenvironment-responsive smart nanoparticles for combination therapy of antiangiogenesis and apoptosis. ACS Nano 7, 2860–2871.

Huang, Y., Jiang, Y., Wang, H., Wang, J., Shin, M.C., Byun, Y., He, H., Liang, Y., Yang, V.C., 2013b. Curb challenges of the "Trojan Horse" approach: smart strategies in achieving effective yet safe cell-penetrating peptide-based drug delivery. Adv. Drug Deliv. Rev. 65, 1299–1315.

Illum, L., 2000. Transport of drugs from the nasal cavity to the central nervous system. Eur. J. Pharm. Sci. 11, 1–18.

Illum, L., 2015. Intranasal delivery to the central nervous system. In: Di, L., Kerns, E.H. (Eds.), Blood–Brain Barrier in Drug Discovery: Optimizing Brain Exposure of CNS Drugs and Minimizing Brain Side Effects for Peripheral Drugs. John Wiley & Sons Inc., Hoboken, NJ, pp. 535–565.

Ji, B., Maeda, J., Higuchi, M., Inoue, K., Akita, H., Harashima, H., Suhara, T., 2006. Pharmacokinetics and brain uptake of lactoferrin in rats. Life Sci. 78, 851–855.

Kabanov, A.V., Batrakova, E.V., Miller, D.W., 2003. Pluronic® block copolymers as modulators of drug efflux transporter activity in the blood–brain barrier. Adv. Drug Deliv. Rev. 55, 151–164.

Khalil, I.A., Kogure, K., Akita, H., Harashima, H., 2006. Uptake pathways and subsequent intracellular trafficking in nonviral gene delivery. Pharmacol. Rev. 58, 32–45.

Kreuter, J., Shamenkov, D., Petrov, V., Ramge, P., Cychutek, K., Koch-Brandt, C., Alyautdin, R., 2002. Apolipoprotein-mediated transport of nanoparticle-bound drugs across the blood–brain barrier. J. Drug Target. 10, 317–325.

Kreuter, J., 2014. Drug delivery to the central nervous system by polymeric nanoparticles: what do we know? Adv. Drug Deliv. Rev. 71, 2–14.

Kubek, M.J., Domb, A.J., Veronesi, M.C., 2009. Attenuation of kindled seizures by intranasal delivery of neuropeptide-loaded nanoparticles. Neurotherapeutics 6, 359–371.

Kumagai, A., Eisenberg, J.B., Pardridge, W., 1987. Absorptive-mediated endocytosis of cationized albumin and a beta-endorphin-cationized albumin chimeric peptide by isolated brain capillaries. Model system of blood–brain barrier transport. J. Biol. Chem. 262, 15214–15219.

Kumar, P., Wu, H., McBride, J.L., Jung, K.-E., Kim, M.H., Davidson, B.L., Lee, S.K., Shankar, P., Manjunath, N., 2007. Transvascular delivery of small interfering RNA to the central nervous system. Nature 448, 39–43.

Kurakhmaeva, K.B., Djindjikhashvili, I.A., Petrov, V.E., Balabanyan, V.U., Voronina, T.A., Trofimov, S.S., Kreuter, J., Gelperina, S., Begley, D., Alyautdin, R.N., 2009. Brain targeting of nerve growth factor using poly(butyl cyanoacrylate) nanoparticles. J. Drug Target. 17, 564–574.

Lee, H.J., Engelhardt, B., Lesley, J., Bickel, U., Pardridge, W.M., 2000. Targeting rat anti-mouse transferrin receptor monoclonal antibodies through blood–brain barrier in mouse. J. Pharmacol. Exp. Ther. 292, 1048–1052.

Lee, J.H., Engler, J.A., Collawn, J.F., Moore, B.A., 2001. Receptor mediated uptake of peptides that bind the human transferrin receptor. Eur. J. Biochem. 268, 2004–2012.

Lee, S.H., Castagner, B., Leroux, J.-C., 2013. Is there a future for cell-penetrating peptides in oligonucleotide delivery? Eur. J. Pharm. Biopharm. 85, 5–11.

Li, S.D., Huang, L., 2008. Pharmacokinetics and biodistribution of nanoparticles. Mol. Pharm. 5, 496–504.

Li, J., Zhou, L., Ye, D., Huang, S., Shao, K., Huang, R., Han, L., Liu, Y., Liu, S., Ye, L., 2011. Choline-derivate-modified nanoparticles for brain-targeting gene delivery. Adv. Mater. 23, 4516–4520.

Liu, L., Guo, K., Lu, J., Venkatraman, S.S., Luo, D., Ng, K.C., Ling, E.-A., Moochhala, S., Yang, Y.-Y., 2008a. Biologically active core/shell nanoparticles self-assembled from cholesterol-terminated PEG–TAT for drug delivery across the blood–brain barrier. Biomaterials 29, 1509–1517.

Liu, L., Venkatraman, S.S., Yang, Y.Y., Guo, K., Lu, J., He, B., Moochhala, S., Kan, L., 2008b. Polymeric micelles anchored with TAT for delivery of antibiotics across the blood–brain barrier. Pept. Sci. 90, 617–623.

Liu, Y., Huang, R., Han, L., Ke, W., Shao, K., Ye, L., Lou, J., Jiang, C., 2009. Brain-targeting gene delivery and cellular internalization mechanisms for modified rabies virus glycoprotein RVG29 nanoparticles. Biomaterials 30, 4195–4202.

Liu, Q., Shao, X., Chen, J., Shen, Y., Feng, C., Gao, X., Zhao, Y., Li, J., Zhang, Q., Jiang, X., 2011. In vivo toxicity and immunogenicity of wheat germ agglutinin conjugated poly(ethylene glycol)-poly(lactic acid) nanoparticles for intranasal delivery to the brain. Toxicol. Appl. Pharmacol. 251, 79–84.

Liu, Z., Jiang, M., Kang, T., Miao, D., Gu, G., Song, Q., Yao, L., Hu, Q., Tu, Y., Pang, Z., 2013. Lactoferrin-modified PEG-co-PCL nanoparticles for enhanced brain delivery of NAP peptide following intranasal administration. Biomaterials 34, 3870–3881.

Liu, Y., Ran, R., Chen, J., Kuang, Q., Tang, J., Mei, L., Zhang, Q., Gao, H., Zhang, Z., He, Q., 2014. Paclitaxel loaded liposomes decorated with a multifunctional tandem peptide for glioma targeting. Biomaterials 35, 4835–4847.

Lochhead, J.J., Thorne, R.G., 2012. Intranasal delivery of biologics to the central nervous system. Adv. Drug Deliv. Rev. 64, 614–628.

Loret, E.P., Vives, E., Ho, P.S., Rochat, H., Van Rietschoten, J., Johnson Jr., W.C., 1991. Activating region of HIV-1 Tat protein: vacuum UV circular dichroism and energy minimization. Biochemistry 30, 6013–6023.

Lu, W., Zhang, Y., Tan, Y.-Z., Hu, K.-L., Jiang, X.-G., Fu, S.-K., 2005. Cationic albumin-conjugated pegylated nanoparticles as novel drug carrier for brain delivery. J. Control. Release 107, 428–448.

Lu, W., Sun, Q., Wan, J., She, Z., Jiang, X.-G., 2006. Cationic albumin–conjugated pegylated nanoparticles allow gene delivery into brain tumors via intravenous administration. Cancer Res. 66, 11878–11887.

Lu, W., Wan, J., She, Z., Jiang, X., 2007. Brain delivery property and accelerated blood clearance of cationic albumin conjugated pegylated nanoparticle. J. Control. Release 118, 38–53.

MacKay, J.A., Deen, D.F., Szoka, F.C., 2005. Distribution in brain of liposomes after convection enhanced delivery; modulation by particle charge, particle diameter, and presence of steric coating. Brain Res. 1035, 139–153.

Mehmood, Y., Tariq, A., Siddiqui, F.A., 2015. Brain targeting drug delivery system: a review. Int. J. Basic Med. Sci. Pharm. 5, 32–40.

Miller, D.S., Bauer, B., Hartz, A.M., 2008. Modulation of P-glycoprotein at the blood–brain barrier: opportunities to improve central nervous system pharmacotherapy. Pharmacol. Rev. 60, 196–209.

Milletti, F., 2012. Cell-penetrating peptides: classes, origin, and current landscape. Drug Discov. Today 17, 850–860.

Moos, T., Morgan, E.H., 2000. Transferrin and transferrin receptor function in brain barrier systems. Cell Mol. Neurobiol. 20, 77–95.

Neuwelt, E., Abbott, N.J., Abrey, L., Banks, W.A., Blakley, B., Davis, T., Engelhardt, B., Grammas, P., Nedergaard, M., Nutt, J., 2008. Strategies to advance translational research into brain barriers. Lancet Neurol. 7, 84–96.

Nirthanan, S., Charpantier, E., Gopalakrishnakone, P., Gwee, M.C., Khoo, H.-E., Cheah, L.-S., Bertrand, D., Kini, R.M., 2002. Candoxin, a novel toxin from *Bungarus candidus*, is a reversible antagonist of muscle (αβγδ) but a poorly reversible antagonist of neuronal α7 nicotinic acetylcholine receptors. J. Biol. Chem. 277, 17811–17820.

Nualart, F., Godoy, A., Reinicke, K., 1999. Expression of the hexose transporters GLUT1 and GLUT2 during the early development of the human brain. Brain Res. 824, 97–104.

O'Brien, M.N., Jones, M.R., Brown, K.A., Mirkin, C.A., 2014. Universal noble metal nanoparticle seeds realized through iterative reductive growth and oxidative dissolution reactions. J. Am. Chem. Soc. 136, 7603–7606.

Pang, Z., Lu, W., Gao, H., Hu, K., Chen, J., Zhang, C., Gao, X., Jiang, X., Zhu, C., 2008. Preparation and brain delivery property of biodegradable polymersomes conjugated with OX26. J. Control. Release 128, 120–127.

Pardridge, W.M., Eisenberg, J., Yang, J., 1985. Human blood–brain barrier insulin receptor. J. Neurochem. 44, 1771–1778.

Pardridge, W.M., Eisenberg, J., Yang, J., 1987. Human blood–brain barrier transferrin receptor. Metabolism 36, 892–895.

Pardridge, W.M., 2002. Drug and gene targeting to the brain with molecular Trojan horses. Nat. Rev. Drug Discov. 1, 131–139.

Pardridge, W., 2003a. Blood–brain barrier genomics and the use of endogenous transporters to cause drug penetration into the brain. Curr. Opin. Drug Discov. Dev. 6, 683–691.

Pardridge, W.M., 2003b. Blood–brain barrier drug targeting: the future of brain drug development. Mol. Interv. 3, 90.

Pardridge, W.M., 2005. Drug and gene targeting to the brain via blood–brain barrier receptor-mediated transport systems. Int. Congr. Ser. 1277, 49–62.

Pardridge, W.M., 2007a. Blood–brain barrier genomics. Stroke 38, 686–690.

Pardridge, W.M., 2007b. Blood–brain barrier delivery. Drug Discov. Today 12, 54–61.

Pardridge, W.M., 2015a. Blood–brain barrier endogenous transporters as therapeutic targets: a new model for small molecule CNS drug discovery. Expert Opin. Ther. Targets 19, 1–14.

Pardridge, W.M., 2015b. Targeted delivery of protein and gene medicines through the blood–brain barrier. Clin. Pharmacol. Ther. 97, 347–361.

Pelkmans, L., Kartenbeck, J., Helenius, A., 2001. Caveolar endocytosis of simian virus 40 reveals a new two-step vesicular-transport pathway to the ER. Nat. Cell Biol. 3, 473–483.

Persidsky, Y., Ramirez, S.H., Haorah, J., Kanmogne, G.D., 2006. Blood–brain barrier: structural components and function under physiologic and pathologic conditions. J. Neuroimmune Pharmacol. 1, 223–236.

Qin, L.-J., Gu, Y.-T., Zhang, H., Xue, Y.-X., 2009. Bradykinin-induced blood–tumor barrier opening is mediated by tumor necrosis factor-α. Neurosci. Lett. 450, 172–175.

Qin, Y., Chen, H., Yuan, W., Kuai, R., Zhang, Q., Xie, F., Zhang, L., Zhang, Z., Liu, J., He, Q., 2011a. Liposome formulated with TAT-modified cholesterol for enhancing the brain delivery. Int. J. Pharm. 419, 85–95.

Qin, Y., Chen, H., Zhang, Q., Wang, X., Yuan, W., Kuai, R., Tang, J., Zhang, L., Zhang, Z., Zhang, Q., 2011b. Liposome formulated with TAT-modified cholesterol for improving brain delivery and therapeutic efficacy on brain glioma in animals. Int. J. Pharm. 420, 304–312.

Qosa, H., Miller, D.S., Pasinelli, P., Trotti, D., 2015. Regulation of ABC efflux transporters at blood–brain barrier in health and neurological disorders. Brain Res. 1628, 298–316.

Rao, K.S., Reddy, M.K., Horning, J.L., Labhasetwar, V., 2008. TAT-conjugated nanoparticles for the CNS delivery of anti-HIV drugs. Biomaterials 29, 4429–4438.

Redzic, Z., 2011. Molecular biology of the blood–brain and the blood-cerebrospinal fluid barriers: similarities and differences. Fluids Barriers CNS 8, 1–25.

Rousselle, C., Smirnova, M., Clair, P., Lefauconnier, J.-M., Chavanieu, A., Calas, B., Scherrmann, J.-M., Temsamani, J., 2001. Enhanced delivery of doxorubicin into the brain via a peptide-vector-mediated strategy: saturation kinetics and specificity. J. Pharmacol. Exp. Ther. 296, 124–131.

van Rooy, I., Mastrobattista, E., Storm, G., Hennink, W.E., Schiffelers, R.M., 2011. Comparison of five different targeting ligands to enhance accumulation of liposomes into the brain. J. Control. Release 150, 30–36.

Ruan, S., Qian, J., Shen, S., Chen, J., Zhu, J., Jiang, X., He, Q., Yang, W., Gao, H., 2014. Fluorescent carbonaceous nanodots for noninvasive glioma imaging after angiopep-2 decoration. Bioconjug. Chem. 25, 2252–2259.

Ruan, S., Yuan, M., Zhang, L., Hu, G., Chen, J., Cun, X., Zhang, Q., Yang, Y., He, Q., Gao, H., 2015. Tumor microenvironment sensitive doxorubicin delivery and release to glioma using angiopep-2 decorated gold nanoparticles. Biomaterials 37, 425–435.

Saar, K., Lindgren, M., Hansen, M., Eiríksdóttir, E., Jiang, Y., Rosenthal-Aizman, K., Sassian, M., Langel, Ü., 2005. Cell-penetrating peptides: a comparative membrane toxicity study. Anal. Biochem. 345, 55–65.

Salahuddin, T.S., Johansson, B.B., Kalimo, H., Olsson, Y., 1988. Structural changes in the rat brain after carotid infusions of hyperosmolar solutions: a light microscopic and immunohistochemical study. Neuropathol. Appl. Neurobiol. 14, 467–482.

Santra, S., Yang, H., Holloway, P.H., Stanley, J.T., Mericle, R.A., 2005a. Synthesis of water-dispersible fluorescent, radio-opaque, and paramagnetic CdS: Mn/ZnS quantum dots: a multifunctional probe for bioimaging. J. Am. Chem. Soc. 127, 1656–1657.

Santra, S., Yang, H., Stanley, J.T., Holloway, P.H., Moudgil, B.M., Walter, G., Mericle, R.A., 2005b. Rapid and effective labeling of brain tissue using TAT-conjugated CdS: Mn/ZnS quantum dots. Chem. Commun. 25, 3144–3146.

Schwarze, S.R., Ho, A., Vocero-Akbani, A., Dowdy, S.F., 1999. In vivo protein transduction: delivery of a biologically active protein into the mouse. Science 285, 1569–1572.

Shah, B., Khunt, D., Bhatt, H., Misra, M., Padh, H., 2015. Application of quality by design approach for intranasal delivery of rivastigmine loaded solid lipid nanoparticles: effect on formulation and characterization parameters. Eur. J. Pharm. Sci. 78, 54–66.

Sharma, G., Modgil, A., Layek, B., Arora, K., Sun, C., Law, B., Singh, J., 2013. Cell penetrating peptide tethered bi-ligand liposomes for delivery to brain in vivo: biodistribution and transfection. J. Control. Release 167, 1–10.

Shi, N., Boado, R.J., Pardridge, W.M., 2001. Receptor-mediated gene targeting to tissues in vivo following intravenous administration of pegylated immunoliposomes. Pharm. Res. 18, 1091–1095.

Stewart, P.A., 2000. Endothelial vesicles in the blood–brain barrier: are they related to permeability? Cell Mol. Neurobiol. 20, 149–163.

Suzuki, Y., Lopez, V., Lönnerdal, B., 2005. Lactoferrin. Cell Mol. Life Sci. 62, 2560–2575.

Tang, B., Fang, G., Gao, Y., Liu, Y., Liu, J., Zou, M., Wang, L., Cheng, G., 2015. Lipid-albumin nanoassemblies co-loaded with borneol and paclitaxel for intracellular drug delivery to C6 glioma cells with P-gp inhibition and its tumor targeting. Asian J. Pharm. Sci. 10, 363–371.

Tuladhar, A., Morshead, C.M., Shoichet, M.S., 2015. Circumventing the blood–brain barrier: local delivery of cyclosporin A stimulates stem cells in stroke-injured rat brain. J. Control. Release 215, 1–11.

Vorbrodt, A., 1989. Ultracytochemical characterization of anionic sites in the wall of brain capillaries. J. Neurocytol. 18, 359–368.

Wadia, J.S., Stan, R.V., Dowdy, S.F., 2004. Transducible TAT-HA fusogenic peptide enhances escape of TAT-fusion proteins after lipid raft macropinocytosis. Nat. Med. 10, 310–315.

Wang, H., Zhao, Y., Wang, H., Gong, J., He, H., Shin, M.C., Yang, V.C., Huang, Y., 2014. Low-molecular-weight protamine-modified PLGA nanoparticles for overcoming drug-resistant breast cancer. J. Control. Release 192, 47–56.

Ward, P.P., Uribe-Luna, S., Conneely, O.M., 2002. Lactoferrin and host defense. Biochem. Cell Biol. 80, 95–102.

Wender, P.A., Galliher, W.C., Goun, E.A., Jones, L.R., Pillow, T.H., 2008. The design of guanidinium-rich transporters and their internalization mechanisms. Adv. Drug Deliv. Rev. 60, 452–472.

Wolburg, H., Wolburg-Buchholz, K., Kraus, J., Rascher-Eggstein, G., Liebner, S., Hamm, S., Duffner, F., Grote, E.-H., Risau, W., Engelhardt, B., 2003. Localization of claudin-3 in tight junctions of the blood–brain barrier is selectively lost during experimental autoimmune encephalomyelitis and human glioblastoma multiforme. Acta Neuropathol. 105, 586–592.

Wolburg, H., Noell, S., Mack, A., Wolburg-Buchholz, K., Fallier-Becker, P., 2009. Brain endothelial cells and the glio-vascular complex. Cell Tissue Res. 335, 75–96.

Wong, H.L., Chattopadhyay, N., Wu, X.Y., Bendayan, R., 2010. Nanotechnology applications for improved delivery of antiretroviral drugs to the brain. Adv. Drug Deliv. Rev. 62, 503–517.

Wu, D., Yang, J., Pardridge, W.M., 1997. Drug targeting of a peptide radiopharmaceutical through the primate blood–brain barrier in vivo with a monoclonal antibody to the human insulin receptor. J. Clin. Invest. 100, 1804.

Xia, H., Gao, X., Gu, G., Liu, Z., Zeng, N., Hu, Q., Song, Q., Yao, L., Pang, Z., Jiang, X., 2011. Low molecular weight protamine-functionalized nanoparticles for drug delivery to the brain after intranasal administration. Biomaterials 32, 9888–9898.

Xie, Y.-L., Lu, W., Jiang, X.-G., 2006. Improvement of cationic albumin conjugated pegylated nanoparticles holding NC-1900, a vasopressin fragment analog, in memory deficits induced by scopolamine in mice. Behav. Brain Res. 173, 76–84.

Xie, F., Yao, N., Qin, Y., Zhang, Q., Chen, H., Yuan, M., Tang, J., Li, X., Fan, W., Zhang, Q., 2012. Investigation of glucose-modified liposomes using polyethylene glycols with different chain lengths as the linkers for brain targeting. Int. J. Nanomed. 7, 163.

Xu, F., Lu, W., Wu, H., Fan, L., Gao, X., Jiang, X., 2009. Brain delivery and systemic effect of cationic albumin conjugated PLGA nanoparticles. J. Drug Target. 17, 423–434.

Yang, Z.-Z., Li, J.-Q., Wang, Z.-Z., Dong, D.-W., Qi, X.-R., 2014. Tumor-targeting dual peptides-modified cationic liposomes for delivery of siRNA and docetaxel to gliomas. Biomaterials 35, 5226–5239.

Yao, H., Wang, K., Wang, Y., Wang, S., Li, J., Lou, J., Ye, L., Yan, X., Lu, W., Huang, R., 2015. Enhanced blood–brain barrier penetration and glioma therapy mediated by a new peptide modified gene delivery system. Biomaterials 37, 345–352.

Ying, X., Wen, H., Lu, W.-L., Du, J., Guo, J., Tian, W., Men, Y., Zhang, Y., Li, R.-J., Yang, T.-Y., 2010. Dual-targeting daunorubicin liposomes improve the therapeutic efficacy of brain glioma in animals. J. Control. Release 141, 183–192.

Zhan, C., Li, B., Hu, L., Wei, X., Feng, L., Fu, W., Lu, W., 2011. Micelle-based brain-targeted drug delivery enabled by a nicotine acetylcholine receptor ligand. Angew. Chem. 123, 5596–5599.

Zhang, Y., Zhu, C., Pardridge, W.M., 2002. Antisense gene therapy of brain cancer with an artificial virus gene delivery system. Mol. Ther. 6, 67–72.

Zhang, Y., Zhang, Y.-F., Bryant, J., Charles, A., Boado, R.J., Pardridge, W.M., 2004. Intravenous RNA interference gene therapy targeting the human epidermal growth factor receptor prolongs survival in intracranial brain cancer. Clin. Cancer Res. 10, 3667–3677.

Zhang, C., Chen, J., Feng, C., Shao, X., Liu, Q., Zhang, Q., Pang, Z., Jiang, X., 2014. Intranasal nanoparticles of basic fibroblast growth factor for brain delivery to treat Alzheimer's disease. Int. J. Pharm. 461, 192–202.

Zheng, G., Zhang, Z., Lockman, P.R., Geldenhuys, W.J., Allen, D.D., Dwoskin, L.P., Crooks, P.A., 2010. bis-Azaaromatic quaternary ammonium salts as ligands for the blood–brain barrier choline transporter. Bioorg. Med. Chem. Lett. 20, 3208–3210.

Zong, T., Mei, L., Gao, H., Cai, W., Zhu, P., Shi, K., Chen, J., Wang, Y., Gao, F., He, Q., 2014. Synergistic dual-ligand doxorubicin liposomes improve targeting and therapeutic efficacy of brain glioma in animals. Mol. Pharm. 11, 2346–2357.

Zorko, M., Langel, Ü., 2005. Cell-penetrating peptides: mechanism and kinetics of cargo delivery. Adv. Drug Deliv. Rev. 57, 529–545.

Chapter 3

The Distribution and Elimination of Nanomaterials in Brain

X. Gao, X. Gu, H. Chen
Shanghai Jiao Tong University School of Medicine, Shanghai, China

Chapter Outline

Brain drug delivery is made difficult by the presence of the blood–brain barrier (BBB), which is formed by tight junctions within the capillary endothelium of the vertebrate brain (Pardridge, 2005a). Finding ways to get diagnostic and therapeutic agents to the brain effectively, safely, and conveniently is becoming more and more important. Recent advances in nanotechnology have provided promising solutions to this challenge. Several nanocarriers such as polymeric nanoparticles, solid lipid nanoparticles, liposomes, metallic nanoparticles, micelles, dendrimers, nanogels, nanoemulsions, and nanosuspensions have been studied for the brain delivery of therapeutics. Many of these nanomedicines can be effectively transported across various in vitro and in vivo BBB models by endocytosis and/or transcytosis, and demonstrated early preclini-

Neurotoxicity of Nanomaterials and Nanomedicine. http://dx.doi.org/10.1016/B978-0-12-804598-5.00003-9

cal success for the management of central nervous system (CNS) conditions such as brain tumors, Alzheimer disease (AD), human immunodeficiency virus encephalopathy, and acute ischemic stroke (Wong et al., 2012). Besides originating from medicine and pharmaceutics, nanomaterials such as combustion-derived nanoparticles (eg, particulate matters, diesel exhaust particles, welding fumes) and manufactured or engineered nanoparticles (eg, titanium dioxide, carbon black, carbon nanotubes, silver, zinc oxide, copper oxide) are two main types of nanoparticulate pollutants that can be absorbed into the CNS following inhalation exposure (Donaldson et al., 2005). These nanoparticles can also reach the brain and accumulate in the CNS as a risk factor of neurotoxicity. There have been increasing reports indicating that inhaled nanomaterials can reach the brain and may be associated with neurodegeneration and neurobehavioral disorders of environmental origin (Oberdorster et al., 2009; Win-Shwe and Fujimaki, 2011). Numerous studies have also reported an adverse effect even toxic effect induced by certain nanocarriers for brain drug delivery (Bertero et al., 2014; Godinho et al., 2014). However, limited work has been done to reveal the fate of nanoparticles in the CNS. The current review will briefly introduce the background of the BBB and the existing pathways for the brain delivery of nanomaterials, review the emerging work related to the distribution and elimination of nanomaterials in the CNS, and discuss the related factors.

1. BLOOD–BRAIN BARRIER

Three kinds of barriers are likely involved in the protection of the CNS: the BBB, which mediates the exchange between the CNS and the blood circulation; the blood–cerebrospinal fluid (CSF) barrier, which is localized in the choroid plexus epithelium in the ventricles; and the ependyma, the obstacle between the CSF and the brain tissue (de Boer and Gaillard, 2007). Among these barriers, the BBB plays a key role not only in CNS protection but also in the hindrance of brain drug delivery (Abbott, 2005; de Boer and Gaillard, 2007).

Brain capillary endothelial cells (BCECs), together with the basal lamina, astrocytes, pericytes, and the end feet of neurons, constitute the neurovascular unit (NVU), which acts as the functional unit of the BBB (Hawkins and Davis, 2005; Del, 2010). BCECs are the central component of the NVU that comprise three key characteristics that hinder the free interaction between the blood and brain interstitial fluid (ISF): tight junctions, a significantly low efficiency of pinocytosis, and the absence of intracellular fenestrations (Reese and Karnovsky, 1967). Surrounding the endothelial cells, pericytes give a support to mediate the development, maintenance, and regulation of BBB (Hawkins and Davis, 2005). Astrocytes also provide a support to the prime development of BBB. The interactions between astrocytes and BCECs regulate vascular proliferation, angiogenesis, and also homeostasis of CNS (Lai and Kuo, 2005). Neurons also play a role in adjusting blood flow and the permeability of BBB. Noradrenergic, serotonergic, cholinergic, and GABA-ergic neurons are found to involve in the communication between the microvascular endothelial cells and the associated astrocytes (Abbott et al., 2006).

Besides the above-mentioned barriers formed by cell–cell interaction, ATP-binding cassette (ABC) transporters also decrease the entry of circulating substances into the CNS by pumping them back to the luminal side of brain capillary (Pinzon-Daza et al., 2013). ABC transporters including P-glycoprotein (P-gp/ABCB1), multidrug resistance-related proteins (MRPs/ABCCs), and breast cancer resistance protein (BCRP/ABCG2) are integral membrane transporters to pump substrates out of the CNS via an ATP-dependent manner.

All in all, the BBB regulates the homeostasis of the brain and protects it from toxic substances from the circulation. On the other hand, it is also an obstacle for brain drug delivery and thus hinders the effective therapy of CNS diseases. More than 98% of small molecule agents and almost all large molecule drugs cannot access the CNS because of the presence of BBB (Pardridge, 2005b). The development of nanotechnology therefore provides a powerful platform that facilitates the brain delivery of therapeutic agents to enable the therapy of brain diseases.

2. EXISTING PATHWAYS FOR THE BRAIN DELIVERY OF NANOMATERIALS

Currently existing pathways for the brain delivery of nanomaterials can be generally divided into three types: entering the CNS across the BBB, nose-to-brain delivery, and direct injection or implantation into the CNS. Among these, entering the CNS across the BBB is the most studied pathway, which can be further classified into several detailed delivery mechanisms.

2.1 Entering the CNS Across the BBB

Transcytosis is a major approach for nanomaterials getting through the BBB into the CNS, among which receptor-mediated transcytosis (RMT), adsorptive-mediated transcytosis (AMT), and carrier-mediated transcytosis (CMT) are the three major mechanisms (Bhaskar et al., 2010). RMT is the most applied strategy for nanocarrier-mediated brain drug delivery. Large molecules are generally transported into the CNS via specific receptors that expressed on the BCECs. Numerous receptors such as transferrin receptor, insulin receptor, low-density lipoprotein receptor (LDLR), and LDLR-related protein 1 (LRP-1) are found to be highly expressed on the endothelial cells (Gabathuler, 2010). These receptors can recognize and capture specific ligands and antibodies. Nanocarriers conjugated to these specific ligands or antibodies forming molecular Trojan horses can achieve the ability to cross the BBB. For instance, liposome surface modified with anti-transferrin receptor antibody can be transported into the brain via transferrin RMT (Huwyler et al., 1997).

Temporarily disrupting the BBB integrity is another potential approach that can promote the penetration of nanocarriers into the CNS. For instance, poly(butyl cyanoacrylate) (PBCA)-based microbubbles (MBs) carrying ultrasmall superparamagnetic iron oxide (USPIO) nanoparticles have been designed to mediate drug delivery across the BBB. USPIO-MB was destroyed on exposure to transcranial ultrasound pulses, resulting in acoustic forces that induce BBB permeability. Meanwhile, USPIO nanoparticles were released and got across the permeabilized BBB (Lammers et al., 2015). This approach has been applied to increase the brain delivery of both small molecular compounds and large biomolecules. In contrast, our group reported that MB-enhanced unfocused ultrasound (MEUUS) could even enhance the brain delivery of nanoparticles without affecting the BBB permeability (Yao et al., 2014).

2.2 Nose-to-Brain Delivery

Intranasal (IN) delivery holds great potential for brain drug delivery, which can bypass the BBB and rapidly deliver therapeutic agents to the CNS. It provides the advantages including a large surface area for absorption, rapid achievement of target drug levels, avoidance of first-pass metabolism, and improvement of drug bioavailability; furthermore, this delivery route is needleless, maximizing patient comfort and compliance. Olfactory nerve, trigeminal nerve, blood vessel, perivascular channels, and lymphatic channels are all possible pathways involved in drug transport from the nasal mucosa to the CNS following IN administration (Dhuria et al., 2010). Nose-to-brain delivery of nanotherapeutics for the therapy of pain, epilepsy, neurodegenerative disease, and infectious disease has been justified in animal models (Ong et al., 2014).

2.3 Direct Injection or Implantation Into the CNS

Direct injection or implantation into the CNS can bypass the BBB and deliver drugs to the target tissue in brain, including intracerebroventricular infusion, intracerebral injection, intrathecal injection, implantation, and convection-enhanced delivery. Polymeric nanocomposite hydrogels had been tested for safe use in the injured spinal cord through intrathecal injection (Baumann et al., 2010). Nanoformulations such as liposome and maghemite nanoparticles have been delivered into the brain via convection-enhanced delivery (Mackay et al., 2005; Perlstein et al., 2008). Although various therapeutic agents have been delivered into the brain via these invasive approaches, the high risk of side effects such as infection, edema, and neuron damage highly restricts their further application (Wahl et al., 1988; Gabathuler, 2010).

3. DISTRIBUTION OF NANOMATERIALS IN BRAIN

Knowledge of the biofate of nanoparticulate therapeutics is essential for maximizing therapeutic effect while minimizing unwanted adverse effects. Biodistribution of nanocarriers can alter the pharmacokinetic and toxicity profiles of the pharmacological agents if they remain trapped in the carriers sufficiently long. Many factors affect the distribution of nanoparticles in brain, including pharmaceutical factors such as particle size, surface charge, shape, targeting ligands, and administration routes, and biological factors such as chronobiology and disease conditions.

3.1 Size

Particle size is a key parameter that determines the kinetics and effects of nanomaterials in the organism and also their penetration within the brain tissue. Generally, larger particles above 2 μm are captured by pulmonary capillary vessels. Small nanoparticles with hydrodynamic diameter less than 5.5 nm are rapidly excreted by the kidney. In the development of therapeutics against CNS diseases, 20 nm, large enough to escape renal glomerular filtration, is thought to be sufficiently small size (Jo et al., 2015). Generally, smaller nanoparticles enter the brain more easily than larger ones. For example, Gao compared the brain drug delivery efficiency of polysorbate 80-coated polybutylcyanoacrylate nanoparticles, finding that nanoparticles at 70 nm delivered more drugs into brain, whereas no significant difference was observed among those at 170, 220, and 345 nm (Gao and Jiang, 2006). In addition, the diffusion of nanoparticles in the brain tissue extracellular space (ECS) was also size dependent. Using the minimally adhesive poly(ethylene glycol) (PEG)-coated particles, it was estimated that human brain tissue ECS has some pores larger than 200 nm, and that more than one-quarter of all pores are ≥100 nm, and 40- and 100 nm, but not 200-nm, nanoparticles, spread rapidly within brain tissue, only if densely coated with PEG. Similar results were observed in rat brain tissue with paclitaxel-loaded biodegradable nanoparticles of similar size (85 nm) and surface properties (Nance et al., 2012).

3.2 Surface Charge

Surface charge also affects the fate of nanoparticles following administration and its distribution into the brain. Generally, positively charged nanoparticles are known to be more easily internalized than neutral and negatively charged nanoparticles. For instance, cationic bovine serum albumin has been modified on the surface of nanoparticles (Lu et al., 2005) and polymersome (Pang et al., 2012) to enhance their brain delivery. Cell-penetrating peptides, typically short cationic sequences capable of being rapidly internalized into cells as well as ferrying attached cargoes to translocate cell membranes, have also been widely applied for drug delivery to the CNS (Xia et al., 2012; Kang et al., 2014). However, the nonspecific interaction of the cationic surface-charged nanoparticles with all biological membranes increases their accumulation and unwanted side effects to the nontargeted organs, and thus highly limits their application. To overcome such dilemma, by taking advantage of the dramatically upregulated expression of matrix metalloproteinases

MMP-2 and MMP-9 in glioblastomas and the powerful transport ability of low-molecular-weight protamine (LMWP), we designed an activatable LMWP and conjugated it to polyethylene glycol-poly(ε-caprolactone) nanoparticles to develop a "smart" drug delivery system for enhanced targeted glioblastoma therapy. The positive charge of LMWP was initially masked by a polyanionic peptide (E10) via an MMP-2/9-cleavable peptide linker sequence. The obtained nanoformulation thus exhibited longer retention in the circulation and much higher and more specific accumulation at the glioma site (Gu et al., 2013).

3.3 Shape

Particle shape has an important impact on the performance of nanoparticles as a carrier of therapeutic agents by affecting their fate not only in the circulation but also in the cell. Rod-shaped nanoparticles have been reported to exhibit higher specific and lower nonspecific accumulation under flow at the target, brain endothelium, compared with their spherical counterparts (Kolhar et al., 2013). Chaturbedy reported shape-directed intracellular compartmentalization of nanoparticles in brain cells and their ability to deliver therapeutic molecules to specific organelles, finding that the iron oxide (Fe_3O_4) nanoparticles with biconcave geometry localized predominantly in the nuclei of the brain cells, whereas those with nanotube geometry were contained mostly in the cytoplasm (Chaturbedy et al., 2015).

3.4 Targeting Ligands

Besides particle size, surface charge, and shape, surface modification with specific molecules (eg, ligand, antibody) that can recognize their targets that locate in the brain capillary endothelium or inside the brain parenchyma plays an important role in modulating the brain distribution of nanomaterials. One promising strategy for overcoming the BBB to deliver biologics is the targeting of endogenous RMT systems that employ vesicular trafficking to transport ligands across the BBB endothelium. For instance, apolipoprotein E and B can specifically bind to LRP-1 that expressed on the BBB. Certain surfactants such as polysorbate 80 and poloxamer 188-coated nanoparticles, which can adsorb apolipoprotein E or B from the blood, exhibited increased permeability into brain (Hekmatara et al., 2009; Gelperina et al., 2010). Insulin receptor is also a candidate receptor that can mediate the transcytosis of nanoparticles. Insulin or an anti-insulin receptor monoclonal antibody (29B4) covalently coupled to the human serum albumin (HSA) nanoparticles enabled the brain delivery of those drugs normally unable to cross the BBB (Ulbrich et al., 2011). In addition, peptides that recognize the transferrin receptor (Prades et al., 2012), simil-opioid peptide (Tosi et al., 2007) and leptin-derived peptide (Tosi et al., 2012), can also confer the permeability of BBB to the nanoparticles.

The human brain contains approximately 100 billion capillaries, which feature an intervessel distance of approximately 40 μm and a total drug transport surface area of roughly 20 m^2 (Chaturbedy et al., 2015). Because of the brain's high vascular density, brain cells are readily accessible to circulating drugs provided that they can cross the BBB. Therefore, those nanoparticles enter the brain via the RMT mechanism generally widely distributed in the CNS. For example, gold nanoparticles conjugated with THRPPMWSPVWP peptide, which targeting the transferrin receptor was detected in various regions of the brain with a diffuse staining pattern observed at 2 h post injection (Prades et al., 2012).

The distribution pattern of these receptors that can mediate BBB crossing in the CNS also affects the distribution profile of the nanoparticles in the brain. For instance, g21 peptide, a ligand of leptin receptor, has been used to enhance the brain delivery of poly-lactide-co-glycolide (PLGA) nanoparticles. The g21 peptide-conjugated nanoparticles seemed to demonstrate higher accumulation in the hypothalamic region than in other brain areas, which was in accordance with the higher level of leptin receptor in this area than the other brain regions (Tosi et al., 2012). In contrast, F3 peptide that specifically binds to nucleolin, which is highly expressed on the surface of both glioma cells and endothelial cells of glioma angiogenic blood vessels, was used to decorate a nanoparticulate drug delivery system to realize glioma cell and neovasculature dual targeting. Compared with unmodified nanoparticles, the resulting F3-modified nanoparticles displayed much higher accumulation at the tumor site and deep penetration into the glioma parenchyma (Hu et al., 2013b).

3.5 Administration Routes

The administration route definitely affects the brain distribution of nanomaterials. Invasive approaches including intracerebroventricular infusion, intracerebral injection, and intracerebral implantation all face a problem that the diffusion of the nanomaterials into the brain parenchyma is poor. The diffusion of intracerebral injection decreases exponentially with distance (Gabathuler, 2010), and extremely high drug concentrations remain at the injection site, which might contribute to neurotoxicity. In contrast, because of the brain's high vascular density, brain cells are readily accessible to circulating drugs that can cross the BBB. Nanoparticles that are intravenously administered and enter the brain via the noninvasive approaches including CMT, RMT, and AMT generally widely distribute in the CNS. For example, lipoprotein-based nanoparticles that can cross the BBB via the LDLR RMT exhibited an extensive and uniform distribution among different brain regions following intravenous administration (Song et al., 2014). In contrast, GM1-modified lipoprotein-like nanoparticle, which was given intranasally, showed higher distribution in the olfactory bulb and the brainstem. This is because intranasally applied nanoparticles were transported into the brain mainly via both the olfactory and the trigeminal nerve pathways (Godinho et al., 2014), and transport via the olfactory pathway accelerates drug delivery to the rostral brain areas, whereas transport via the trigeminal nerve pathway facilitates drug delivery directly to the caudal brain regions (Huang et al., 2015). Focused ultrasound especially in combination with MB could noninvasively enhance the delivery of therapeutic agents into the CNS locally (Treat et al., 2007; Raymond et al., 2008; Vykhodtseva et al., 2008; Howles et al., 2010; Jordao et al., 2010; Aslund et al., 2015). In contrast, MEUUS was explored as a feasible modality for improving the extensive brain delivery of poly (ethylene glycol)–poly (lactic acid) (PLA) nanoparticles. It was showed that following MEUUS treatment, the nanoparticle signals were found to be widely distributed in the brain tissues. Such technique might serve as a powerful platform especially for the application of nanoparticles in the diagnosis of brain diseases (Yao et al., 2014).

3.6 Chronobiology

Chronobiology is another important factor that can influence the uptake of nanoparticles into the brain. For instance, after intravenous injection of dalargin-loaded polysorbate 80-coated PBCA nanoparticles, the pain reaction of mice exhibited a circadian phase-dependent manner, which was because the endo- and exocytotic nanoparticle transport activity of the BCECs or within the brain was enhanced during the rest time or suppressed during the active phase (Kreuter, 2015).

3.7 Disease Conditions

Besides, disease condition largely affects the brain distribution of nanomaterials. First, the permeability of the BBB may undergo significant changes under various pathological conditions, including stroke, AD, Parkinson disease, ischemia, and brain tumors (Guan et al., 2013; Banks, 2012). This often leads to the compromise of the integrity of the BBB and thus enhancement of BBB permeability for nanomaterials. For instance, in the case of brain tumors, it is generally acknowledged that the BBB remains intact at the early stage, but become damaged when the tumor reaches a volume large enough, in which blood–brain tumor barrier (BBTB) forms. With the further deterioration of the tumors, both BBB and BBTB become impaired, endothelial gaps form in the microvessels of brain tumors, where enhanced permeability and retention effect may facilitate the accumulation of nanomaterials in the tumor tissue (Zhan and Lu, 2012). Damaged BBB was also found along the progress of AD, which might become an advantage for drug delivery. For example, the increased BBB permeability, reduced expression of efflux transporter, and decreased CSF reabsorption can increase the retention of drug in the CNS. On the contrary, the decreased cerebral blood flow along with the disease progress could greatly reduce brain drug distribution (Mcdade et al., 2014), especially for those drugs with high BBB permeability (Banks, 2012). Second, the expression of certain target moieties or receptors often changes under disease conditions. Taking glioblastoma multiforme as an example, the expression of certain receptors such as neuropilin (Hu et al., 2013a), nucleolin (Guo et al., 2011), and heparan sulfate sulfated polysaccharide (Guo et al., 2011) was elevated in both the neovascular endothelial cells and tumor cells, providing ideal targets for glioma-targeting drug delivery, which would induce specific accumulation of nanotherapeutics in the glioma site.

4. ELIMINATION OF NANOMATERIALS IN BRAIN

Numerous studies regarding nanoparticulate brain delivery system focus on how to increase drug accumulation in the whole brain or in a specific area of the brain. However, the elimination of nanoparticles in the CNS, which is closely related to the efficacy and safety of brain drug delivery, has been hardly studied. Four routes are likely involved in the clearance of nanoparticles in the brain: (1) extracellular degradation mediated by metabolizing enzymes, (2) capture by neurons or glial cells followed by intracellular degradation, (3) entering CSF circulation mediated by the convective bulk flow and then being cleared from the CNS to the bloodstream or the cervical lymphatics, and (4) brain-to-blood efflux. Metabolic enzymes such as aminopeptidase and hyaluronidases that exist in the ECS of the brain parenchyma could account for the extracellular degradation of nanoparticles such as PEGylated hyaluronic acid nanocarrier (Choi et al., 2011). In addition, cellular uptake of nanomaterials had been found in both neurons (Barandeh et al., 2012) and glial cells (Maysinger et al., 2007), therefore intracellular degradation especially in lysosome could be another mechanism for the brain clearance of nanomaterials. Microglia, serving as the gatekeeper of the CNS, could be the major cell type that mediates the intracellular degradation of nanomaterials in the CNS (Nunes et al., 2012). Besides, a paravascular pathway, namely glymphatic pathway, that facilitates CSF flow through the brain parenchyma was found and suggested to be one of the key pathways in the clearance of interstitial endogenous and exogenous solutes, such as amyloid β and dextrans with various molecular weights (Iliff et al., 2012). Functional lymphatic vessels were also found in the CNS lining the dural sinuses, being able to carry both fluid and immune cells from the CSF, and connecting to the deep cervical lymph nodes (Louveau et al., 2015). Such lymphatic system could also be a pathway that mediated the clearance of nanomaterials from brain. In addition, brain-to-blood efflux mediated by P-gp, MRPs, and BCRP plays an important role in the brain clearance of not only agents with small molecular weight such as paclitaxel (Cannon et al., 2012) but also large molecules like amyloid beta (Paganetti et al., 2014). Such brain-to-blood efflux may also involve in the brain elimination of nanomaterials.

Factors that affect the elimination of nanomaterials from the brain also include pharmaceutical factors such as deformability and biodegradability of the matrix, targeting ligand, size, surface charge, and administration routes, and biological factors such as conscious state and disease conditions.

4.1 Deformability and Biodegradability of the Matrix

Deformability and biodegradability are the key factors determining particle stability in the biological system, which also definitely affect the elimination of nanomaterials in brain. Deformable micelle (Sonali et al., 2015), and biodegradable liposome (Boado and Pardridge, 2011), reconstituted lipoproteins (Song et al., 2014; Huang et al., 2015), PBCA or poly(isohexyl cyanoacrylate), PLA or its copolymer (lactide-co-glycolide) (PLGA), and HSA, appear to be the proper candidate nanoparticle matrix that can be eliminated from the brain by deformation or degradation (Wohlfart et al., 2012). In contrast, for nondegradable or hardly degradable nanomaterials such as gold nanoparticles, silver nanoparticles (van der Zande et al., 2012), and cerium oxide particles (Geraets et al., 2012), clearance via the CSF flow or brain-to-blood efflux could be the major elimination mechanism (Prades et al., 2012).

4.2 Size and Surface Charge

Size and surface charge are important design parameters that can be tailored for purposes of directing particle distribution in brain, which may also affect the elimination of nanoparticles in brain. Previous work demonstrated that dextran with lower molecular weight were cleared via the CSF flow quicker than those with higher molecular weight (Iliff et al., 2012); therefore it is possible that the CSF-mediated clearance of nanomaterials would also be size dependent.

4.3 Targeting Ligands

Targeting ligands that mediate the brain delivery of nanomaterials can also affect their clearance from the CNS. Biomimetic nanocarrier composed of 1,2-dimyristoyl-*sn*-glycero-3-phosphocholine and ApoA-I that target the scavenger receptor class B type I receptor leads to lysosome escape and direct cytosolic drug delivery (Zhang et al., 2009). In contrast, ApoE-functionalized liposome that targets the low density lipoprotein receptor was likely to be cleared by intracellular degradation in the lysosome (Bana et al., 2014).

4.4 Conscious State

Sleep has a critical function in ensuring metabolic homeostasis. Natural sleep or anesthesia had been shown to associate with a 60% increase in the interstitial space, and resulted in a striking increase in convective exchange of CSF with ISF, and thus increased the rate of Aβ clearance during sleep (Xie et al., 2013). Thus, sleep and anesthesia might also accelerate the brain clearance of certain nanomaterials.

4.5 Disease Conditions

Metabolic enzymes such as hyaluronidases are known to increase in various malignant tumors including brain tumors. Thus the degradation of PEG-conjugated hyaluronic acid nanoparticles occurred at the tumor site, and this unique degradation pathway for hyaluronic acid allows its use as a carrier for tumor-targeting drug delivery (Choi et al., 2011).

4.6 Brian Regional Distribution

The brain regional distribution of membrane-bound pyroglutamate aminopeptidase was not homogeneous: a 10-fold difference was observed between regions. The highest activity was detected in the olfactory bulb, whereas the lowest was in the cervical part of the spinal cord. The differential distribution of this enzyme is consistent with the hypothesis that it is responsible for extracellular degradation of neuroactive peptides (Vargas et al., 1987). Such site-specific distribution profile of metabolic enzyme in the brain can also induce site-specific degradation of certain nanomaterials.

5. CONCLUSIONS

Nanomaterials have significant potential for drug delivery to the CNS. Over the past 40 years, this technology has undergone rapid expansion and is now poised for clinical translation. However, limited work has been done to reveal the fate of nanomaterials in the CNS. Here, we introduce emerging work in this area and highlight the importance of both pharmaceutical factors such as particle size, surface charge, deformability, and biodegradability of the matrix, targeting ligands, and administration routes, and biological factors such as chronobiology and disease conditions that determine the distribution and elimination of nanomaterials in brain. We emphasize that for the safe and efficient application of nanotechnology in medicine, especially for the treatment of brain disorders, more investigations on the fate of nanomaterial in brain should be conducted.

ACKNOWLEDGMENTS

This contribution was supported by National Natural Science Foundation of China (No. 81573382).

REFERENCES

Abbott, N.J., Ronnback, L., Hansson, E., 2006. Astrocyte-endothelial interactions at the blood–brain barrier. Nat. Rev. Neurosci. 7, 41–53.

Abbott, N.J., 2005. Dynamics of CNS barriers: evolution, differentiation, and modulation. Cell Mol. Neurobiol. 25, 5–23.

Aslund, A.K., Berg, S., Hak, S., Morch, Y., Torp, S.H., Sandvig, A., Wideroe, M., Hansen, R., de Lange, D.C., 2015. Nanoparticle delivery to the brain – by focused ultrasound and self-assembled nanoparticle-stabilized microbubbles. J. Control. Release 220, 287–294.

Bana, L., Minniti, S., Salvati, E., Sesana, S., Zambelli, V., Cagnotto, A., Orlando, A., Cazzaniga, E., Zwart, R., Scheper, W., Masserini, M., Re, F., 2014. Liposomes bi-functionalized with phosphatidic acid and an ApoE-derived peptide affect Aβ aggregation features and cross the blood-brain-barrier: implications for therapy of Alzheimer disease. Nanomedicine 10, 1583–1590.

Banks, W.A., 2012. Drug delivery to the brain in Alzheimer's disease: consideration of the blood–brain barrier. Adv. Drug Deliv. Rev. 64, 629–639.

Barandeh, F., Nguyen, P.L., Kumar, R., Iacobucci, G.J., Kuznicki, M.L., Kosterman, A., Bergey, E.J., Prasad, P.N., Gunawardena, S., 2012. Organically modified silica nanoparticles are biocompatible and can be targeted to neurons in vivo. PLoS One 7, e29424.

Baumann, M.D., Kang, C.E., Tator, C.H., Shoichet, M.S., 2010. Intrathecal delivery of a polymeric nanocomposite hydrogel after spinal cord injury. Biomaterials 31, 7631–7639.

Bertero, A., Boni, A., Gemmi, M., Gagliardi, M., Bifone, A., Bardi, G., 2014. Surface functionalisation regulates polyamidoamine dendrimer toxicity on blood–brain barrier cells and the modulation of key inflammatory receptors on microglia. Nanotoxicology 8, 158–168.

Bhaskar, S., Tian, F., Stoeger, T., Kreyling, W., de la Fuente, J.M., Grazu, V., Borm, P., Estrada, G., Ntziachristos, V., Razansky, D., 2010. Multifunctional nanocarriers for diagnostics, drug delivery and targeted treatment across blood–brain barrier: perspectives on tracking and neuroimaging. Part. Fibre Toxicol. 7, 3.

Boado, R.J., Pardridge, W.M., 2011. The Trojan horse liposome technology for nonviral gene transfer across the blood–brain barrier. J. Drug Deliv. 2011, 296151.

de Boer, A.G., Gaillard, P.J., 2007. Strategies to improve drug delivery across the blood–brain barrier. Clin. Pharmacokinet. 46, 553–576.

Cannon, R.E., Peart, J.C., Hawkins, B.T., Campos, C.R., Miller, D.S., 2012. Targeting blood–brain barrier sphingolipid signaling reduces basal P-glycoprotein activity and improves drug delivery to the brain. Proc. Natl. Acad. Sci. USA 109, 15930–15935.

Chaturbedy, P., Kumar, M., Salikolimi, K., Das, S., Sinha, S.H., Chatterjee, S., Suma, B.S., Kundu, T.K., Eswaramoorthy, M., 2015. Shape-directed compartmentalized delivery of a nanoparticle-conjugated small-molecule activator of an epigenetic enzyme in the brain. J. Control. Release 217, 151–159.

Choi, K.Y., Yoon, H.Y., Kim, J.H., Bae, S.M., Park, R.W., Kang, Y.M., Kim, I.S., Kwon, I.C., Choi, K., Jeong, S.Y., Kim, K., Park, J.H., 2011. Smart nanocarrier based on PEGylated hyaluronic acid for cancer therapy. ACS Nano 5, 8591–8599.

Del, Z.G., 2010. The neurovascular unit in the setting of stroke. J. Intern. Med. 267, 156–171.

Dhuria, S.V., Hanson, L.R., Frey, W.N., 2010. Intranasal delivery to the central nervous system: mechanisms and experimental considerations. J. Pharm. Sci. 99, 1654–1673.

Donaldson, K., Tran, L., Jimenez, L.A., Duffin, R., Newby, D.E., Mills, N., Macnee, W., Stone, V., 2005. Combustion-derived nanoparticles: a review of their toxicology following inhalation exposure. Part. Fibre Toxicol. 2, 10.

Gabathuler, R., 2010. Approaches to transport therapeutic drugs across the blood–brain barrier to treat brain diseases. Neurobiol. Dis. 37, 48–57.

Gao, K., Jiang, X., 2006. Influence of particle size on transport of methotrexate across blood–brain barrier by polysorbate 80-coated polybutylcyanoacrylate nanoparticles. Int. J. Pharm. 310, 213–219.

Gelperina, S., Maksimenko, O., Khalansky, A., Vanchugova, L., Shipulo, E., Abbasova, K., Berdiev, R., Wohlfart, S., Chepurnova, N., Kreuter, J., 2010. Drug delivery to the brain using surfactant-coated poly(lactide-co-glycolide) nanoparticles: influence of the formulation parameters. Eur. J. Pharm. Biopharm. 74, 157–163.

Geraets, L., Oomen, A.G., Schroeter, J.D., Coleman, V.A., Cassee, F.R., 2012. Tissue distribution of inhaled micro- and nano-sized cerium oxide particles in rats: results from a 28-day exposure study. Toxicol. Sci. 127, 463–473.

Godinho, B.M., Mccarthy, D.J., Torres-Fuentes, C., Beltran, C.J., Mccarthy, J., Quinlan, A., Ogier, J.R., Darcy, R., O'Driscoll, C.M., Cryan, J.F., 2014. Differential nanotoxicological and neuroinflammatory liabilities of non-viral vectors for RNA interference in the central nervous system. Biomaterials 35, 489–499.

Gu, G., Xia, H., Hu, Q., Liu, Z., Jiang, M., Kang, T., Miao, D., Tu, Y., Pang, Z., Song, Q., Yao, L., Chen, H., Gao, X., Chen, J., 2013. PEG-co-PCL nanoparticles modified with MMP-2/9 activatable low molecular weight protamine for enhanced targeted glioblastoma therapy. Biomaterials 34, 196–208.

Guan, J., Pavlovic, D., Dalkie, N., Waldvogel, H.J., O'Carroll, S.J., Green, C.R., Nicholson, L.F., 2013. Vascular degeneration in Parkinson's disease, Brain Pathol. 23, 154–164.

Guo, J., Gao, X., Su, L., Xia, H., Gu, G., Pang, Z., Jiang, X., Yao, L., Chen, J., Chen, H., 2011. Aptamer-functionalized PEG-PLGA nanoparticles for enhanced anti-glioma drug delivery. Biomaterials 32, 8010–8020.

Hawkins, B.T., Davis, T.P., 2005. The blood–brain barrier/neurovascular unit in health and disease. Pharmacol. Rev. 57, 173–185.

Hekmatara, T., Bernreuther, C., Khalansky, A.S., Theisen, A., Weissenberger, J., Matschke, J., Gelperina, S., Kreuter, J., Glatzel, M., 2009. Efficient systemic therapy of rat glioblastoma by nanoparticle-bound doxorubicin is due to antiangiogenic effects. Clin. Neuropathol. 28, 153–164.

Howles, G.P., Qi, Y., Johnson, G.A., 2010. Ultrasonic disruption of the blood–brain barrier enables in vivo functional mapping of the mouse barrel field cortex with manganese-enhanced MRI. Neuroimage 50, 1464–1471.

Hu, Q., Gao, X., Gu, G., Kang, T., Tu, Y., Liu, Z., Song, Q., Yao, L., Pang, Z., Jiang, X., Chen, H., Chen, J., 2013a. Glioma therapy using tumor homing and penetrating peptide-functionalized PEG-PLA nanoparticles loaded with paclitaxel. Biomaterials 34, 5640–5650.

Hu, Q., Gu, G., Liu, Z., Jiang, M., Kang, T., Miao, D., Tu, Y., Pang, Z., Song, Q., Yao, L., Xia, H., Chen, H., Jiang, X., Gao, X., Chen, J., 2013b. F3 peptide-functionalized PEG-PLA nanoparticles co-administrated with tLyp-1 peptide for anti-glioma drug delivery. Biomaterials 34, 1135–1145.

Huang, M., Hu, M., Song, Q., Song, H., Huang, J., Gu, X., Wang, X., Chen, J., Kang, T., Feng, X., Jiang, D., Zheng, G., Chen, H., Gao, X., 2015. GM1-modified lipoprotein-like nanoparticle: multifunctional nanoplatform for the combination therapy of Alzheimer's disease. ACS Nano 9, 10801–10816.

Huwyler, J., Yang, J., Pardridge, W.M., 1997. Receptor mediated delivery of daunomycin using immunoliposomes: pharmacokinetics and tissue distribution in the rat. J. Pharmacol. Exp. Ther. 282, 1541–1546.

Iliff, J.J., Wang, M., Liao, Y., Plogg, B.A., Peng, W., Gundersen, G.A., Benveniste, H., Vates, G.E., Deane, R., Goldman, S.A., Nagelhus, E.A., Nedergaard, M., 2012. A paravascular pathway facilitates CSF flow through the brain parenchyma and the clearance of interstitial solutes, including amyloid beta. Sci. Transl. Med. 4, 111r–147r.

Jo, D.H., Kim, J.H., Lee, T.G., Kim, J.H., 2015. Size, surface charge, and shape determine therapeutic effects of nanoparticles on brain and retinal diseases. Nanomedicine 11, 1603–1611.

Jordao, J.F., Ayala-Grosso, C.A., Markham, K., Huang, Y., Chopra, R., Mclaurin, J., Hynynen, K., Aubert, I., 2010. Antibodies targeted to the brain with image-guided focused ultrasound reduces amyloid-beta plaque load in the TgCRND8 mouse model of Alzheimer's disease. PLoS One 5, e10549.

Kang, T., Gao, X., Chen, J., 2014. Harnessing the capacity of cell-penetrating peptides for drug delivery to the central nervous system. Curr. Pharm. Biotechnol. 15, 220–230.

Kolhar, P., Anselmo, A.C., Gupta, V., Pant, K., Prabhakarpandian, B., Ruoslahti, E., Mitragotri, S., 2013. Using shape effects to target antibody-coated nanoparticles to lung and brain endothelium. Proc. Natl. Acad. Sci. USA 110, 10753–10758.

Kreuter, J., 2015. Influence of chronobiology on the nanoparticle-mediated drug uptake into the brain. Pharmaceutics 7, 3–9.

Lai, C.H., Kuo, K.H., 2005. The critical component to establish in vitro BBB model: Pericyte. Brain Res. Brain Res. Rev. 50, 258–265.

Lammers, T., Koczera, P., Fokong, S., Gremse, F., Ehling, J., Vogt, M., Pich, A., Storm, G., van Zandvoort, M., Kiessling, F., 2015. Theranostic USPIO-loaded microbubbles for mediating and monitoring blood–brain barrier permeation. Adv. Funct. Mater. 25, 36–43.

Louveau, A., Smirnov, I., Keyes, T.J., Eccles, J.D., Rouhani, S.J., Peske, J.D., Derecki, N.C., Castle, D., Mandell, J.W., Lee, K.S., Harris, T.H., Kipnis, J., 2015. Structural and functional features of central nervous system lymphatic vessels. Nature 523, 337–341.

Lu, W., Zhang, Y., Tan, Y.Z., Hu, K.L., Jiang, X.G., Fu, S.K., 2005. Cationic albumin-conjugated pegylated nanoparticles as novel drug carrier for brain delivery. J. Control. Release 107, 428–448.

Mackay, J.A., Deen, D.F., Szoka, F.J., 2005. Distribution in brain of liposomes after convection enhanced delivery; modulation by particle charge, particle diameter, and presence of steric coating. Brain Res. 1035, 139–153.

Maysinger, D., Behrendt, M., Lalancette-Hebert, M., Kriz, J., 2007. Real-time imaging of astrocyte response to quantum dots: in vivo screening model system for biocompatibility of nanoparticles. Nano Lett. 7, 2513–2520.

Mcdade, E., Kim, A., James, J., Sheu, L.K., Kuan, D.C., Minhas, D., Gianaros, P.J., Ikonomovic, S., Lopez, O., Snitz, B., Price, J., Becker, J., Mathis, C., Klunk, W., 2014. Cerebral perfusion alterations and cerebral amyloid in autosomal dominant Alzheimer disease. Neurology 83 (8), 710–717.

Nance, E.A., Woodworth, G.F., Sailor, K.A., Shih, T.Y., Xu, Q., Swaminathan, G., Xiang, D., Eberhart, C., Hanes, J., 2012. A dense poly(ethylene glycol) coating improves penetration of large polymeric nanoparticles within brain tissue. Sci. Transl. Med. 4, 119r–149r.

Nunes, A., Bussy, C., Gherardini, L., Meneghetti, M., Herrero, M.A., Bianco, A., Prato, M., Pizzorusso, T., Al-Jamal, K.T., Kostarelos, K., 2012. In vivo degradation of functionalized carbon nanotubes after stereotactic administration in the brain cortex. Nanomedicine 7, 1485–1494.

Oberdorster, G., Elder, A., Rinderknecht, A., 2009. Nanoparticles and the brain: cause for concern? J. Nanosci. Nanotechnol. 9, 4996–5007.

Ong, W.Y., Shalini, S.M., Costantino, L., 2014. Nose-to-brain drug delivery by nanoparticles in the treatment of neurological disorders. Curr. Med. Chem. 21, 4247–4256.

Paganetti, P., Antoniello, K., Devraj, K., Toni, N., Kieran, D., Madani, R., Pihlgren, M., Adolfsson, O., Froestl, W., Schrattenholz, A., Liebner, S., Havas, D., Windisch, M., Cirrito, J.R., Pfeifer, A., Muhs, A., 2014. Increased efflux of amyloid-beta peptides through the blood–brain barrier by muscarinic acetylcholine receptor inhibition reduces pathological phenotypes in mouse models of brain amyloidosis. J. Alzheimers Dis. 38, 767–786.

Pang, Z., Gao, H., Chen, J., Shen, S., Zhang, B., Ren, J., Guo, L., Qian, Y., Jiang, X., Mei, H., 2012. Intracellular delivery mechanism and brain delivery kinetics of biodegradable cationic bovine serum albumin-conjugated polymersomes. Int. J. Nanomed. 7, 3421–3432.

Pardridge, W.M., 2005a. The blood–brain barrier: bottleneck in brain drug development. NeuroRx 2, 3–14.

Pardridge, W.M., 2005b. The blood–brain barrier and neurotherapeutics. NeuroRx 2, 1–2.

Perlstein, B., Ram, Z., Daniels, D., Ocherashvilli, A., Roth, Y., Margel, S., Mardor, Y., 2008. Convection-enhanced delivery of maghemite nanoparticles: increased efficacy and MRI monitoring. Neurooncology 10, 153–161.

Pinzon-Daza, M.L., Campia, I., Kopecka, J., Garzon, R., Ghigo, D., Riganti, C., 2013. Nanoparticle- and liposome-carried drugs: new strategies for active targeting and drug delivery across blood–brain barrier. Curr. Drug Metab. 14, 625–640.

Prades, R., Guerrero, S., Araya, E., Molina, C., Salas, E., Zurita, E., Selva, J., Egea, G., Lopez-Iglesias, C., Teixido, M., Kogan, M.J., Giralt, E., 2012. Delivery of gold nanoparticles to the brain by conjugation with a peptide that recognizes the transferrin receptor. Biomaterials 33, 7194–7205.
Raymond, S.B., Treat, L.H., Dewey, J.D., Mcdannold, N.J., Hynynen, K., Bacskai, B.J., 2008. Ultrasound enhanced delivery of molecular imaging and therapeutic agents in Alzheimer's disease mouse models. PLoS One 3, e2175.
Reese, T.S., Karnovsky, M.J., 1967. Fine structural localization of a blood–brain barrier to exogenous peroxidase. J. Cell Biol. 34, 207–217.
Sonali, Agrawal, P., Singh, R.P., Rajesh, C.V., Singh, S., Vijayakumar, M.R., Pandey, B.L., Muthu, M.S., 2015. Transferrin receptor-targeted vitamin E TPGS micelles for brain cancer therapy: preparation, characterization and brain distribution in rats. Drug Deliv. 1–11.
Song, Q., Huang, M., Yao, L., Wang, X., Gu, X., Chen, J., Chen, J., Huang, J., Hu, Q., Kang, T., Rong, Z., Qi, H., Zheng, G., Chen, H., Gao, X., 2014. Lipoprotein-based nanoparticles rescue the memory loss of mice with Alzheimer's disease by accelerating the clearance of amyloid-beta. ACS Nano 8, 2345–2359.
Tosi, G., Costantino, L., Rivasi, F., Ruozi, B., Leo, E., Vergoni, A.V., Tacchi, R., Bertolini, A., Vandelli, M.A., Forni, F., 2007. Targeting the central nervous system: in vivo experiments with peptide-derivatized nanoparticles loaded with Loperamide and Rhodamine-123. J. Control. Release 122, 1–9.
Tosi, G., Badiali, L., Ruozi, B., Vergoni, A.V., Bondioli, L., Ferrari, A., Rivasi, F., Forni, F., Vandelli, M.A., 2012. Can leptin-derived sequence-modified nanoparticles be suitable tools for brain delivery? Nanomedicine 7, 365–382.
Treat, L.H., Mcdannold, N., Vykhodtseva, N., Zhang, Y., Tam, K., Hynynen, K., 2007. Targeted delivery of doxorubicin to the rat brain at therapeutic levels using MRI-guided focused ultrasound. Int. J. Cancer 121, 901–907.
Ulbrich, K., Knobloch, T., Kreuter, J., 2011. Targeting the insulin receptor: nanoparticles for drug delivery across the blood–brain barrier (BBB). J. Drug Target. 19, 125–132.
Vargas, M., Mendez, M., Cisneros, M., Joseph-Bravo, P., Charli, J.L., 1987. Regional distribution of the membrane-bound pyroglutamate amino peptidase-degrading thyrotropin-releasing hormone in rat brain. Neurosci. Lett. 79, 311–314.
Vykhodtseva, N., Mcdannold, N., Hynynen, K., 2008. Progress and problems in the application of focused ultrasound for blood–brain barrier disruption. Ultrasonics 48, 279–296.
Wahl, M., Unterberg, A., Baethmann, A., Schilling, L., 1988. Mediators of blood–brain barrier dysfunction and formation of vasogenic brain edema. J. Cereb. Blood Flow Metab. 8, 621–634.
Win-Shwe, T.T., Fujimaki, H., 2011. Nanoparticles and neurotoxicity. Int. J. Mol. Sci. 12, 6267–6280.
Wohlfart, S., Gelperina, S., Kreuter, J., 2012. Transport of drugs across the blood–brain barrier by nanoparticles. J. Control. Release 161, 264–273.
Wong, H.L., Wu, X.Y., Bendayan, R., 2012. Nanotechnological advances for the delivery of CNS therapeutics. Adv. Drug Deliv. Rev. 64, 686–700.
Xia, H., Gao, X., Gu, G., Liu, Z., Hu, Q., Tu, Y., Song, Q., Yao, L., Pang, Z., Jiang, X., Chen, J., Chen, H., 2012. Penetratin-functionalized PEG-PLA nanoparticles for brain drug delivery. Int. J. Pharm. 436, 840–850.
Xie, L., Kang, H., Xu, Q., Chen, M.J., Liao, Y., Thiyagarajan, M., O'Donnell, J., Christensen, D.J., Nicholson, C., Iliff, J.J., Takano, T., Deane, R., Nedergaard, M., 2013. Sleep drives metabolite clearance from the adult brain. Science 342, 373–377.

Yao, L., Song, Q., Bai, W., Zhang, J., Miao, D., Jiang, M., Wang, Y., Shen, Z., Hu, Q., Gu, X., Huang, M., Zheng, G., Gao, X., Hu, B., Chen, J., Chen, H., 2014. Facilitated brain delivery of poly(ethylene glycol)-poly(lactic acid) nanoparticles by microbubble-enhanced unfocused ultrasound. Biomaterials 35, 3384–3395.

van der Zande, M., Vandebriel, R.J., Van Doren, E., Kramer, E., Herrera, R.Z., Serrano-Rojero, C.S., Gremmer, E.R., Mast, J., Peters, R.J., Hollman, P.C., Hendriksen, P.J., Marvin, H.J., Peijnenburg, A.A., Bouwmeester, H., 2012. Distribution, elimination, and toxicity of silver nanoparticles and silver ions in rats after 28-day oral exposure. ACS Nano 6, 7427–7442.

Zhan, C., Lu, W., 2012. The blood–brain/tumor barriers: challenges and chances for malignant gliomas targeted drug delivery. Curr. Pharm. Biotechnol. 13, 2380–2387.

Zhang, Z., Cao, W., Jin, H., Lovell, J.F., Yang, M., Ding, L., Chen, J., Corbin, I., Luo, Q., Zheng, G., 2009. Biomimetic nanocarrier for direct cytosolic drug delivery. Angew. Chem. 48, 9171–9175.

Chapter 4

Current Perspective on Nanomaterial-Induced Adverse Effects: Neurotoxicity as a Case Example

J. Li[1], F.L. Martin[1,2]
[1]*Lancaster University, Lancaster, United Kingdom;* [2]*University of Central Lancashire, Preston, United Kingdom*

Chapter Outline

1. NANOTOXICOLOGY

The term "nanoparticles" (NPs) is used to define a particle <100 nm in diameter. With rapid technological developments, nanotechnology is no longer just confined to the scientist's bench. Now it is more common to find nanomaterials in

Neurotoxicity of Nanomaterials and Nanomedicine. http://dx.doi.org/10.1016/B978-0-12-804598-5.00004-0

commercial or industrial products. They are increasingly found in plasticwares, clothing, cosmetics, electrical appliances, and even food products (Becheri et al., 2008; Serpone et al., 2007). Their applications extend into the biomedical and healthcare fields, particularly in medical imaging and diagnosis, pharmaceuticals, drug delivery, and clinical therapy (Bakry et al., 2007; Bianco et al., 2005; Gooding, 2005; Katz and Willner, 2004; Liu et al., 2009). Because of the increasing production of synthetic NPs, the potential release of NPs in the environment is expected to dramatically increase, as well as occupational and public exposure in the coming years. To realize proposed benefits, significant research has been performed over the past decades so as to determine whether the value of nanotechnology could be realized without any adverse impacts on human health. This has been termed nanotoxicology and focuses on gaining a thorough understanding of the relationship between properties such as particle size, surface area and reactivity, dose (concentration), and material composition and the potential toxicity of NPs. In relation to this, it is also proposed that nanotoxicology will develop and implement "nanospecific" protocols to investigate and gain the knowledge necessary to determine the potential toxicity of different NPs.

2. NEUROTOXICOLOGY

The nervous system consists of a highly complex organization with unique structural and functional characteristics, containing many cell types with various functions. The study of its (dys)functionality, ie, neurotoxicology, constitutes an important issue in neurosciences. The field of neurotoxicology has emerged from the integration of toxicology, pharmacology, psychopharmacology, and experimental psychology, and involves the study of changes in the function and/or structure of the nervous system as a result of chemical exposure or other environmental influences, as well as an interpretation of the consequences and adverse effects (Barbosa et al., 2015). Neurotoxicity can be defined as any adverse effect on the chemistry, structure, or function of the nervous system, during development or at maturity, induced by chemical, biological, or physical influences. A large number of compounds have shown capability in causing neurotoxicity, including metals, industrial chemicals, solvents, natural toxins, pharmaceutical drugs, and pesticides. The nervous system is particularly sensitive to toxic insults because of certain intrinsic characteristics, such as dependence on aerobic metabolism for a constant oxygen supply, the presence of axonal transport, or the processes of neurotransmission. A neurotoxic effect can be the direct alteration of the neuronal structure or activity, or it can be the result of a cascade of effects due to glial activation and glial–neuronal interactions. In addition, a neurotoxic effect can manifest immediately or years after the insult. Neurotoxicity can be permanent or reversible, and it can affect the whole nervous system, or only parts of it.

In a general perspective, neurotoxicity can be observed by neurotoxicants in pathways including causing neuropathy, thus damaging the whole neuron; targeting the axon, causing axonopathy; inducing myelopathy; and affecting neurotransmission (Costa et al., 2008). A number of chemicals may cause toxicity that result in the loss of neurons (neuronopathy), either by necrosis or by apoptosis. Such neuronal loss is irreversible and may result in a global encephalopathy; when only subpopulations of neurons are affected, it results in the loss of particular functions. Such chemical toxicants can cause neuronal cell death by a variety of mechanisms, including disruption of the cytoskeleton, induction of oxidative stress (OS), Ca^{2+} overload, or damaging mitochondria. In addition, a large amount of neurotoxic chemicals mainly target the axon, and cause axonopathy. The axon degenerates, and with it the myelin sheath surrounding the axon, whereas the cell body remains intact. However, axonal degeneration may lead to neuronal death. Other chemicals may target myelin, causing intramyelinic edema or demyelination. Although in this case neurons are structurally unaffected, their functions are altered. Such toxicants causing intramyelinic edema can lead to the formation of vacuoles and spongiosis in the brain. Moreover, a number of neurotoxicants are able to interfere with neurotransmission. They can inhibit the release of neurotransmitters, such as the botulin toxin, which inhibits acetylcholine release, and act as agonists or antagonists of specific receptors, which blocks muscarinic receptors and cause interference with signal transduction processes. Effects resulting from an interference with synaptic functioning are usually reversible. However, of toxicological relevance, they may lead to severe acute toxicity or even death.

3. BRAIN AS THE TARGET OF NPs

The blood–brain barrier (BBB) (Fig. 4.1) is a term used to describe a series of unique properties associated with the blood vessels in the brain, which tightly

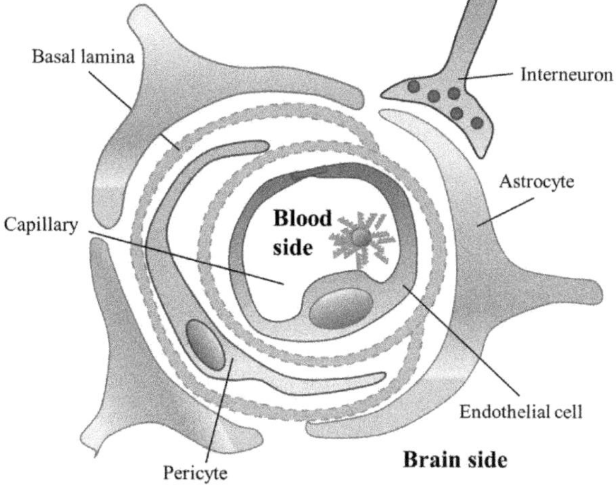

FIGURE 4.1 A brief illustration of the structure of blood–brain barrier.

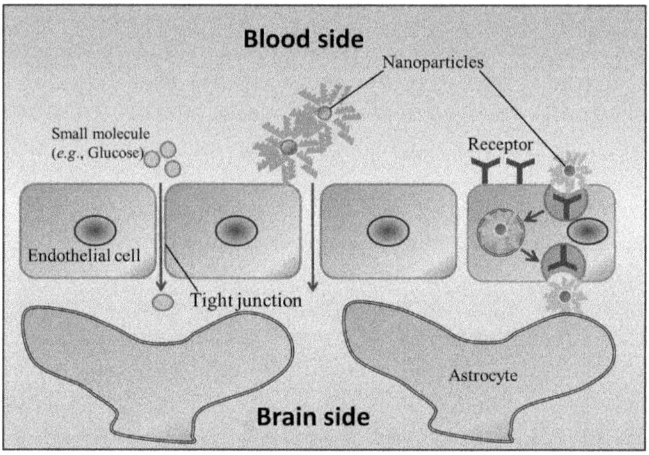

FIGURE 4.2 Paracellular pathways in the blood–brain barrier.

regulate the movement of molecules, ions, and cells between the blood and the neural tissue. The highly-regulated BBB provides stringent control of the extracellular environment in neural tissues, which is critical for proper neuronal function that requires precise ionic concentrations in its surroundings. The existence of the BBB limits the entry of many substances into the brain. In addition, the BBB protects the CNS from injury and disease by limiting the entry of toxins, pathogens, and the body's own immune system into the neural tissue. Most of the properties of the BBB in limiting the passage of compounds into the neural tissue are derived from the endothelial cells, which form highly polarized cells held together by tight junctions and greatly limit the movement of molecules and ions between cells. Consequently, the passage of molecules through the BBB generally requires the existence of active transport systems. Furthermore, astrocytes have long been shown to play a major role in regulating junction and transport properties of the BBB in capillary endothelial cells. Therefore, neurotoxicants may disrupt the BBB either by damaging the endothelial cells or by interfering in the interdependence of astrocyte and endothelial cells. In addition, increasing evidence points to a similarly important role for pericytes in barrier development and maintenance.

Systemic NPs can overcome this physical barrier (Fig. 4.2) and enter into the brain, or enter into the brain via nerve endings of the olfactory bulb (Kreuter, 2001). Generally, the specific mechanisms for most NPs targeting the brain have yet to be elucidated. It is proposed that NPs could reach the brain through their uptake by sensory nerve endings embedded in airway epithelia, followed by axonal translocation to CNS structures. Thus, NPs can be taken up by the nerve endings of the olfactory bulb and translocated to the CNS. This pathway has been studied primarily using carbon, Au, or MnO_2 NPs in experimental inhalation models (Oberdörster et al., 2005). Another portal of entry for NPs into

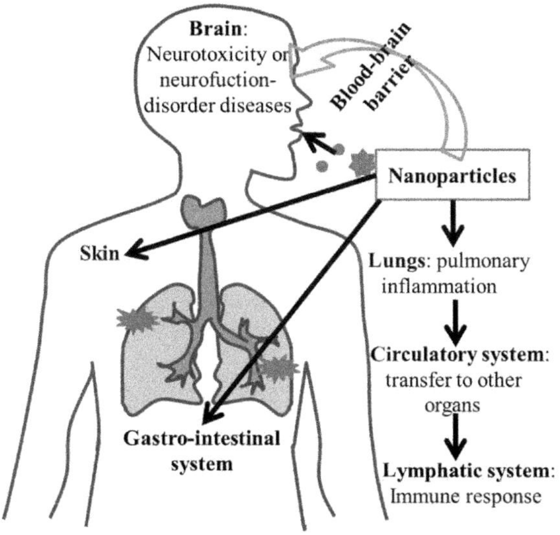

FIGURE 4.3 Pathway of nanoparticles into body following exposure.

brain is from systemic circulation. This has been extensively studied in drug delivery, as an approach to deliver drugs to the brain (Kreuter, 2004). Nanosized particles can enter and penetrate some organs such as the lungs, intestine, and skin (Fig. 4.3). Some can penetrate into the deepest layers of the skin (dermis). Their penetration depends on their size and NP surface features. In addition, the digestive (gastrointestinal) tract has a close relationship with the environment. Owing to the high loading of NPs, macromolecule absorption will increase, so that they can cross through the digestive tract and enter into the circulatory system. The effects on inflammatory and immunological systems may include OS or preinflammatory cytotoxin activity in the lungs, liver, heart, and brain. The physicochemical properties of NPs at different surfactant concentrations, stabilizers, and amyloid-affinity agents could influence the transport mechanism. In the brain, NPs may induce inflammation, apoptosis, and OS by releasing various mediators from microglia and astrocyte. Depending on the production of toxic (eg, NO, excitatory neurotransmitters) or antitoxic mediators (eg, anti-inflammatory cytokines, neurotrophins), it may lead to neurodegeneration or neuroregeneration.

4. TOXICITY OF NANOPARTICLES

4.1 Toxicity of Carbon-Based Nanoparticles

Carbon-based nanoparticles (CNPs) are man-made materials mainly consisting of carbon with a tiny size scale of ≤100 nm. Among CNPs, fullerene and carbon nanotubes (CNTs) are the two most widely used and well-developed types of CNPs in terms of research investigations and current applications in consumer products. Graphene, as a novel material, has only emerged in recent times. Although they consist of only one element, the carbon NP types are complex, and they vary in size, shape, surface charge, production method, and physicochemical properties. These factors may affect their potential toxicity both in vitro and in vivo. Our investigations show that C_{60} and CNTs can exert toxicity in MCF-7 cells or zebrafish at environmentally relevant low doses (Li et al., 2013, 2015). It is also of significant interest that such NPs can alter the methylome of target cells (Li et al., 2016); this underlying and apparently "non-toxic" phenomenon could conceivably alter the developmental plasticity of tissues such as the brain. We speculate on whether this could give rise to idiopathic end points such as autism.

4.1.1 Fullerenes

Fullerene consists of carbon molecules arranged into a spherical shape resembling a geodesic dome (McHedlov-Petrossyan, 2013). It is commonly identified as molecules of C_N, N being the number of carbon atoms in the fullerene, eg, C_{60}, C_{70}, and C_{84}. By far, C_{60} fullerenes are the most studied and widely used, and especially they are employed to serve as platforms for producing advanced materials. Fullerenes exhibit enormous electron affinity (eg, the high electron-acceptor ability and polarizability), and they are used in versatile applications (eg, electronics, optics, fuel cells) (Jensen et al., 1996). A number of cosmetic products, such as face creams that contain C_{60} NPs, have been introduced into the marketplace.

After 6-h exposure to C_{60}, toxicity tests on alveolar macrophages suggest no significant adverse effects. However, C_{60} was found to aggregate along the plasma membrane. Furthermore, it has been shown that C_{60} is likely to pass into the cells and accumulate in lysosomes and cytoplasm, and even interact with the nuclear membrane and penetrate the nucleus. C_{60} may exert genotoxicity in bacteria and give rise to DNA damage (Matsuda et al., 2011). However, it has also been reported that C_{60} did not show genotoxicity either in vitro or in vivo (Shinohara et al., 2009). In addition, it has been shown that in aquatic environments, C_{60} could induce OS in the fish brain and result in lipid peroxidation (Oberdörster, 2004).

4.1.2 Carbon Nanotubes

CNTs are the most promising NPs among various engineered types. CNTs are cylindrical molecules composed solely of carbon atoms, which is distinct from carbon fibers. There are two main forms of manufactured CNTs, the single-walled CNTs (SWCNTs) and multiwalled CNTs (MWCNTs). In term of structure, the SWCNTs are one-layered graphitic cylinders with diameters of a few nanometers, whereas MWCNTs have two or more concentric cylinders with outer diameters commonly between 30 and 50 nm. They consist of extensive sp^2 carbon atoms arranged in fused benzene rings. CNTs exhibit several unique physicochemical properties and afford a wide range of potential functionality. Their extraordinary structures give them exceptional material properties, which contribute to their applications in composite materials, sensors, hydrogen-storage fuel cells, and various environmental utilizations (Mauter and Elimelech, 2008).

In contrast to contradicting toxicity observations for C_{60}, SWCNTs have consistently been shown to generate toxic effects both in vitro and in vivo. In their toxicity assessment, SWCNTs rank as the most toxic among the three CNP types (Jia et al., 2005). When cells were exposed to SWCNTs, toxic effects were found both in a dose-dependent and time-dependent manner (Nel et al., 2009). It is also determined that SWCNTs are able to induce OS in biological systems (Ren and Zhong, 2010). In aquatic environments, it is reported that SWCNTs are a respiratory toxicant in fish, involving gill irritation and brain injury by SWCNTs (Smith et al., 2007). Studies on MWCNTs have yielded results similar to those of SWCNTs. Studies indicate that CNTs have the ability to induce pulmonary toxicity (Kayat et al., 2011), and tests on mice show that they can induce inflammation (Crouzier et al., 2010). However, for MWCNTs, those that are larger in size (especially in length) seem to be more toxic (Lowry et al., 2012).

4.1.3 Graphene

Graphene is defined as a two-dimensional crystal composed of monolayers of carbon atoms arranged in a honeycombed network with six-membered rings (Geim and Novoselov, 2007). It exhibits exceptional properties, including doubled external surface area, easy modification, planar support for biomaterials, ballistic transporting capacity, chemical inertness, mechanical strength, high thermal conductivity, and optical transmittance, compared with fullerenes and CNTs (Hu and Zhou, 2013). Intrinsic graphene and graphene oxide (GO) are the two main materials of interest to scientists. Intrinsic graphene is characterized as a zero-gap semiconductor with super hydrophobicity, whereas GO with O_2-containing groups exhibit a great transporting gap and well-polydispersed properties at room temperature (Stoller et al., 2008). The availability of several types of O_2-containing functional groups on the basal plane and the sheet edge allows GO to interact with a wide range of organic and inorganic materials in noncovalent, covalent, and/or ionic manners, so that functional hybrids and composites with unusual properties can be readily synthesized (Hu and Zhou, 2013). These excellent physicochemical properties of graphene have resulted in a wide range of applications in electrochemical devices, energy storage, catalysis, cell imaging, photochemotherapy, drug delivery, biosensors, contamination purification, and extraction devices for chemical, biological, or environmental samples (Allen et al., 2010; Chavez-Valdez et al., 2013).

The majority of toxicity research on graphene has focused on interactions with nucleic acids, proteins/enzymes, or cells in vitro. Previous investigations reported that GO does not penetrate A549 cells and has no obvious cytotoxicity, but GO can cause a dose-dependent OS in cells and induce a slight loss of cell viability at high concentrations (Chang et al., 2011). These effects are dose- and size-related. It was also found that graphene could induce cytotoxic effects and mitochondrial injury in human neuronal cells after 4 and 24h at a dose of 10μg/mL (Zhang et al., 2010). In addition, graphene and GO nanosheets exerted toxicity in both gram-positive and gram-negative bacteria (Akhavan and Ghaderi, 2010). In vivo studies show that GO or GO with surface modifications can accumulate in the lung (Wang et al., 2011; Zhang et al., 2011).

4.2 Metal-Based Nanoparticles

4.2.1 Silver Nanoparticles

One of the interesting properties of silver nanoparticles (Ag NPs) is its antibacterial and antifungal activity. As a result, the use of nanoparticulate silver is one of the fastest growing areas of commercial NP applications (Sondi and Salopek-Sondi, 2004). With their excellent antibacterial properties, Ag NPs have been widely applied in food services, building materials, textiles, medical instruments, personal care products, and washing machines, as well as room sprays, deodorants, wall paints, and laundry detergents; they are also used for indoor air purification and water detoxification (Jain and Pradeep, 2005; Marambio-Jones and Hoek, 2010; Tolaymat et al., 2010). Despite the numerous papers reporting the beneficial antimicrobial effects of Ag nanomaterials, a relatively modest number of studies have attempted to elucidate the mechanisms by which they exert this antimicrobial activity. As a result, the mechanisms are not widely understood or agreed upon. For antimicrobial activity, proposed mechanisms include the release of Ag ions, as well as generation of reactive oxygen species (ROS) (Hwang et al., 2008; Smetana et al., 2008). Given the vast amount of products with incorporating benefits of Ag nanomaterials, the more widespread will become the potential risk for human and ecosystem health. The main routes of human exposure are the respiratory system, gastrointestinal system, and skin (Chen and Schluesener, 2008).

In addition to toxicity to microorganisms (bacteria, fungi, virus, and algae), there is significant evidence highlighting the toxicity that silver NPs can exert to higher organisms. In vitro studies show that Ag NPs reduce mitochondrial function and increase membrane leakage of mouse spermatogonial stem cells and rat liver cells (Braydich-Stolle et al., 2005; Hussain et al., 2005). Another study in mammalian cells reported that Ag NPs at a dose of 50μg/mL induced apoptosis in 43.4% of fibroblast cells, whereas 100μg/mL generated necrosis in 40.2% of the cancer cells (Hsin et al., 2008). An in vivo study in rats showed that Ag NPs reduce lung function and produce inflammatory pulmonary lesions (Sung et al., 2008).

4.2.2 Gold Nanoparticles

Although colloidal gold nanoparticles (Au NPs) have been used since ancient times for their optical properties, in particular for staining glass, their uses for biological applications have primarily arisen since 2005. This coincided with the development of both biotechnology and nanotechnology, which today allows one to control synthesis and functionalization of materials on the nanometer scale (Sperling et al., 2008). Au NPs are being more and more applied to diagnostics in vivo, photothermal therapy, and targeted delivery of drugs, genetic material, and antigens; as well as being used as a therapeutic agent (eg, in the treatment of tumors) (Boisselier and Astruc, 2009; Dykman and Khlebtsov, 2014).

Although bulk gold has been deemed "safe," nanoscale particles of Au still require examination for biocompatibility and environmental impact if they are to be manufactured for large-scale usage. Cells exposed to Au NPs will incorporate these particles and they are then stored intracellularly in perinuclear compartments, vesicular structures close to the cell nucleus (Patra et al., 2007). It has been reported that the cellular response following exposure is size dependent, in that 1.4-nm particles cause predominantly rapid cell death by necrosis within 12 h, whereas closely related particles 1.2 nm in diameter induce predominantly programmed cell death, ie, apoptosis (Pan et al., 2007). As citrate/Au NPs are composed of an inert material, it was found that as a result of intracellular NP presence in human dermal fibroblasts, actin stress fibers disappeared, thereby inducing major adverse effects on cell viability, including dramatically altering extracellular matrix–forming properties such as cell spreading and adhesion, cell growth, and protein synthesis (Pernodet et al., 2006).

4.2.3 Metal Oxide Nanoparticles

Metal oxide NPs are extensively used in a considerable number of applications in food, materials, chemical, and biological sciences. They are a class of engineered nanomaterials with widespread applications; their special properties of small size and large specific surface area endow them with high chemical reactivity and intrinsic toxicity. Their health effects in living creatures, especially on the nervous system, have raised concerns. Metal oxide toxicity can also be induced by dissolved metal ions from the oxides. Investigations show that the toxicity of soluble NPs was from the soluble metal ions released from NP dissolution before or after the NPs entered the neural cells (Brunner et al., 2006). Considering their unique physicochemical properties, including small size effect, large specific surface area, and high biological surface reactivity, NPs might induce the neurotoxicological behavior and effects in organisms. Metal oxide NPs are capable of translocating along the olfactory nerve pathway to the brain after intranasal instillation and accumulating in the olfactory bulb, cortex, and cerebellum. Moreover, NP deposition in the brain can stimulate OS, inflammatory responses, and pathological changes. These observations have provided evidence that metal oxide NPs can reach the brain and cause tissue damage.

Among several metal oxide NPs, titanium dioxide (TiO_2) is the most common compound of titanium that has found a variety of uses in everyday life. TiO_2 is a white, odorless, water-insoluble material that was believed to have low toxicity. It is a relatively stable, nonflammable material that is found naturally in the form of various ores such as rutile, anatase, and brookite. TiO_2 possesses certain physiochemical properties that make it useful for multiple applications. With properties of corrosion resistance, biocompatibility, mechanical strength, whitening property, and opacity, along with photocatalytic, optical, and electrical activity, TiO_2 is an attractive material for large-scale applications. Nanoparticulate TiO_2 has major uses in developing various products such as cosmetics, foodstuffs, toothpaste, sun blocks, printing ink, car materials, rubber, cleaning products, materials for industrial photocatalytic applications including solar cells, and catalysts for remediation of organic matter in wastewater (Linsebigler et al., 1995). Recent toxicological studies indicate harmful effects of TiO_2 NPs in biological systems, which is of major concern (Nel et al., 2006).

Like TiO_2, another metal-based NP is zinc oxide (ZnO), which has broad uses and applications. ZnO is also white, thermally stable, and a naturally occurring material. It can be used to develop sunscreens, biosensors, food additives, cement, rubber, ceramics, pigments, plastic, catalysts, and electronic materials (Becheri et al., 2008). ZnO toxicity has been demonstrated both in vitro and in vivo in various mammalian cells. Studies in mice have reported that oral exposure of ZnO NPs causes a dose-related damaging effect in the stomach, liver, kidney, pancreas, and spleen (Wang et al., 2007). It is considered that dissolved Zn^{2+} from the NPs is responsible for this toxicity.

Manganese is a major component in the production of different types of steel and cast iron, and has found several other uses in everyday life. Oxides of manganese, such as manganese oxide (MnO), are used in colored glass, ceramics, paints, textile printing, fertilizers, and food supplements and additives. As manganese is known to cause neurotoxicity, toxicity studies associated with manganese-containing nanomaterials provide a useful test case in the evaluation of nanomaterial toxicity. Investigations show that after nasal exposure to manganese oxide NPs, the concentration of manganese in the olfactory bulb, striatum, frontal, and other brain regions is increased. It was concluded that the olfactory neuronal pathway is efficient for translocating inhaled Mn oxide as solid ultrafine particles (UFPs) to the central nervous system and that this can result in inflammatory changes (Elder et al., 2006).

4.3 Quantum Dots

Quantum dots (QDs) are nanometer-scale crystalline semiconductors consisting of chemical elements from groups III to V or groups II to VI of the periodic table with sizes ranging from 1 to 10 nm (Alivisatos, 1996). A typical QD is composed of one kind of semiconductor material as the packaged core and another as the shell, where the core determines the optical property of the QD and the shell strengthens photostability. Colloidal semiconductor QDs are precisely size and shape controlled by the duration, temperature, and ligand molecules used in the synthesis; this process yields QDs possessing a property of composition- and size-dependent absorption and emission. Absorption of a photon with energy above the semiconductor band gap energy results in the creation of an electron–hole pair (or exciton), whereas the absorption has an increased probability at higher energies (ie, shorter wavelengths) and results in a broadband absorption spectrum by contrast to standard fluorophores. Thus they have been extensively applied in biology and medicine as advanced fluorescent probes because of their superior optical properties, such as broad absorption and narrow emission spectra, strong fluorescent intensity, and resistance to photobleaching.

With the rapid development of QDs, there have been huge technological innovations in biotechnology whereby QDs have become useful tools for fundamental biological research as well as the diagnostic or therapeutic strategies. As the biological organism is a critical system, the potential toxic effects of QDs need to be well understood before they can be approved for use in living organisms. It was found that CdSe-core QDs are indeed acutely toxic under certain conditions (Derfus et al., 2004). Specifically, it was reported that the cytotoxicity of QDs is modulated by processing parameters during synthesis, exposure to ultraviolet light, and surface coatings. In addition, it has been suggested that the cytotoxicity correlates with the liberation of free Cd^{2+} ions due to deterioration of the CdSe lattice. A similar process occurs in the cadmium-based QDs (Chen et al., 2012). Some other studies have found that QDs containing different chemical components could elicit severe adverse effects, such as OS injury and immune system damage in the lung, liver, and kidney (Sadaf et al., 2012; Tang et al., 2013; Wu and Tang, 2014).

4.4 Particulate Matter

Other than the manufactured or engineered NPs introduced earlier, combustion-derived NPs also need to be considered (eg, particulate matter (PM), diesel exhaust particles, welding fumes). Although these two types of NPs might be of the same size, there are obvious differences between them. Combustion-derived NPs are polydispersed, chemically complex in nature, and in soluble or poorly soluble forms, their toxicity may result from physicochemical characteristics on their surface. In contrast to the combustion-derived NPs, manufactured or engineered NPs are monodispersed, with a precise chemical nature, solid mainly spherical and may be uniform. Increased vehicular traffic and other combustion processes have resulted in a significant increase in ambient PM over the past two decades. Regarding combustion-derived NPs, several epidemiological studies demonstrate that exposure to elevated levels of PM in urban air is associated with adverse health effects in humans, including an increase in morbidity and mortality related to pulmonary and cardiovascular diseases in susceptible populations (Win-Shwe and Fujimaki, 2011).

Both in vitro and in vivo studies of the health effects of PM have identified the generation of OS as one of the major mechanisms by which air pollution particles exert adverse biological effects. Among particles of different sizes, it has also been established that UFPs, which have an aerodynamic size of ≤100 nm, are potentially the most dangerous because of their small size, large surface area, deep penetration and ability to be retained in the lung, and high content of redox-cycling organic chemicals. PM that includes NPs can trigger free radical activity on the surface of the particles. Free radicals or OS may damage lipids, nucleic acids, and proteins at the site of particle deposition and at translocation sites, and the brain is particularly vulnerable to OS because of its high energy demand, low level of antioxidants, and high cellular content of lipids and proteins (Dellinger et al., 2001). PM is capable of generating acute airway inflammation that can lead to asthma flares after a sudden surge in ambient $PM_{2.5}$ [mass per cubic meter of air of particles with a size (diameter) generally <2.5 μm] levels. Acute exacerbations are characterized by increased symptom scores as well as the requirement for more frequent medication and hospitalization (Dockery et al., 1993).

5. MECHANISMS OF NANOTOXICITY

The mechanisms underlying the toxicity of NPs have been well studied, and the generation of ROS (Fig. 4.4) is considered to be a predominant mechanism in nanotoxicity, which may lead to subsequent formation of OS in cells or tissues (Gonzalez et al., 2008). During intracellular metabolism, a small amount of O_2 escapes during the process of ATP synthesis, where a sequence of coupled proton and electron transfer reactions have not been reduced completely; this

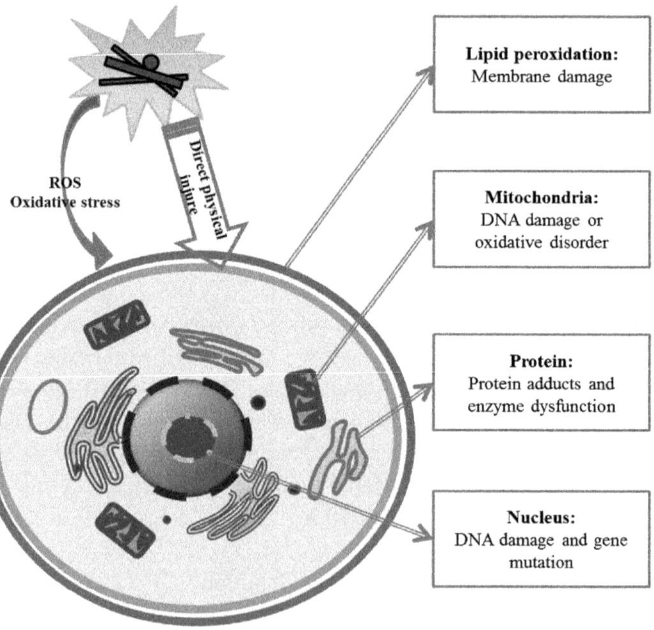

FIGURE 4.4 A brief illustration of the mechanisms underlying nanotoxicity. *ROS*, reactive oxygen species.

further causes the production of superoxide anion radicals (O_2^-) or other oxygen-containing radicals. Consequently, ROS, including O_2^-, hydroxyl radicals (•OH), singlet oxygen ($^1O_2$), and hydrogen peroxide (H_2O_2), are then produced as by-products of intracellular oxidative metabolism, which usually occurs in the mitochondria (Halliwell, 1991; Turrens, 2003). In some cases, overproduction of ROS occurs in cells and induces OS, resulting in cells failing to maintain normal physiological redox-regulated functions (Simon et al., 2000). Dysfunction in cells will then cause oxidative modification of proteins to generate protein radicals, lipid peroxidation, DNA-strand breaks, and even modification to nucleic acids (Bagchi et al., 1995; Cabiscol et al., 2010; Griveau et al., 1995). Furthermore, it will impact on the modulation of gene expression through activation of redox-sensitive transcription factors, and modulation of inflammatory responses through signal transduction, finally leading to cell death and genotoxic effects (Shi et al., 2004; Xia et al., 2006). The level of ROS generated by NPs depends on the physicochemical properties of the NPs, including their small size, large surface area, and high reactivity. In addition, the direct interaction between NPs and cellular constituents at the nano–bio interface is regarded as a physical mechanism (non-OS effects) in the nanotoxicity process resulting from the size and surface properties of the NPs (Elsaesser and Howard, 2012). It could lead to membrane disruption and disturb the transport processes in membranes, and even conduct the NP itself into the cell to interfere directly with cellular function (Shvedova et al., 2012). These effects would finally lead to a number of follow-up biological responses. These cellular responses can occur pre- or postinternalization of particles.

ROS generation is associated with several neurodegenerative disorders such as Parkinson's disease, Alzheimer's disease, and Huntington's disease (Matés et al., 1999). The brain is especially vulnerable to OS damage because of its high content of easily peroxidizable unsaturated fatty acids, high O_2 consumption rate, and relative paucity of antioxidant enzymes compared with other organs (Skaper et al., 1999). Therefore, brain could be continuously exposed to ROS during exposure to NPs. In addition, CNS OS is largely mediated by the microglia, a macrophage-like, phagocytic cell that is normally inactive unless confronted by potential damage and exogenous stimuli (eg, xenobiotics, chemicals, particles). Generally, the neurotoxicity of PM or engineered NPs may be induced initially via the generation of ROS, which could cause OS, and consequently upregulate mitogen-activated protein (MAP) kinases to activate the MAP signaling pathway. Then, inflammation-related cytokines are highly expressed in the brain, causing brain injury.

6. RELEASE OF NANOPARTICLES INTO ENVIRONMENT

The health risk of a substance is determined by both effect and exposure. Knowing how the engineered nanomaterials behave in the environment can help us understand their potential impact on human health. Given the increasing production of nanomaterials of all types, their presence in the environment has increased; subsequent effects on ecosystem health are becoming an increasing concern. Incorporating NPs into more products means that they will ultimately further increase in the environment (Fig. 4.5). NPs enter the environment through various routes. It is determined that potential release of NPs to the environment possibly occurs during their production and the manufacturing of NP-incorporated products (Bello et al., 2009). Direct release is possible from accidental cases or spillage from equipment during manufacture. There could be indirect release via exhaust emissions and wastewater discharge. Normal mechanical use of NP consumer products can equate to a lifetime in years. During use and disposal, NP nanocomposites are exposed to a wide range of harsh environmental conditions, which is likely to cause the release of NPs from composites. For example, during use of a CNT composite-framed tennis racket, some CNTs may be released whenever the racket frame is scratched by abrasion on the court, or after disposal in a landfill or by incineration. In a landfill, the matrix may undergo hydrolytic degradative and, to a lesser extent, photodegradative, release into the environment (Petersen et al., 2011). Furthermore, fire or ultraviolet radiation exposure during long-term use can also accelerate the release of NPs from products (Wohlleben et al., 2011). However, there is a knowledge gap regarding the fate and transport of any NPs within engineered systems, including wastewater treatment facilities. NPs synthesized from silver, copper, fullerenes, and titanium, cerium, or silicon oxides have already been detected in sludge at the site of wastewater treatment plants (Benn and Westerhoff, 2008; Ganesh et al., 2010; Kiser et al., 2010). Unfortunately, removal of the NPs in wastewater treatment plants is not clearly understood. In addition, similarly to other pollutants, some of NP release occurs from non-point sources (eg,

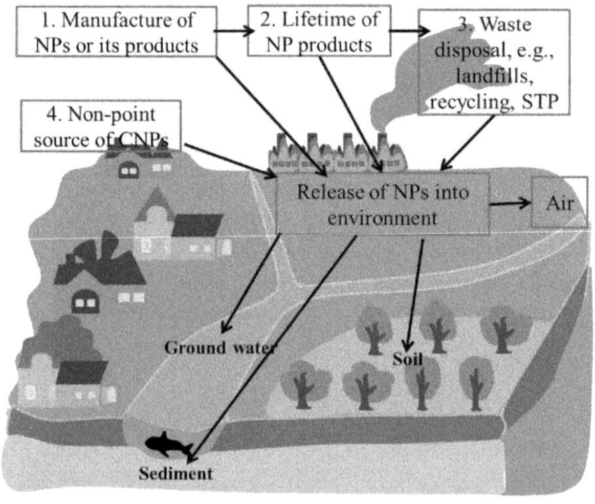

FIGURE 4.5 Potential pathway in release of nanoparticles (NPs) into environment. *CNP*, carbon-based nanoparticle; *STP*, sewage treatment plant.

using cosmetics or sunscreens containing NPs; textiles with nanomaterials), which is difficult to control.

Generally, the possible release routes include discharge during material production, incorporation of NPs into products, the life cycle of NP-containing products, as well as disposal processes at sewage treatment plants, wastewater incineration plants, landfills, and recycling (Gottschalk and Nowack, 2011) (Fig. 4.1). Thus, it is likely that humans will be exposed to NPs via inhalation, ingestion, or dermal absorption in occupational, environmental, and consumer settings (Johnston et al., 2013). All these potential release pathways pose a serious concern from material performance and environmental health.

7. FACTORS CONTRIBUTING TO NANOTOXICITY

Distinguished from common chemical toxicants, NPs require a comprehensive characterization of their physicochemical properties, including size distribution, shape, surface area, crystallinity, purity, agglomeration state, surface charge, and solubility. Among all the parameters of NPs that is relevant to their toxicity, size is the most important, which is critical for determining their interactions with living systems. Researchers have concluded that particle size can clearly have a dramatic effect on the way in which an organism responds on exposure. Previous study shows that NP-mediated cellular response is size dependent (Jiang et al., 2008). Size can govern where and how the body reacts to the particulates (exposure). Size is a key factor in whether the body can remove foreign particles or not. The lung is designed to clear PM from the upper airways via the ciliary system. Phagocytosis by macrophages and giant cells is a common mechanism by which the body's innate immune system attempts to clear particles in the order of a few microns or less. Particle size can potentially influence direct mechanisms and extent of toxicity (cytotoxicity, necrosis, and mutagenicity). In addition, size directly affects the surface-to-mass ratio (specific surface area) and can have dramatic effects on surface reactivity and solubility of particulate systems. The NP surface area is another important factor in toxicity, as the interaction of NPs with biological systems takes place at their surface. A reduction in the size of nanosized particles results in an increase in particle surface area. Therefore more chemical molecules may attach to this surface, which would enhance its reactivity and result in an increase in its toxic effects. Many studies on the absorption of NPs from the mucus have examined these effects. After absorption, NPs reach the blood stream and then spread through the tissue.

Chemical components of the particle surface have important effects on NPs as they can react with metals. Iron can be affected by NPs, which increases the induction of ROS in a free cell system. Surface modification of NPs can reduce toxicity. It has been shown that the toxicity of super paramagnetic iron oxide NPs can be reduced by coating them with pullulan (a polysaccharide polymer consisting of maltotriose units). Toxicity and other responses depend on the prescribed dosage and substances used. High exposures of NPs, small or big, can be harmful to health.

8. INTERACTION OF NANOPARTICLES WITH OTHER CHEMICALS IN THE ENVIRONMENT

Owing to their high adsorption capacity, NPs can be used for biomedical applications (Bakry et al., 2007; Bianco et al., 2005) or environmental remediation (Mauter and Elimelech, 2008). However, when NPs are released to the environment, this inherent property can represent a double-edged sword. Especially in aquatic environments, NPs will interact with other components, including not only dissolved organic matter, but also hydrophobic organic contaminants and inorganic ions, including heavy metals (Aiken et al., 2011; Nowack and Bucheli, 2007; Zhang et al., 2009). The interaction of NPs with toxic compounds can amplify as well as alleviate the toxicity of both the NPs and/or such compounds, which is still largely obscure. Investigations into such binary effects of NPs with other compounds arose from the "Trojan horse" effect postulated by Limbach et al. (2007), who reported that NPs may facilitate the cellular intake of absorbed toxic molecules. However, when this concept is extended to the environmental context, the biological effects of the mixtures are difficult to predict. It is more challenging to deal with mixtures of NPs with chemical compounds rather than mixtures of common chemical pollutants, as the interaction of contaminants with NPs depends on NP characteristics, including their size, composition, morphology, structure, and status of agglomeration/aggregation (Christian et al., 2008).

Engineered NPs have been widely applied in many areas, and they exhibit a strong affinity for organic contaminants because of their excellent absorption capacity. Adsorption of hydrophobic organic compounds (HOCs) by CNPs may impact on their toxicity and affect the fate, transformation, and transport of HOCs in the environment (Yang et al., 2006). Investigations into the absorption of chemical compounds show that it is a complicated process (Pan and Xing, 2008), which would in turn be more complex in real environmental conditions. Although there are a lot of studies being carried out on the toxicity of CNPs, few studies have investigated their interactive effects with other pollutants or biological organisms. Such reports on interactive toxicity are often quite controversial. A few investigations determined that CNPs could enhance the toxicity of pollutants with coexposure. Investigation into the exposure of phenanthrene in the presence of SWCNTs showed that the SWCNTs act as a contaminant carrier, which could enhance the bioavailability of phenanthrene (Su et al., 2013). Other studies also highlighted the same fact with C_{60} (Baun et al., 2008; Lehto et al., 2014). It was determined in vitro that coexposure of arsenium and C_{60} could facilitate the uptake of arsenium in zebrafish hepatocytes (Azevedo Costa et al., 2012). Another study showed that C_{60} could enhance the toxicity of benzo[*a*]pyrene (B[*a*]P) (Ferreira et al., 2014). However, this is not always the case. It was found that when 17α-ethinylestradiol (EE_2) is delivered to cultured cells together with carboxyl-functionalized (cf) SWCNTs, the bioactivity of adsorbed EE_2 on cf-SWCNTs was significantly inhibited (Song et al., 2014). C_{60} was also found to reduce the bioavailability of EE_2 in zebrafish during coexposure (Park et al., 2010, 2011). In addition, MWCNTs may reduce the bioavailability and the toxicity of the biocide triclocarban (Simon et al., 2014). TiO_2 NPs can act as carrier and facilitate bioaccumulation of phenanthrene in marine bivalve, ark shell (Tian et al., 2014). In contrast, another study showed that TiO_2 NPs served as a guardian against environmental carcinogen B[*a*]P (Dhasmana et al., 2014).

9. SAFETY CONSIDERATIONS

Although numerous in vivo and in vitro studies have provided evidence of the toxic effects of various types of NPs, our understanding of the potential health and safety issues regarding NPs still lags behind the rapid commercialization of nanomaterials. The benefits of nanomaterials must be weighed against their potential toxic effects. One major problem is the lacking information on the possible adverse health effects caused by exposure to different nanomaterials. As a consequence, a regulatory risk assessment framework is a long way from being implemented. Understanding the toxic effects of manufactured or engineered NPs would help in the development of safety guidelines to promote nanotechnology for applications without hazard. In addition to physiological safety considerations associated with the NPs themselves, practical safety issues will also have to be considered to properly protect health. Detailed handling and administration protocols will have to be developed for the safe use of NPs because there is growing evidence that inhaled particles can result in multiple adverse effects, including nonspecific CNS effects, translocation from the lungs to the circulation, and localized or systemic inflammatory responses. Recent observations of real-world levels of NPs being capable of altering the methylome of target cells raises the risk that apparently "nontoxic" effects could influence development plasticity (Li et al., 2016). Taking safety considerations into account, it is necessary to understand the fundamental molecular and cellular mechanisms involved in NPs crossing the BBB, and design well-controlled and appropriate preclinical and clinical translational studies. In addition, the use of nanotechnologies for safely, efficiently, and specifically delivering drugs and other molecules across the BBB may prove to be one of the highest-impact contributions to clinical neuroscience.

10. REDUCING EXPOSURE AND NEUROTOXICITY

There is the potential that the possible range of adverse effects of NPs will be reminiscent of the asbestos story. Very little is known about NP toxicity and assays to ascertain their effects remain limited or developmental (Li et al., 2016). Through a number of routes (inhalation, dermal absorption), NPs can pass into the systemic circulation. In circulation, they can bypass the BBB (Kostarelos et al., 2007). How NPs interact with biological systems, especially at the "nanophysical–biology interface," remains poorly understood (Kuzma,

2007). Because of safety uncertainties, the Control of Substances Hazardous to Health Regulations (COSHH) 2002 requires adoption of the "precautionary principles" where the toxicity is unknown (Calliess and Stockhaus, 2012); this states that an assessment should be carried out for all work involving nanomaterials, with risk management procedures applied (Health and Safety Executive, 2005). With emerging evidence of toxicity, regulators are further investigating risks to human health and the adverse effects likely to arise due to increasing exposure with further exploitation of nanotechnology. Thus there is an urgent need to define a hazard profile and risk management strategies need to incorporate a fine balance between overregulation and maintaining a competitive edge (Doak et al., 2012). Slowing this process up is insufficient toxicological hazard information for nanomaterials.

There are also no known techniques to create appropriate dose metrics for NPs as there is uncertainty as to what these would be (in fact, there remains debate as to the true definition for an NP). Hence, there remain no criteria to develop health surveillance strategies. Based on laboratory evidence, there is increasing concern about the possible hazards posed by exposure to NPs, but it is too soon for long-term effects to manifest themselves. Therefore a novel approach to detect early NP-induced effects on a mechanism-based model is overdue. As such, creating a framework to minimize neurotoxicity as a consequence of NP exposure remains a long way off. There is an urgent need for further research into NP-mediated mechanisms of action to develop and validate early biomarkers of NP-induced effects. In a real-world scenario with low environmental levels, it would be likely to take several years to identify NP-induced neurotoxicity effects (especially in populations) and then to implement strategies to reduce or remediate causative candidate exposures.

REFERENCES

Aiken, G.R., Hsu-Kim, H., Ryan, J.N., 2011. Influence of dissolved organic matter on the environmental fate of metals, nanoparticles, and colloids. Environ. Sci. Technol. 45, 3196–3201.

Akhavan, O., Ghaderi, E., 2010. Toxicity of graphene and graphene oxide nanowalls against bacteria. ACS Nano 4, 5731–5736.

Alivisatos, A.P., 1996. Semiconductor clusters, nanocrystals, and quantum dots. Science 271, 933–937.

Allen, M.J., Tung, V.C., Kaner, R.B., 2010. Honeycomb carbon: a review of graphene. Chem. Rev. 110, 132–145.

Azevedo Costa, C.L., Chaves, I.S., Ventura-Lima, J., Ferreira, J.L.R., Ferraz, L., de Carvalho, L.M., Monserrat, J.M., 2012. In vitro evaluation of co-exposure of arsenium and an organic nanomaterial (fullerene, C60) in zebrafish hepatocytes. Comp. Biochem. Physiol. C 155, 206–212.

Bagchi, D., Bagchi, M., Hassoun, E.A., Stohs, S.J., 1995. In vitro and in vivo generation of reactive oxygen species, DNA damage and lactate dehydrogenase leakage by selected pesticides. Toxicology 104, 129–140.

Bakry, R., Vallant, R.M., Najam-ul-Haq, M., Rainer, M., Szabo, Z., Huck, C.W., Bonn, G.K., 2007. Medicinal applications of fullerenes. Int. J. Nanomed 2, 639.

Barbosa, D.J., Capela, J.P., de Lourdes Bastos, M., Carvalho, F., 2015. In vitro models for neurotoxicology research. Toxicol. Res. 4, 801–842.

Baun, A., Sørensen, S.N., Rasmussen, R.F., Hartmann, N.B., Koch, C.B., 2008. Toxicity and bioaccumulation of xenobiotic organic compounds in the presence of aqueous suspensions of aggregates of nano-C60. Aquat. Toxicol. 86, 379–387.

Becheri, A., Dürr, M., Nostro, P.L., Baglioni, P., 2008. Synthesis and characterization of zinc oxide nanoparticles: application to textiles as UV-absorbers. J. Nanopart. Res. 10, 679–689.

Bello, D., Wardle, B., Yamamoto, N., Guzman deVilloria, R., Garcia, E., Hart, A., Ahn, K., Ellenbecker, M., Hallock, M., 2009. Exposure to nanoscale particles and fibers during machining of hybrid advanced composites containing carbon nanotubes. J. Nanopart. Res. 11, 231–249.

Benn, T.M., Westerhoff, P., 2008. Nanoparticle silver released into water from commercially available sock fabrics. Environ. Sci. Technol. 42, 4133–4139.

Bianco, A., Kostarelos, K., Prato, M., 2005. Applications of carbon nanotubes in drug delivery. Curr. Opin. Chem. Biol. 9, 674–679.

Boisselier, E., Astruc, D., 2009. Gold nanoparticles in nanomedicine: preparations, imaging, diagnostics, therapies and toxicity. Chem. Soc. Rev. 38, 1759–1782.

Braydich-Stolle, L., Hussain, S., Schlager, J.J., Hofmann, M.-C., 2005. In vitro cytotoxicity of nanoparticles in mammalian germline stem cells. Toxicol. Sci. 88, 412–419.

Brunner, T.J., Wick, P., Manser, P., Spohn, P., Grass, R.N., Limbach, L.K., Bruinink, A., Stark, W.J., 2006. In vitro cytotoxicity of oxide nanoparticles: comparison to asbestos, silica, and the effect of particle solubility. Environ. Sci. Technol. 40, 4374–4381.

Cabiscol, E., Tamarit, J., Ros, J., 2010. Oxidative stress in bacteria and protein damage by reactive oxygen species. Int. Microbiol. 3, 3–8.

Calliess, C., Stockhaus, H., 2012. Precautionary principle and nanomaterials: REACH revisited. J. Eur. Environ. Plan. Law 9, 113–135.

Chang, Y., Yang, S.-T., Liu, J.-H., Dong, E., Wang, Y., Cao, A., Liu, Y., Wang, H., 2011. In vitro toxicity evaluation of graphene oxide on A549 cells. Toxicol. Lett. 200, 201–210.

Chavez-Valdez, A., Shaffer, M.S.P., Boccaccini, A.R., 2013. Applications of graphene electrophoretic deposition. A review. J. Phys. Chem. B 117, 1502–1515.

Chen, X., Schluesener, H.J., 2008. Nanosilver: a nanoproduct in medical application. Toxicol. Lett. 176, 1–12.

Chen, N., He, Y., Su, Y., Li, X., Huang, Q., Wang, H., Zhang, X., Tai, R., Fan, C., 2012. The cytotoxicity of cadmium-based quantum dots. Biomaterials 33, 1238–1244.

Christian, P., Von der Kammer, F., Baalousha, M., Hofmann, T., 2008. Nanoparticles: structure, properties, preparation and behaviour in environmental media. Ecotoxicology 17, 326–343.

Costa, L.G., Giordano, G., Guizzetti, M., Vitalone, A., 2008. Neurotoxicity of pesticides: a brief review. Front. Biosci. 1240–1249.

Crouzier, D., Follot, S., Gentilhomme, E., Flahaut, E., Arnaud, R., Dabouis, V., Castellarin, C., Debouzy, J.C., 2010. Carbon nanotubes induce inflammation but decrease the production of reactive oxygen species in lung. Toxicology 272, 39–45.

Dellinger, B., Pryor, W.A., Cueto, R., Squadrito, G.L., Hegde, V., Deutsch, W.A., 2001. Role of free radicals in the toxicity of airborne fine particulate matter. Chem. Res. Toxicol. 14, 1371–1377.

Derfus, A.M., Chan, W.C.W., Bhatia, S.N., 2004. Probing the cytotoxicity of semiconductor quantum dots. Nano Lett. 4, 11–18.

Dhasmana, A., Sajid Jamal, Q.M., Mir, S.S., Bhatt, M.L.B., Rahman, Q., Gupta, R., Siddiqui, M.H., Lohani, M., 2014. Titanium dioxide nanoparticles as guardian against environmental carcinogen benzo[alpha]pyrene. PLoS One 9, e107068.

Doak, S., Manshian, B., Jenkins, G., Singh, N., 2012. In vitro genotoxicity testing strategy for nanomaterials and the adaptation of current OECD guidelines. Mutat. Res. 745, 104–111.

Dockery, D.W., Pope, C.A., Xu, X., Spengler, J.D., Ware, J.H., Fay, M.E., Ferris, B.G., Speizer, F.E., 1993. An association between air pollution and mortality in six U.S. cities. N. Engl. J. Med. 329, 1753–1759.

Dykman, L.A., Khlebtsov, N.G., 2014. Uptake of engineered gold nanoparticles into mammalian cells. Chem. Rev. 114, 1258–1288.

Elder, A., Gelein, R., Silva, V., Feikert, T., Opanashuk, L., Carter, J., Potter, R., Maynard, A., Ito, Y., Finkelstein, J., Oberdörster, G., 2006. Translocation of inhaled ultrafine manganese oxide particles to the central nervous system. Environ. Health Perspect. 114, 1172–1178.

Elsaesser, A., Howard, C.V., 2012. Toxicology of nanoparticles. Adv. Drug Deliv. Rev. 64, 129–137.

Ferreira, J.L.R., Lonné, M.N., França, T.A., Maximilla, N.R., Lugokenski, T.H., Costa, P.G., Fillmann, G., Antunes Soares, F.A., de la Torre, F.R., Monserrat, J.M., 2014. Co-exposure of the organic nanomaterial fullerene C60 with benzo[a]pyrene in *Danio rerio* (zebrafish) hepatocytes: evidence of toxicological interactions. Aquat. Toxicol. 147, 76–83.

Ganesh, R., Smeraldi, J., Hosseini, T., Khatib, L., Olson, B.H., Rosso, D., 2010. Evaluation of nanocopper removal and toxicity in municipal wastewaters. Environ. Sci. Technol. 44, 7808–7813.

Geim, A.K., Novoselov, K.S., 2007. The rise of graphene. Nat. Mater. 6, 183–191.

Gonzalez, L., Lison, D., Kirsch-Volders, M., 2008. Genotoxicity of engineered nanomaterials: a critical review. Nanotoxicology 2, 252–273.

Gooding, J.J., 2005. Nanostructuring electrodes with carbon nanotubes: a review on electrochemistry and applications for sensing. Electrochim. Acta 50, 3049–3060.

Gottschalk, F., Nowack, B., 2011. The release of engineered nanomaterials to the environment. J. Environ. Monit. 13, 1145–1155.

Griveau, J.F., Dumont, E., Renard, P., Callegari, J.P., Le Lannou, D., 1995. Reactive oxygen species, lipid peroxidation and enzymatic defence systems in human spermatozoa. J. Reprod. Fertil. 103, 17–26.

Halliwell, B., 1991. Reactive oxygen species in living systems: source, biochemistry, and role in human disease. Am. J. Med. 91, S14–S22.

Health and Safety Executive, 2005. Control of Substances Hazardous to Health: The Control of Substances Hazardous to Health Regulations 2002 (As Amended): Approved Code of Practice and Guidance. HSE Books, Sudbury.

Hsin, Y.-H., Chen, C.-F., Huang, S., Shih, T.-S., Lai, P.-S., Chueh, P.J., 2008. The apoptotic effect of nanosilver is mediated by a ROS- and JNK-dependent mechanism involving the mitochondrial pathway in NIH3T3 cells. Toxicol. Lett. 179, 130–139.

Hu, X., Zhou, Q., 2013. Health and ecosystem risks of graphene. Chem. Rev. 113, 3815–3835.

Hussain, S.M., Hess, K.L., Gearhart, J.M., Geiss, K.T., Schlager, J.J., 2005. In vitro toxicity of nanoparticles in BRL 3A rat liver cells. Toxicol. In Vitro 19, 975–983.

Hwang, E.T., Lee, J.H., Chae, Y.J., Kim, Y.S., Kim, B.C., Sang, B.-I., Gu, M.B., 2008. Analysis of the toxic mode of action of silver nanoparticles using stress-specific bioluminescent bacteria. Small 4, 746–750.

Jain, P., Pradeep, T., 2005. Potential of silver nanoparticle-coated polyurethane foam as an antibacterial water filter. Biotechnol. Bioeng. 90, 59–63.

Jensen, A.W., Wilson, S.R., Schuster, D.I., 1996. Biological applications of fullerenes. Bioorg. Med. Chem. 4, 767–779.

Jia, G., Wang, H., Yan, L., Wang, X., Pei, R., Yan, T., Zhao, Y., Guo, X., 2005. Cytotoxicity of carbon nanomaterials: single-wall nanotube, multi-wall nanotube, and fullerene. Environ. Sci. Technol. 39, 1378–1383.

Jiang, W., Kim, B.Y.S., Rutka, J.T., Chan, W.C.W., 2008. Nanoparticle-mediated cellular response is size-dependent. Nat. Nano 3, 145–150.

Johnston, H., Pojana, G., Zuin, S., Jacobsen, N.R., Møller, P., Loft, S., Semmler-Behnke, M., McGuiness, C., Balharry, D., Marcomini, A., Wallin, H., Kreyling, W., Donaldson, K., Tran, L., Stone, V., 2013. Engineered nanomaterial risk. Lessons learnt from completed nanotoxicology studies: potential solutions to current and future challenges. Crit. Rev. Toxicol. 43, 1–20.

Katz, E., Willner, I., 2004. Biomolecule-functionalized carbon nanotubes: applications in nanobio-electronics. ChemPhysChem 5, 1084–1104.

Kayat, J., Gajbhiye, V., Tekade, R.K., Jain, N.K., 2011. Pulmonary toxicity of carbon nanotubes: a systematic report. Nanomed. Nanotechnol. Biol. Med. 7, 40–49.

Kiser, M.A., Ryu, H., Jang, H., Hristovski, K., Westerhoff, P., 2010. Biosorption of nanoparticles to heterotrophic wastewater biomass. Water Res. 44, 4105–4114.

Kostarelos, K., Lacerda, L., Pastorin, G., Wu, W., Wieckowski, S., Luangsivilay, J., Godefroy, S., Pantarotto, D., Briand, J.P., Muller, S., Prato, M., Bianco, A., 2007. Cellular uptake of functionalized carbon nanotubes is independent of functional group and cell type. Nat. Nanotechnol. 2, 108–113.

Kreuter, J., 2001. Nanoparticulate systems for brain delivery of drugs. Adv. Drug Deliv. Rev. 47, 65–81.

Kreuter, J., 2004. Influence of the surface properties on nanoparticle-mediated transport of drugs to the brain. J. Nanosci. Nanotechnol. 4, 484–488.

Kuzma, J., 2007. Moving Forward Responsibly: Oversight for the Nanotechnology-Biology Interface Nanotechnology and Occupational Health. Springer, pp. 165–182.

Lehto, M., Karilainen, T., Róg, T., Cramariuc, O., Vanhala, E., Tornaeus, J., Taberman, H., Jänis, J., Alenius, H., Vattulainen, I., Laine, O., 2014. Co-exposure with fullerene may strengthen health effects of organic industrial chemicals. PLoS One 9, e114490.

Li, J., Strong, R., Trevisan, J., Fogarty, S.W., Fullwood, N.J., Jones, K.C., Martin, F.L., 2013. Dose-related alterations of carbon nanoparticles in mammalian cells detected using biospectroscopy: potential for real-world effects. Environ. Sci. Technol. 47, 10005–10011.

Li, J., Ying, G.-G., Jones, K.C., Martin, F.L., 2015. Real-world carbon nanoparticle exposures induce brain and gonadal alterations in zebrafish (*Danio rerio*) as determined by biospectroscopy techniques. Analyst 140, 2687–2695.

Li, J., Tian, M., Cui, L., Dwyer, J., Fullwood, N.J., Shen, H., Martin, F.L., 2016. Low-dose effects induced by carbon-based nanoparticles in A549 lung cells determined by biospectroscopy are associated with increases in genomic methylation. Sci. Rep. 6, 20207.

Limbach, L.K., Wick, P., Manser, P., Grass, R.N., Bruinink, A., Stark, W.J., 2007. Exposure of engineered nanoparticles to human lung epithelial cells: influence of chemical composition and catalytic activity on oxidative stress. Environ. Sci. Technol. 41, 4158–4163.

Linsebigler, A.L., Lu, G., Yates, J.T., 1995. Photocatalysis on TiO_2 surfaces: principles, mechanisms, and selected results. Chem. Rev. 95, 735–758.

Liu, Z., Tabakman, S., Welsher, K., Dai, H., 2009. Carbon nanotubes in biology and medicine: in vitro and in vivo detection, imaging and drug delivery. Nano Res. 2, 85–120.

Lowry, G.V., Gregory, K.B., Apte, S.C., Lead, J.R., 2012. Transformations of nanomaterials in the environment. Environ. Sci. Technol. 46, 6893–6899.

Marambio-Jones, C., Hoek, E.M.V., 2010. A review of the antibacterial effects of silver nanomaterials and potential implications for human health and the environment. J. Nanopart. Res. 12, 1531–1551.

Matés, J.M., Pérez-Gómez, C., De Castro, I.N., 1999. Antioxidant enzymes and human diseases. Clin. Biochem. 32, 595–603.

Matsuda, S., Matsui, S., Shimizu, Y., Matsuda, T., 2011. Genotoxicity of colloidal fullerene C60. Environ. Sci. Technol. 45, 4133–4138.

Mauter, M.S., Elimelech, M., 2008. Environmental applications of carbon-based nanomaterials. Environ. Sci. Technol. 42, 5843–5859.

McHedlov-Petrossyan, N.O., 2013. Fullerenes in liquid media: an unsettling intrusion into the solution chemistry. Chem. Rev. 113, 5149–5193.

Nel, A., Xia, T., Mädler, L., Li, N., 2006. Toxic potential of materials at the nanolevel. Science 311, 622–627.

Nel, A.E., Madler, L., Velegol, D., Xia, T., Hoek, E.M.V., Somasundaran, P., Klaessig, F., Castranova, V., Thompson, M., 2009. Understanding biophysicochemical interactions at the nano-bio interface. Nat. Mater. 8, 543–557.

Nowack, B., Bucheli, T.D., 2007. Occurrence, behavior and effects of nanoparticles in the environment. Environ. Pollut. 150, 5–22.

Oberdörster, G., Oberdörster, E., Oberdörster, J., 2005. Nanotoxicology: an emerging discipline evolving from studies of ultrafine particles. Environ. Health Perspect. 113, 823–839.

Oberdörster, E., 2004. Manufactured nanomaterials (fullerenes, C60) induce oxidative stress in the brain of juvenile largemouth bass. Environ. Health Perspect. 112, 1058–1062.

Pan, B., Xing, B., 2008. Adsorption mechanisms of organic chemicals on carbon nanotubes. Environ. Sci. Technol. 42, 9005–9013.

Pan, Y., Neuss, S., Leifert, A., Fischler, M., Wen, F., Simon, U., Schmid, G., Brandau, W., Jahnen-Dechent, W., 2007. Size-dependent cytotoxicity of gold nanoparticles. Small 3, 1941–1949.

Park, J.-W., Henry, T.B., Menn, F.-M., Compton, R.N., Sayler, G., 2010. No bioavailability of 17α-ethinylestradiol when associated with nC60 aggregates during dietary exposure in adult male zebrafish (*Danio rerio*). Chemosphere 81, 1227–1232.

Park, J.-W., Henry, T.B., Ard, S., Menn, F.-M., Compton, R.N., Sayler, G.S., 2011. The association between nC60 and 17α-ethinylestradiol (EE2) decreases EE2 bioavailability in zebrafish and alters nanoaggregate characteristics. Nanotoxicology 5, 406–416.

Patra, H.K., Banerjee, S., Chaudhuri, U., Lahiri, P., Dasgupta, A.K., 2007. Cell selective response to gold nanoparticles. Nanomed. Nanotechnol. Biol. Med. 3, 111–119.

Pernodet, N., Fang, X., Sun, Y., Bakhtina, A., Ramakrishnan, A., Sokolov, J., Ulman, A., Rafailovich, M., 2006. Adverse effects of citrate/gold nanoparticles on human dermal fibroblasts. Small 2, 766–773.

Petersen, E.J., Zhang, L., Mattison, N.T., O'Carroll, D.M., Whelton, A.J., Uddin, N., Nguyen, T., Huang, Q., Henry, T.B., Holbrook, R.D., Chen, K.L., 2011. Potential release pathways, environmental fate, and ecological risks of carbon nanotubes. Environ. Sci. Technol. 45, 9837–9856.

Ren, L., Zhong, W., 2010. Oxidation reactions mediated by single-walled carbon nanotubes in aqueous solution. Environ. Sci. Technol. 44, 6954–6958.

Sadaf, A., Zeshan, B., Wang, Z., Zhang, R., Xu, S., Wang, C., Cui, Y., 2012. Toxicity evaluation of hydrophilic CdTe quantum dots and CdTe@SiO2 nanoparticles in mice. J. Nanosci. Nanotechnol. 12, 8287–8292.

Serpone, N., Dondi, D., Albini, A., 2007. Inorganic and organic UV filters: their role and efficacy in sunscreens and suncare products. Inorg. Chim. Acta 360, 794–802.

Shi, H., Hudson, L.G., Liu, K.J., 2004. Oxidative stress and apoptosis in metal ion-induced carcinogenesis. Free Radic. Biol. Med. 37, 582–593.

Shinohara, N., Matsumoto, K., Endoh, S., Maru, J., Nakanishi, J., 2009. In vitro and in vivo genotoxicity tests on fullerene C60 nanoparticles. Toxicol. Lett. 191, 289–296.

Shvedova, A.A., Pietroiusti, A., Fadeel, B., Kagan, V.E., 2012. Mechanisms of carbon nanotube-induced toxicity: focus on oxidative stress. Toxicol. Appl. Pharmacol. 261, 121–133.

Simon, H.U., Haj-Yehia, A., Levi-Schaffer, F., 2000. Role of reactive oxygen species (ROS) in apoptosis induction. Apoptosis 5, 415–418.

Simon, A., Maletz, S.X., Hollert, H., Schäffer, A., Maes, H.M., 2014. Effects of multiwalled carbon nanotubes and triclocarban on several eukaryotic cell lines: elucidating cytotoxicity, endocrine disruption, and reactive oxygen species generation. Nanoscale Res. Lett. 396.

Skaper, S.D., Floreani, M., Ceccon, M., Facci, L., Giusti, P., 1999. Excitotoxicity, oxidative stress, and the neuroprotective potential of melatonin. Ann. NY Acad. Sci. 890, 107–118.

Smetana, A.B., Klabunde, K.J., Marchin, G.R., Sorensen, C.M., 2008. Biocidal activity of nanocrystalline silver powders and particles. Langmuir 24, 7457–7464.

Smith, C.J., Shaw, B.J., Handy, R.D., 2007. Toxicity of single walled carbon nanotubes to rainbow trout, (*Oncorhynchus mykiss*): respiratory toxicity, organ pathologies, and other physiological effects. Aquat. Toxicol. 82, 94–109.

Sondi, I., Salopek-Sondi, B., 2004. Silver nanoparticles as antimicrobial agent: a case study on *E. coli* as a model for Gram-negative bacteria. J. Colloid Interface Sci. 275, 177–182.

Song, M., Wang, F., Zeng, L., Yin, J., Wang, H., Jiang, G., 2014. Co-exposure of carboxyl-functionalized single-walled carbon nanotubes and 17α-ethinylestradiol in cultured cells: effects on bioactivity and cytotoxicity. Environ. Sci. Technol. 48, 13978–13984.

Sperling, R.A., Rivera Gil, P., Zhang, F., Zanella, M., Parak, W.J., 2008. Biological applications of gold nanoparticles. Chem. Soc. Rev. 37, 1896–1908.

Stoller, M.D., Park, S., Zhu, Y., An, J., Ruoff, R.S., 2008. Graphene-based ultracapacitors. Nano Lett. 8, 3498–3502.

Su, Y., Yan, X., Pu, Y., Xiao, F., Wang, D., Yang, M., 2013. Risks of single-walled carbon nanotubes acting as contaminants-carriers: potential release of phenanthrene in Japanese Medaka (*Oryzias latipes*). Environ. Sci. Technol. 47, 4704–4710.

Sung, J.H., Ji, J.H., Yoon, J.U., Kim, D.S., Song, M.Y., Jeong, J., Han, B.S., Han, J.H., Chung, Y.H., Kim, J., Kim, T.S., Chang, H.K., Lee, E.J., Lee, J.H., Yu, I.J., 2008. Lung function changes in Sprague-Dawley rats after prolonged inhalation exposure to silver nanoparticles. Inhal. Toxicol. 20, 567–574.

Tang, S., Cai, Q., Chibli, H., Allagadda, V., Nadeau, J.L., Mayer, G.D., 2013. Cadmium sulfate and CdTe-quantum dots alter DNA repair in zebrafish (*Danio rerio*) liver cells. Toxicol. Appl. Pharmacol. 272, 443–452.

Tian, S., Zhang, Y., Song, C., Zhu, X., Xing, B., 2014. Titanium dioxide nanoparticles as carrier facilitate bioaccumulation of phenanthrene in marine bivalve, ark shell (*Scapharca subcrenata*). Environ. Pollut. 192, 59–64.

Tolaymat, T.M., El Badawy, A.M., Genaidy, A., Scheckel, K.G., Luxton, T.P., Suidan, M., 2010. An evidence-based environmental perspective of manufactured silver nanoparticle in syntheses and applications: a systematic review and critical appraisal of peer-reviewed scientific papers. Sci. Total Environ. 408, 999–1006.

Turrens, J.F., 2003. Mitochondrial formation of reactive oxygen species. J. Physiol. 552, 335–344.

Wang, B., Feng, W., Wang, M., Wang, T., Gu, Y., Zhu, M., Ouyang, H., Shi, J., Zhang, F., Zhao, Y., Chai, Z., Wang, H., Wang, J., 2007. Acute toxicological impact of nano- and submicro-scaled zinc oxide powder on healthy adult mice. J. Nanopart. Res. 10, 263–276.

Wang, K., Ruan, J., Song, H., Zhang, J., Wo, Y., Guo, S., Cui, D., 2011. Biocompatibility of graphene oxide. Nanoscale Res. Lett. 6, 1–8.

Win-Shwe, T.-T., Fujimaki, H., 2011. Nanoparticles and neurotoxicity. Int. J. Mol. Sci. 12, 6267.

Wohlleben, W., Brill, S., Meier, M.W., Mertler, M., Cox, G., Hirth, S., von Vacano, B., Strauss, V., Treumann, S., Wiench, K., Ma-Hock, L., Landsiedel, R., 2011. On the lifecycle of nanocomposites: comparing released fragments and their in-vivo hazards from three release mechanisms and four nanocomposites. Small 7, 2384–2395.

Wu, T., Tang, M., 2014. Toxicity of quantum dots on respiratory system. Inhal. Toxicol. 26, 128–139.

Xia, Q., Yin, J.J., Cherng, S.-H., Wamer, W.G., Boudreau, M., Howard, P.C., Fu, P.P., 2006. UVA photoirradiation of retinyl palmitate—formation of singlet oxygen and superoxide, and their role in induction of lipid peroxidation. Toxicol. Lett. 163, 30–43.

Yang, K., Zhu, L., Xing, B., 2006. Adsorption of polycyclic aromatic hydrocarbons by carbon nanomaterials. Environ. Sci. Technol. 40, 1855–1861.

Zhang, Y., Chen, Y., Westerhoff, P., Crittenden, J., 2009. Impact of natural organic matter and divalent cations on the stability of aqueous nanoparticles. Water Res. 43, 4249–4257.

Zhang, Y., Ali, S.F., Dervishi, E., Xu, Y., Li, Z., Casciano, D., Biris, A.S., 2010. Cytotoxicity effects of graphene and single-wall carbon nanotubes in neural phaeochromocytoma-derived PC12 cells. ACS Nano 4, 3181–3186.

Zhang, X., Yin, J., Peng, C., Hu, W., Zhu, Z., Li, W., Fan, C., Huang, Q., 2011. Distribution and biocompatibility studies of graphene oxide in mice after intravenous administration. Carbon 49, 986–995.

Chapter 5

Toxicity of Titanium Dioxide Nanoparticles on Brain

L.Q. Shao[1], B. Song[1,2]
[1]Southern Medical University, Guangzhou, China; [2]Guizhou Provincial People's Hospital, Guiyang, China

Chapter Outline

1. INTRODUCTION

As nanotechnology is advancing rapidly, more concerns on health risks about exposure to nanoparticles (NPs) have been arising. The titanium dioxide (TiO_2) particles possess low toxicity and are widely applied in biomedical applications for their excellent biocompatibility (Wu et al., 2014a; Catauro et al., 2015;

Neurotoxicity of Nanomaterials and Nanomedicine. http://dx.doi.org/10.1016/B978-0-12-804598-5.00005-2

Logan et al., 2015). However, when the size of TiO_2 NPs is diminished to the nanoscale (1–100 nm), the bioactivity and physicochemical properties of nanosized TiO_2 are apparently different from that of their bulk analogue. As a consequence, the understanding about the risk assessments of NPs lags behind their rapid advancement and widespread applications (Newman et al., 2009; Montazer and Pakdel, 2011; Ma et al., 2014; Martirosyan and Schneider, 2014). TiO_2 NPs-based products are widely used as well, which could unintentionally lead to human exposure and environmental pollution. The TiO_2 NPs might be potentially absorbed mainly through inhalation, ingestion, and skin penetrations into the circulation of human beings. Then they may be redistributed into other tissues, which could induce impairments on organs. As concerns about unintentional exposure of human beings to NPs are arising, an increasing number of studies have been performed to investigate the potential toxic effects of TiO_2 NPs. Several in vivo researches have assessed the biosafety of TiO_2 NPs. The main administration routes in in vivo studies include inhalation (Gustafsson et al., 2014), intratracheal (Shinohara et al., 2014b) or nasal instillation (Wang et al., 2008b), oral gavage and dermal exposure (Auttachoat et al., 2014), intragastric feeding (Hong et al., 2015), intraperitoneal (Chen et al., 2009), and intravenous injection (Shinohara et al., 2014b).

The central nervous system (CNS), including the brain and spinal cord, is the extremely important nervous system for human beings. Several in vivo studies have investigated the biodistribution of TiO_2 NPs. The main pathways for TiO_2 NPs to be transported into the brain included (1) the blood–brain barrier (BBB) pathway (Lockman et al., 2004), (2) the olfactory nerve translocation pathway (Wang et al., 2008a), and (3) the placental barrier pathway (Tsyganova et al., 2014). Once the TiO_2 NPs have been transported into the brain regions, major cells in the brain would be affected by NPs. Reactive oxygen species (ROS), apoptosis, and inflammation would be induced by TiO_2 NPs, which may lead to cell death and disturb the brain functions, or even induce neurodegenerative disease. However, current knowledge about neurotoxicity induced by TiO_2 NPs is insufficient and more detailed and standardized researches are needed. Therefore, in this chapter, we discuss the applications of TiO_2 NPs, the main routes of TiO_2 NPs into the brain, the main mechanisms underlying neurotoxicity of TiO_2 NPs, and major factors that might influence the neurotoxicity of TiO_2 NPs.

2. APPLICATIONS OF TIO_2 NANOPARTICLES

TiO_2 NPs, at least with one dimension in the range of 1–100 nm, are widely used in many fields, such as paints, lacquers, laminates, food wraps, foils, transparent plastics, papers, textiles, cosmetics, food products, medicines, pharmaceuticals, and toothpaste. Owing to their catalytic properties they are also used in self-cleaning tiles, windows, antifogging car mirrors, in the purification of water, sewage, combustion gases, as a antimicrobial material in decontamination, as catalyst in organic synthesis, and as a photoactive material in solar cells (Shi et al., 2013). In situ deposition of TiO_2 NPs on polyester fabric can improve ultraviolet (UV) protection and exhibit photocatalytic solution discoloration and good washing fastness properties (Khan et al., 2015). TiO_2 NPs-coated papers show much great stability to UV illumination (Cheng et al., 2014). Fluorine-doped anatase TiO_2 NPs are synthesized by hydrothermal method and applied in dye-sensitized solar cells as excellent photoelectrode material (Yu et al., 2015).

In medical applications, Nano Ag/TiO_2 coating bracket not only has an antibacterial effect but also has good biocompatibility; therefore, it can satisfy the clinical request of orthodontic treatment (Zhang et al., 2015). A complex ceramic coating on Ti plates, containing TiO_2 nanotubes obtained by anodizing, multiwalled carbon nanotubes functionalized with —COOH groups and hydroxyapatite, has an optimal microstructure and morphology, showing a proper surface for osteoblast adhesion, viability, and proliferation. It signifies a progress in the field of the implant biomaterials (Prodana et al., 2015). TiO_2 nanofilm may have potential application in the diathermy rehabilitation treatment of orthopedic metal implants (Hua et al., 2015). Wu et al. (2014b) have summarized recent advances in the research of nanostructured TiO_2-based biomaterials pertaining to bone tissue engineering (nanostructured TiO_2 NPs-based bone implant materials, modulation of cell and tissue behavior), intravascular stents, drug delivery systems (antibacterial agent delivery, anticancer drug delivery, gene therapy carriers), and biosensors (enzymatic TiO_2-based biosensors, nonenzymatic TiO_2 NPs-based biosensors).

3. MAIN ROUTES OF TIO_2 NANOPARTICLES INTO THE BRAIN

The CNS, including the brain and the spinal cord, is the extremely important nervous system for human beings. The main functions of brain include (1) transferring, storing, and processing information; (2) generating numerous psychological activities; (3) commanding and controlling behaviors of humans. A great number of researches have focused on the biodistribution of TiO_2 NPs when they are absorbed through several administration routes. After rodents were treated with TiO_2 NPs, the NPs would accumulate in brain areas and the TiO_2 NPs concentrations in the brain were enhanced gradually after repeated exposure. The main pathways for TiO_2 NPs to be transported into the brain include (1) the BBB pathway (Lockman et al., 2004), (2) the olfactory nerve translocation pathway (Wang et al., 2008a), and (3) the placental barrier pathway (Tsyganova et al., 2014).

3.1 Translocation of TiO_2 Nanoparticles From the Blood to the Brain

The BBB is an effectively protective structure, which is mainly composed of endothelial cells, astrocytes, and pericytes (Ballabh et al., 2004). The endothelial cells are connected with each other through complicated tight junctions, whereas the connections are supported by the astrocytes and pericytes. Only specific substances with small size or low molecular weight could be allowed to pass through the BBB. In other word, the BBB is capable of protecting the healthy and functional brain from being influenced by harmful chemicals, toxins, and drugs in the circulatory system. However, TiO_2 NPs are able to pass through the BBB and enter into the brain due to their unique physicochemical characteristics (Roney et al., 2005; Barbu et al., 2009; Dominguez et al., 2014). On the other hand, the permeability of the BBB can be changed by NPs, which could assist in influx of exogenous substances into the brain. As a result, NPs might induce inflammation, edema, and cell injury or even cell death in brain zones.

An in vivo study (Li et al., 2010) demonstrated that after TiO_2 NPs were repeatedly administered to mice through intratracheal instillations every day for 4 weeks, the ratio of brain/body weight in the experimental group was significantly downregulated compared with that in the control. Moreover, histopathological changes, inflammatory cell aggregation and cell necrosis were present in the brain zones. The content of TiO_2 NPs in brain, measured by inductively coupled plasma mass spectrometry (ICP-MS), was significantly increased in TiO_2 NPs-treated mice. These results suggested that (1) after mice were exposed to TiO_2 NPs in a chronic way, those NPs were transported into the blood, and then they passed through the BBB and finally accumulated in the brain; (2) TiO_2 NPs that were transported into the brain could induce impairments.

TiO_2 NPs could not only pass through the BBB, but also affect the integrity of the BBB. The toxic effects of TiO_2 NPs on the BBB were investigated based on an in vitro BBB model. The model, mimicking the specific characteristics of in vivo one (Brun et al., 2012), was composed of rat primary endothelial cells and astrocytes. After acute or long-term exposure of the BBB model to TiO_2 NPs, the expression of P-glycoprotein, claudin 5, caveolin-1, and caveolin-2, which can regulate the integrity of BBB, was significantly downregulated. These findings suggested that direct harmful impacts of TiO_2 NPs on BBB integrity were induced during the acute and long-term exposure. Besides, authors revealed that the mRNA expression of CXC chemokines, CC chemokines, a disintegrin and metalloproteinase 17, CC chemokine ligand 2, transforming growth factor β1, intercellular adhesion molecule, and vascular cell adhesion molecule increased, paralleled by decreased ATP-binding cassette transporters mRNA expression after exposure. The upregulated expression of those target genes was related with the decreased permeability of BBB (Candelario-Jalil et al., 2009; Hosking et al., 2009; Mestre et al., 2011; D'Aversa et al., 2013; Fang et al., 2013; Louboutin and Strayer, 2013; Baello et al., 2014), which could facilitate the transportation of other exogenous substances into brain. These findings meant that in addition to direct impairments, indirect harmful impacts (inflammatory effects) of TiO_2 NPs on BBB integrity also occurred.

3.2 Axonal Translocation of TiO_2 Nanoparticles From the Nose to the Brain

Axonal transport is defined as the process by which proteins and other substances synthesized in the neurosome are transported to the nerve endings through cytoskeleton (De Vos et al., 2008). However, some low-molecular-weight or small-size substances such as NPs can be taken up by the nerve ending, and then be transported to the neurosome. This process is called retrograde axonal transport (Perlson et al., 2010). The olfactory and trigeminal nerve endings are present in the nasal areas. Once NPs are instilled through nose, they can enter the circulation system and pass through the BBB, or they can bypass the BBB to be transported into the brain along the axons. The second direct pathway is the major route for NPs to be transported to the brain zones after intranasal instillation.

Wang et al. (2008b) confirmed that after female mice were exposed to TiO_2 NPs through intranasal instillation, the TiO_2 NPs concentration was significantly increased in the brain compared with that in the control. Wang et al. (2008a) discovered that when female mice were treated with TiO_2 NPs through intranasal instillation every other day, the Ti content was detected in the brains after 30 days exposure. The TiO_2 NP contents, determined by synchrotron radiation X-ray fluorescence analysis and ICP-MS, were significantly increased in the hippocampus area, olfactory bulb, cerebellum, and cerebral cortex. These findings suggested that the TiO_2 NPs could be transported to the mice brain through the olfactory bulb after intranasal instillation. The same research group conducted another study (Wang et al., 2008c), which also detected enhanced TiO_2 NP contents in hippocampus, olfactory bulb, cerebellum, and cerebral cortex by ICP-MS. These findings indicated that there was a high TiO_2 NP content in hippocampus, so this brain area would be easily affected by TiO_2 NPs. It is generally accepted that the hippocampus is mainly in charge of memory and learning (Bird and Burgess, 2008; Howland and Wang, 2008). Therefore, impairments to this area might probably induce neurodegenerative diseases, such as Alzheimer disease (AD). Another study (Ze et al., 2014a) also discovered that after female mice were exposed to TiO_2 NPs through nasal administration for 90 consecutive days, the TiO_2 NPs were detected in the brain and this accumulation can induce brain impairments.

3.3 Translocation Into the Brain of Offspring Through the Placental Barrier

Placental barrier, composed of both maternal and fetal tissue, is another internal barrier that can protect the development of embryo (Chu et al., 2010). It could protect the fetus from being affected by harmful substances in maternal blood circulation, whereas the fetus could get nutrients and oxygen from the mother via the placenta. However, a great number of studies (Di Bona et al., 2014; Semmler-Behnke et al., 2014) have already revealed that after pregnant mice/rats were exposed to exogenous substances, such as NPs, those substances could be detected in the brain of the fetus, and then they can perturb the homeostasis of brain or even lead to neuronal death. Those harmful impacts on fetus brain are related with psychiatric disorders such as autism, schizophrenia, and depression in their later life (O'Connor and Paley, 2009; Ekblad et al., 2010). As a consequence, those findings suggested that placental barrier plays an important part in fetal growth and development.

In a study by Mohammadipour et al. (2014), pregnant Wistar rats were administered TiO_2 NPs intragastrically daily from gestational day 2 to 21. The TiO_2 NPs concentration in the hippocampus of 1-day-old neonates, determined by ICP-MS, was significantly upregulated as compared with that in the control group. Yamashita et al. (2011) also detected silica (70 nm) and TiO_2 (35 nm) NPs in the placenta and fetal brain after the pregnant mice were injected intravenously with these NPs, which led to pregnancy complication.

In another study (Takahashi et al., 2010), the authors discovered that when pregnant mice were treated with TiO_2 NPs, the levels of dopamine and its metabolites were increased in some regions of the fetal brain on postnatal day 21. Another study (Umezawa et al., 2012) adopted microarray to assess gene expression changes in brains of male fetus and pups after pregnant mice were exposed to TiO_2 NPs. Data showed that the gene expression related with the dopamine neuron system was altered. Shimizu et al. (2009) also used microarray in their study, and pregnant mice were administered TiO_2 NPs. By analyzing the gene expression alternations in brain of male fetus and pups, data obtained revealed that the expression of genes related with oxidative stress, neurotransmitters, and psychiatric diseases was dysregulated. The neurobehavioral performance of the offspring might be moderately altered because of maternal exposure to TiO_2 NPs (Hougaard et al., 2010). Similarly, Cui et al. (2014) discovered that when Sprague–Dawley rats were injected subcutaneously with TiO_2 NPs, the antioxidant ability of pups' brain was impaired. Although these researches did not measure the contents of TiO_2 NPs in the brain directly, data collected indirectly demonstrated that the TiO_2 NPs in the maternal circulation system would affect the development of the embryo. Moreover, TiO_2 NPs could impair brain development and finally lead to brain disorders in later life.

4. BIODISTRIBUTION AFTER DIFFERENT ADMINISTRATION ROUTES AND ELIMINATION RATE OF TiO_2 NANOPARTICLES FROM THE BRAIN

4.1 Biodistribution

When TiO_2 NPs are absorbed into circulation, they would redistribute to second organs. However, different administration routes may lead to different biodistribution status. Several researches have been performed to study the biodistribution of TiO_2 NPs after administration (Table 5.1) (Song et al., 2015). In a study by Fabian et al. (2008), when rats were treated with TiO_2 NPs by intravenous

TABLE 5.1 Biodistribution of TiO_2 Nanoparticles After Rat/Mice Were Administered by Different Routes (Song et al., 2015)

Crystal Type	Animal	Administration	Parameters/Dose	Biodistribution	References
Both anatase and rutile forms (70/30)	Male Wistar rats	Intravenous injection	20–30 nm; no surface coating; 5 mg/kg body weight (BW); single injection	Liver, spleen, lung, and kidney detected; blood cells, plasma, brain, or lymph nodes not detected	Fabian et al. (2008)
Both anatase and rutile forms (80/20)	Balb/c female mice	Subcutaneous injection	Hydrodynamic diameters ranging from 114 to 122 nm; 5600 mg/kg BW; two consecutive days	Liver, lymph node, and spleen detected; brain not detected	Patri et al. (2009)
		Intravenous injection	Hydrodynamic diameters ranging from 114 to 122 nm; 560 mg/kg BW; two consecutive days	Lung, liver, lymph node, spleen, and kidney detected; brain not detected	
Rutile	Male mice	Intravenous injection	Primary particle diameter 15 nm, secondary particle size 120 nm; 1813 μg/animal	Liver, kidney, blood detected; brain not detected	Sugibayashi et al. (2008)
Rutile	Female mice	Intranasal instillation	80 nm; 50 mg/kg BW; every other day for 30 days	Lung and brain detected; liver, heart, and spleen not detected	Wang et al. (2008b)
Anatase			155 nm; 50 mg/kg BW; every other day for 30 days		
Degussa P25	Rats	Intravenous administration	21 nm; spherical; 0.95 mg/kg BW; single injection	Liver (highest), spleen, lung, kidney, heart, and blood detected; brain not detected	Shinohara et al. (2014a)
Anatase 80:20 Rutile	Sprague–Dawley rats	Oral administration	21 nm; spherical; 260.4, 520.8, and 1041.5 mg/kg/day BW; every day for 13 weeks (7 days/week)	Low absorption in organs	Cho et al. (2013)
Anatase	Male Kunming mice	Inhalation exposure	20 nm; steady concentration ($6.34 \pm 0.22\ mg/m^3$); 8 h per day for 3 weeks	Lungs (highest), liver, blood, and urine increased, brain and kidney not detected	Yin et al. (2014)
Anatase	Male Wistar rats	Intraperitoneal injection	30 nm; 20 mg/kg; every 2 days for 20 days	Liver (highest), lung, and brain detected	Ben Younes et al. (2015)

injection, TiO_2 NPs were detected in liver, spleen, lung, and kidney in addition to blood cells, plasma, brain, and lymph nodes. The female mice were exposed to TiO_2 NPs by intravenous or subcutaneous injection. The TiO_2 NPs were detected by energy dispersive X-ray spectroscopy in lung, liver, lymph node, spleen, and kidney from intravenously administered mice but only in liver, lymph node, and spleen of subcutaneously administered mice. The content of NPs was not elevated in brain (Patri et al., 2009). Another study (Sugibayashi et al., 2008) also did not detect TiO_2 NPs in the brain of male mice except in blood and liver after intravenous injection. However, after hairless mice were treated with TiO_2 NPs by dermal exposure for 60 days, significant pathological alterations were present in the skin and liver and the NPs were also detected in brain without pathological changes (Wu et al., 2009). Wang et al. (2008b) studied the biodistribution of TiO_2 NPs after female mice were treated with NPs by intranasal instillation every other day for 30 days. The biochemical parameters of liver, spleen, heart, and serum were not affected by NPs as compared with the control group, whereas the concentration of NPs was apparently enhanced in lung and brain regions. Another study investigated the biodistribution of TiO_2 NPs after rats were repeatedly orally administered for 13 weeks. Even in the highest dosage group, the Ti content in brain was minimal with no statistical significance. Geraets et al. (2014) compared the different distribution of TiO_2 NPs in rats after oral and intravenous administration. The data obtained demonstrated that the TiO_2 NP concentrations were not detectable in tissues, including the brain after oral administration. However, the TiO_2 NP contents were detected in liver, spleen, kidney, lung, heart, brain, thymus, and reproductive organs after intravenous injection. Another study (Ben Younes et al., 2015) also discovered that after male rats were injected intraperitoneally with TiO_2 NPs every 2 days for 20 days, the TiO_2 NP contents were highest in liver, followed by lung and brain.

Undoubtedly, conclusions drawn from the above-mentioned in vivo researches might not be convincing, because the translocations of TiO_2 NPs into the brain would be influenced by several parameters, such as administration routes, size, and dosage, which would be discussed in later chapters.

4.2 Elimination

Although TiO_2 NPs are capable of entering into brain regions through specific routes by a variety of delivery ways, the capability of excretion would keep the brain from being affected by NPs. But few researches about the elimination of TiO_2 NPs from brain were published. In a study by Cho et al. (2013), when Sprague–Dawley rats were administered TiO_2 NPs by intravenous injection for 5 consecutive days, the contents of NPs were determined on day 2/day 6, day 14, day 30, and day 90 after administration. The TiO_2 NP contents were detected in liver (the highest), spleen, kidney, lung, heart, brain, thymus, and reproductive organs on day 2/day 6. During the observation period, TiO_2 NP concentrations in feces and urine demonstrated no increase as compared with the control group. Also, TiO_2 NP contents in brain demonstrated no detectable alterations.

It can be inferred from those results that (1) the excretion of TiO_2 NPs from the brain is limited and (2) even a negligible dosage of TiO_2 NPs could be accumulated in the brain with minimal elimination, so chronic or long-term exposure might be potentially risky for brain health. TiO_2 NPs are thought to be less toxic. However, in the long term, repeated exposure and durative existence of TiO_2 NPs in the brain regions made it necessary to re-evaluate their biosafety. Therefore, more in vivo researches are needed to further investigate the parameters of TiO_2 NPs that might influence their elimination rates from the brain, with the purpose of lowering the health risks of TiO_2 NPs to the brain.

5. MAIN MECHANISMS UNDERLYING NEUROTOXICITY OF TIO_2 NANOPARTICLES

Once the TiO_2 NPs are translocated into the brain through the above-mentioned pathways, they may accumulate in the brain regions. Because of their slow elimination rates, those NPs could remain in the brain zones for a long period and the TiO_2 NP contents would gradually increase with repeated exposure. This would induce pathological changes, such as inflammation, immunological response, edema, cell injury, and cell necrosis, which would ultimately lead to brain dysfunctions, including neurodegenerative diseases and psychiatric disorders. Generally, the neurons and glial cells are the main cell types in the brain. Therefore, the toxic effects of TiO_2 NPs on these cells would lead to impairments of the brain as a consequence (Tables 5.2 and 5.3) (Song et al., 2015). The main mechanisms underlying neurotoxicity of TiO_2 NPs are illustrated as follows.

5.1 Oxidative Stress

ROS are highly reactive molecules that consist of an oxygen atom. For example, ROS might include the superoxide radical (O_2^-) or H_2O_2 (Rang and Boonstra, 2014). Under physiological conditions, ROS are present in every cell, being produced by the mitochondrial and cytoplasmic oxidation processes. A low or moderate ROS concentration is necessary to maintain normal physiological processes. However, an excessive production of ROS caused by oxidative stress would be extremely harmful to cells. The brain is particularly susceptible to oxidative stress because of its high oxygen consumption, its weak antioxidative ability, and the terminal-differentiation feature of neurons (Li et al., 2013). Oxidative stress has been established as a crucial contributor to acute brain injury-related neurodegenerative diseases, such as DNA damage, which can impair the viability of cerebral cells (Smith et al., 2013). An increasing number of studies show that ROS play a pivotal role in neurodegenerative diseases, such as AD and Parkinson disease (PD). Moreover, ROS can lead to neuronal apoptosis,

TABLE 5.2 Toxic Effects of TiO_2 Nanoparticles on Brain in In Vivo Studies (Song et al., 2015)

Crystal Type	Animals	Administration	Parameters/Dose	Main Findings	References
Rutile 80 nm Anatase 155 nm	Female mice	Nasal instillation	500 μg; every other day for 30 days	Ti contents in the brain (olfactory bulb, hippocampus, cerebellum, cerebral cortex) increased; cell lost; GFAP-positive cell increased; CAT increased; SOD increased; MDA, protein carbonyls, AChE activities increased; glutamic acid, NO increased	Wang et al. (2008a)
Anatase bulk	Female mice	Delivered to the abdominal cavity	5 nm; 5, 10, 50, 100, 150 mg/kg BW; every day for 14 days	The coefficients of the brain to body weight decreased with increased dosage; Ti contents in brain increased; O_2, H_2O_2, MDA, NOS, NO increased; antioxidative enzymes (SOD, CAT, APx, GSH-Px) decreased; nonenzymatic antioxidant contents (AsA/DAsA, GSH/GSSG) and Glu contents, AChE activity decreased	Ma et al. (2010)
Anatase	Male mice	Intranasal administration	5–6 nm; 2.5, 5, 10 mg/kg BW; every day for 90 days	Ti contents increased without daily behavioral changes; overproliferation of spongiocytes observed; levels of O_2, H_2O_2, MDA, protein carbonyl and 8-OHdG contents increased; the mRNA expressions of p38, JNK, NF-κB, Nrf-2, and HO-1 increased	Ze et al. (2013)
Anatase	Sprague–Dawley rats (male and female)	Subcutaneous injection	5 nm; 500 μL (1 μg/μL) on GD 6, 9, 12, 15, and 18	The levels of CAT, GSH-PX, T-AOC decreased; MDA increased; 8-OHdG increased; experimental group spending less time exploring the novel object; the immobility increased; consumed less sucrose water	Cui et al. (2014)
Rutile 80 nm Anatase 155 nm	Female mice	Intranasal instillation	500 μg; every other day for 30 days; evaluated at 2, 10, 20, and 30 days of postinstillation time points	Ti contents highest in hippocampus, followed by olfactory bulb, cerebellum, cerebral cortex; activities of GSH-Px, GST, SOD, and GSH level not changed. MDA increased; TNF-α and IL-1β in brain increased	Wang et al. (2008c)
Anatase	Female mice	Intranasal administration	5–6 nm; 2.5, 5, 10 mg/kg BW; every day for 90 days	Brain indices lower than the control; histopathological changes in mouse hippocampus; TLR2 and TLR4 genes and their protein levels increased; mRNA and protein levels of TNF-α, IKK1, IKK2, NF-κB, NF-κBP52, NF-κBP65, NIK, and IL-1β increased with decreased level of IκB; Spatial recognition memory and locomotor activity affected	Ze et al. (2014b)

Rutile	Male C57BL/6 mice	Intraperitoneal injection	Fine (<1 μm), ultrafine (21 nm); 40 mg/kg BW; one injection 30 min after LPS or vehicle injection	mRNA expression levels of IL-1β, TNF-α, and iNOS in cortex and hippocampus upregulated by ultrafine TiO_2 in the LPS-stimulated group; the protein level of IL-1β increased by ultrafine TiO_2 in the LPS-stimulated group; ROS production increased by ultrafine TiO_2 in LPS-treated group; the expression of OX-42 enhanced by ultrafine TiO_2 in the LPS-stimulated group	Shin et al. (2010)
Anatase	Female mice	Intragastric administration	6.5 nm; 5, 10, 50 mg/kg BW; every day for 60 days	The contents of titanium in the hippocampus increased; shrinkage of nucleus and shrinkage of nuclear membrane and chromatin marginalization; classical DNA ladder; the mRNA and protein levels of caspase-9, caspase-3, Bax, cytochrome c upregulated with downregulated Bcl-2; caspase-8 not affected; levels of O_2 and H_2O_2 increased; the activities of SOD, CAT, APx, and GSH-Px downregulated; the ratios of AsA to DAsA and GSH to GSSGG decreased; the percentage of duration to explore the novel arm reduced	Hu et al. (2011)
Anatase	Female mice	Nasal administration	5–6 nm; 2.5, 5, 10 mg/kg BW; every day for 90 days	TiO_2 NPs-treated mice learned the training task more slowly than the control group; TiO_2 NPs-treated mice showed downregulation of LTP amplitudes of fEPSP; mRNA and protein levels of NR2A, NR2B, CREB-1, CREB-2, FosB/DFosB, CaMKIV decreased; protein level of pCREB and mRNA level of CaMKIV decreased	Ze et al. (2014c)
Anatase	Female mice	Intragastric administration	5 nm; 5, 10, 50 mg/kg BW; every day for 60 days	Ti concentrations in the brain upregulated; the concentrations of Ca, Na elevated with decreased levels of Mg, K, Zn, Fe; reduction in the activities of Na^+/K^+-ATPase, Ca^{2+}-ATPase, Ca^{2+}/Mg^{2+}-ATPase of brain. The levels of Ach, Glu, and NO elevated; the contents of NE, DA, DOPAC, 5-HT, and 5-HIAA reduced	Hu et al. (2010)
Anatase	Pregnant mice	Subcutaneous injection	2570 nm; 100 μg, injection on GD 6, 9, 12, 15	The expressions of genes related with cell death, apoptosis, oxidative stress, inflammation, and neurotransmitters changed	Shimizu et al. (2009)
Rutile	Pregnant BALB/c mice	Intravenous injection	35 nm; 0.8 mg, injections on GD 16, 17	TiO_2 NPs-treated mice had lower uterine weights and smaller fetuses, leading to fetal resorption and retarded fetal growth	Yamashita et al. (2011)
Anatase	Pregnant Wistar rats	Intragastric administration	10 nm; 100 mg/kg BW, every day from GD 2 to GD 21	Ti contents elevated in the hippocampus and the number of Ki-67-positive cells in the hippocampus reduced in the experimental group; learning and memory in offspring (PD21, PD60) disrupted	Mohammadipour et al. (2014)

5-HIAA, 5-hydroxyindoleacetic acid; *5-HT*, 5-hydroxytryptamine; *8-OHdG*, 8-hydroxydeoxyguanosine; AChE, acetylcholinesterase; *APx*, ascorbate peroxidase; *AsA/DAsA*, ascorbic acid/oxidized AsA; *BW*, body weight; *CAT*, catalase; *DA*, dopamine; *DOPAC*, dihydroxy-phenylacetic acid; *GD*, gestational day; *GFAP*, glial fibrillary acidic protein; *GSH/GSSG*, glutathione/oxidized glutathione; *GSH-Px*, glutathione peroxidase; *HO-1*, heme oxygenase; *IL*, interleukin; *JNK*, Jun N-terminal kinase; *MDA*, malonaldehyde; *NO*, nitric oxide; *NOS*, nitric oxide synthase; *Nrf-2*, nuclear factor E2-related factor 2; *NSC*, neural stem cell; *ROS*, reactive oxygen species; *SOD*, superoxide dismutase; *TNF*, tumor necrosis factor.

TABLE 5.3 Toxic Effects of TiO_2 Nanoparticles on Brain in In Vitro Studies (Song et al., 2015)

Crystal Type	Cell Type	Parameters/Dose	Main Findings	References
Unknown	Primary microglia from Sprague–Dawley rats and PC12	20 nm; 0.25 mg/mL or 0.5 mg/mL; 24 h or 48 h	For primary microglia, NO production elevated, accompanied by increased mRNA and protein levels of iNOS and the mRNA expressions of MCP-1 and MIP-1α enhanced; NF-κB binding activity increased; the mRNA expression of Th inhibited. For PC12 cells, marked cytotoxicity of PC12 after cells incubation with supernatant of TiO_2 NPs-treated microglia	Xue et al. (2012)
Anatase	PC12	Average 21 nm (range from 20 to 50 nm); 1, 10, 50, 100 μg/mL; incubated for 6, 12, 24, 48 h	The viability of cells decreased in all experimental groups except the 1 μg/mL; the percentage of DCF-positive cells elevated; increased ratio of PC12 apoptosis	Liu et al. (2010a)
Anatase and rutile	PC12	20 nm; 25, 50, 100, 200 μg/mL for 24 h	Apparent cytotoxicity; LDH leakage; DCF fluorescence intensity increased; the levels of GSH and SOD decreased; the level of MDA elevated; reduction in mitochondrial membrane potential; cell apoptosis and necrosis; the elevated percentage of G2/M phase cells; the protein levels of p-JNK, JNK, p-c-Jun, Jun, p-P53, p53, p21 GADD45, bcl-2, bax disrupted	Wu et al. (2010)
Anatase (96%)	C6 U373	40–200 nm; 2.5, 5, 10, 20, 40 μg/mL; 24, 48 or 96 h	The proliferation of C6 and U373 depressed; morphology of both cell lines changed; TiO_2 NPs internalized by both cell lines; reduction in immune-location of F-actin fibers; changes in cytoskeleton; apoptosis	Marquez-Ramirez et al. (2012)

Unknown	C6 U373	50 nm; 20 μg/cm^2 for 2, 4, 6, 24, 48, 72 h	Elevation in H2DCFDA in both cell lines; imbalance in the mRNA levels of GPx, SOD, and catalase; fluorescence of cis-parinaric acid and Rh123 downregulated; MitoTracker Green FM staining elevated	Huerta-Garcia et al. (2014)
Rutile coated by SiO_2	Mouse NSCs line C17.2	80–100 nm; 50, 100, 150, 200, 250 μg/mL exposed for 12, 24, 36, 48, 60, 72 h or 7 days	TiO_2 NPs internalized; distribution in the cytoplasma; inhibition on cellular proliferation; β-tublin-positive cells; expressions of nine proteins changed; mRNA level of Cx43 elevated with the reduction in PKCε	Liu et al. (2010b)
Unknown	Human cerebral endothelial cells (HCECs)	21 nm; 2 mg/mL; 0.12, 0.6, 3, 15, 75 μg/cm^2 for 4, 24, 48, or 72 h	Internalized by HCECs; significant cytotoxicity; ROS production; marked DNA damage; the levels of activated cathepsin D and LC3-II upregulated	Kenzaoui et al. (2012)
Anatase	Primary hippocampal neurons	5 nm; 5, 15, 30, 40, 50 μg/mL for 6, 12, 24, or 48 h	Cell viabilities reduced; LDH activities elevated; ultrastructure of cells altered; apoptotic rate enhanced; reduction of MMP; elevation of cytoplasmic Ca^{2+}; apoptotic cytokine levels disturbed	Sheng et al. (2015)
Anatase(S) Anatase (80%)+rutile(D)	Human SHSY5Y neuronal cells	25 nm; 20, 40, 60, 80, 100, 120, 140, 160 μg/mL for 3, 6, 24 h	No cytotoxicity; TiO_2 NPs internalized; changed cell cycle in TiO_2-S; elevated percentage of apoptotic cells; genotoxicity; no oxidative damage	Valdiglesias et al. (2013)
Anatase	Human neural stem cell line	80 nm; 0.01, 0.1, 1 mg/mL for 7 days	Cellular aggregation; changed morphology; no change with mitochondrial activity; elevated protein levels of Nestin (stem cell marker) and neurofilament heavy polypeptide (N-FH; neuron marker); gene level of the high mobility group AT-hook 1 (HMGA1) enhanced	Fujioka et al. (2014)

DCF, dihydrodichlorofluorescein; *GSH*, glutathione; *HCEC*, human cerebral endothelial cell; *JNK*, Jun N-terminal kinase; *LDH*, lactate dehydrogenase; *MDA*, malonaldehyde; *MMP*, mitochondrial membrane potential; *NF-κB*, nuclear factor κB; *NO*, nitric oxide; *NPs*, nanoparticles; *NSC*, neural stem cell; *ROS*, reactive oxygen species; *SOD*, superoxide dismutase.

which is a key mechanism in brain development (Sorce and Krause, 2009). Furthermore, ROS can regulate neuronal ion channels, kinases, and transcription factors (Thomas et al., 2007; Schaeppi et al., 2008). ROS generated from the NADPH oxidase 2 (Nox2) system also can induce long-term memory dysfunction (Massaad and Klann, 2011).

ROS are capable of causing membrane impairment that influences the structure and function of inner proteins, lipid denaturation, and structural alteration of DNA. DNA oxidation is of the utmost concern because it can lead to mutations and alterations in gene expression. Owing to the lack of DNA repair enzymes, mitochondrial DNA appears to be more susceptible to ROS-induced mutations. Protein oxidation may cause the aggregation of insoluble protein, which serves as the molecular basis of some diseases, especially neurodegenerative pathologies (Brieger et al., 2012). There is an increasing amount of evidence indicating that NPs can cause cytotoxicity because of their large surface area, which is capable of inducing ROS, a key factor that generates cellular stress and disease pathogenesis (Nel et al., 2006). However, the impact of NP-induced ROS on the brain is unclear.

A study revealed that after female mice were injected with TiO_2 NPs for 14 consecutive days, TiO_2 NPs were detected in the mice brain and the activities of superoxide dismutase (SOD), catalase (CAT), ascorbate peroxidase (APx), glutathione peroxidase (GSH-Px) and the ratio of ascorbic acid (AsA)/oxidized AsA (DAsA), and glutathione (GSH)/oxidized glutathione (GSSG) in mice brain were significantly downregulated as compared with that in the control group. Meanwhile, the experimental group exhibited higher levels of O_2, H_2O_2, malonaldehyde (MDA), nitric oxide synthase, and nitric oxide than the control group (Ma et al., 2010). Similarly, Hu et al. (2011) discovered that after female mice were treated with TiO_2 NPs every day for 60 days, the expressions of O_2 and H_2O_2 were markedly enhanced and the activities of SOD, CAT, APx, and GSH-Px were significantly downregulated. Meanwhile, the ratios of AsA/DAsA and GSH/GSSG were decreased as well. Another research (Ze et al., 2013) also demonstrated that when male mice were exposed to TiO_2 NPs for 90 consecutive days, the content of TiO_2 NPs in brain was significantly increased and the mRNA expression of genes regulating oxidative stress, such as p38, Jun N-terminal kinase (JNK), nuclear factor (NF)-κB, nuclear factor E2-related factor 2, and heme oxygenase-1, was apparently enhanced in the exposed group. At the same time, the levels of O_2, H_2O_2, MDA, protein carbonyl, and 8-hydroxydeoxyguanosine in brains of treated group were significantly promoted. These results indicated that the antioxidant capabilities of the brain were impaired by TiO_2 NPs, which might lead to excessive oxidative stress.

Prenatal exposure to TiO_2 NPs might affect the development of brain in the fetus. Cui et al. (2014) found that when pregnant rats were exposed to TiO_2 NPs, the antioxidant capabilities of the brain in their pups were affected. The maternal exposure to TiO_2 NPs could induce oxidative stress to the nucleic acids in the fetal brain. Moreover, their offspring exhibited poor performance in behavioral tests due to impairments to the brain.

In an in vitro research, TiO_2 NPs can decrease the viability of PC12 cells and the percentage of dihydrodichlorofluorescein (DCF)-positive cells was significantly increased. However, pretreating PC12 cells with ROS scavenger can ameliorate the harmful effects induced by TiO_2 NPs (Liu et al., 2010a). In addition to the increased percentage of DCF-positive cells, the expression of GSH and SOD was decreased and the levels of lactate dehydrogenase (LDH) and MDA were significantly promoted, after PC12 cells were incubated with TiO_2 NPs (Wu et al., 2010). The gene expression of antioxidant enzymes in C6 and U373 cell lines was decreased and oxidation of lipids was promoted by TiO_2 NPs treatment (Huerta-Garcia et al., 2014).

5.2 Apoptosis and Autophagy

Apoptosis, which can be called programmed cell death, is defined as the genetically determined elimination of cells, in which caspase activation plays a pivotal role. However, autophagy, which is a highly regulated cellular process for eliminating long-lived proteins and damaged organelle components via the lysosomal mechanism, has attracted considerable attention recently. Generally, autophagy can be simply classified into three major types: (1) microautophagy, (2) macroautophagy, and (3) identified-chaperone-mediated autophagy. It is reported that macroautophagy (usually referred to as autophagy) is the major degradative pathway among the three subtypes of autophagy (Hara et al., 2006; Levine and Kroemer, 2008; Mizushima et al., 2008). Meanwhile, autophagy also serves as a dynamic recycling system by which new building blocks and energy for cellular renovation and homeostasis are produced (Mizushima and Komatsu, 2011). Increasing evidences disclosed that autophagy alterations might be the main mechanisms of neurodegenerative diseases, such as AD, PD, and Huntington's disease (Cheung and Ip, 2011; Nixon, 2013). Moreover, neurodegeneration and an elevated level of protein aggregation have been detected in mice with neuron-specific knockdown of key autophagy proteins (ie, Atg5, Atg7, and beclin-1) (Podila and Brown, 2013). The cell autophagy process can be induced by various forms of stress, such as interactions with various types of NPs. Thus, NPs are considered to be a novel category of autophagy inducers (Zhang et al., 2010).

In addition to impaired antioxidant capabilities in TiO_2 NPs-treated female mice brains, Hu et al. (2011) disclosed that the levels of apoptosis-related genes in mice brains were altered by TiO_2 NPs. The mRNA and protein expressions of caspase-9, caspase-3, Bax, and cytochrome c were significantly promoted, whereas Bcl-2 level was markedly downregulated. However, the caspase-8 level was not affected. These findings indicated that apoptosis induced by TiO_2 NPs might probably depend on the intrinsic pathway. Furthermore, the spatial recognition memory of mice was impaired due to the apoptosis of neurons in hippocampus in the experimental groups. Besides, TiO_2 NPs can induce mitochondria-mediated apoptosis in brains of rats as well (Meena et al., 2015).

Apoptosis was induced after PC12 cells were treated with TiO_2 NPs (Liu et al., 2010a; Wu et al., 2010). After primary hippocampal neurons were exposed to TiO_2 NPs, apoptotic cytokine levels, including cytochrome c, caspase-3, Bax, and caspase-12, were markedly increased and the apoptotic rate was apparently increased (Sheng et al., 2015). Besides PC12 cells, the percentage of apoptotic SH-SY5Y neuronal cells was increased by TiO_2 NPs as well (Valdiglesias et al., 2013). Moreover, apoptosis can be induced by TiO_2 NPs in both C6 (rat glial cell) and U373 (human glial cell) (Marquez-Ramirez et al., 2012).

After human cerebral endothelial cells were exposed to TiO_2 NPs, the level of LC3-II was significantly upregulated, which indicated that TiO_2 NPs can induce autophagy. Meanwhile, these findings indicated that the integrity of BBB can be impaired by TiO_2 NPs, which might facilitate TiO_2 NPs to be transported into brain (Kenzaoui et al., 2012).

5.3 Immune Mechanism

Once NPs enter the brain, they immediately encounter a complex environment of resident microglial immune cells and neurons. Microglia, accounting for ~20% of the glial cells in the brain, are a type of glial cells, which are the resident innate immune cells in the brain and the predominant regulators of neuroinflammation (Block et al., 2012). As macrophages located in the brain, the microglial cells are the most likely to respond when NPs enter the brain. Once activated, they are able to change in form and function.

When PC12 cells were incubated with supernatant of TiO_2 NPs-treated primary microglia for 48 h, the viability of PC12 cells was significantly reduced (Xue et al., 2012). In another study (Wang et al., 2008c), TiO_2 NPs can lead to enhanced expression of tumor necrosis factor (TNF)-α and interleukin (IL)-1β in brains of female mice. Ze et al. (2014b) also revealed that TiO_2 NPs induced upregulated expression of Toll-like receptor (TLR) 2, TLR-4, TNF-α, NF-κB, and IL-1β in mouse brains. Moreover, the spatial recognition memory and locomotor activity were impaired because of this inflammatory response.

5.4 Activated Signaling Pathways

NPs can impair or pass through the cellular membrane into the cell. In addition to ROS, NPs can change the configuration and conformation of biological macromolecules (including membrane lipid, protein, and nucleus), resulting in their corresponding dysfunction. Certain cell-signaling pathways can be activated upon NPs interaction, such as the proinflammatory cytokines referred to as mitogen-activated protein kinase and NF-κB cascades.

Only few studies illustrated this mechanism. One study showed that the protein levels of JNK, p-c-jun, and jun of PC12 cells were disrupted by TiO_2 NPs, which indicated that the JNK pathway was implicated in cytotoxicity of PC12 cells induced by TiO_2 NPs (Wu et al., 2010).

Based on the mechanisms mentioned earlier, oxidative stress and apoptosis might be the major neurotoxic mechanisms of TiO_2 NPs. However, other mechanisms should not be ignored, such as autophagy. Therefore, more studies are needed to further investigate the mechanisms of nanoneurotoxicity.

6. MAJOR FACTORS INFLUENCING THE NEUROTOXICITY OF TIO_2 NANOPARTICLES

The neurotoxic effects of TiO_2 NPs could be modulated by its peculiar physicochemical characteristics, administration routes, dosage, and so on. Therefore, although the above-mentioned studies are all focused on the harmful impacts of TiO_2 NPs on the brain, different conclusions have been obtained. As a consequence, for the purpose of assessing their neurotoxicity in a standard way, it is vital to discuss the major factors that might influence the neurotoxicity of TiO_2 NPs.

6.1 Crystal Type

TiO_2 NPs, unlike other NPs, have two crystal types: the anatase and the rutile (Chen and Mao, 2007). Both possess subtly different physicochemical characteristics, which lead to different toxic effects. It was reported that the toxicity of the anatase form was more severe than that of rutile (Gitrowski et al., 2014; Lin et al., 2014; Clemente et al., 2015). However, some studies did not draw the same conclusions (Numano et al., 2014; Sekar et al., 2014). Concerning the TiO_2 NPs, how different crystal forms of TiO_2 NPs affect neurotoxicity is still unclear. In the study by Wang et al. (2008a), female mice were treated with anatase or rutile TiO_2 NPs. The levels of glial fibrillary acidic protein, MDA, acetylcholinesterase activity, and glutamic acid in the anatase group were higher than that in rutile group. In another report of the same research group (Wang et al., 2008c), the elevated levels of GSH-Px, GST, GSH, and SOD in rutile group were significantly higher than in the control and anatase groups at 10 days after female mice were treated with TiO_2 NPs. But the levels of IL-1β and TNF-α in the anatase group were apparently higher than those in the control and rutile groups after 30 days exposure. The rutile TiO_2 NPs did not induce elevation in the levels of IL-1β and TNF-α.

When PC12 cells were exposed to anatase or rutile TiO_2 NPs, the reduction in cell viability, mitochondrial membrane potential (MMP), and levels of GSH and SOD in the anatase group were remarkably higher than that in the rutile group. The levels of LDH and MDA, caspase 3 activity, and the percentage of necrosis in the anatase group were increased, which were higher than that in the rutile group (Wu et al., 2010). However, an inconsistent conclusion was drawn from this study that the effects of both TiO_2-S (100% anatase NPs) and TiO_2-D (80% anatase+20% rutile) on the brain were similar with no statistical difference (Valdiglesias et al., 2013).

6.2 Size of Nanoparticles

Several researches have revealed that the dimension of nanomaterial is another vital factor that can influence the nanoneurotoxicity (Chen et al., 2015). When it comes to the effects of TiO_2 NPs on the brain, little is known about how different sizes influence the neurotoxicity of TiO_2 NPs. Data collected from current studies only compared the toxicity of micro-TiO_2 with that of nanosized-TiO_2. Micro-TiO_2 was not detected in the brain regions. Thus it was demonstrated to have no toxic effects (Ma et al., 2010; Shin et al., 2010; Wu et al., 2010; Zhang et al., 2011; Fujioka et al., 2014). Therefore, for fully understanding the neurotoxicity of TiO_2 NPs, more studies are needed to investigate the effects of TiO_2 NPs on the brain with different sizes at the nanoscale level.

6.3 Shape and Surface Modification

The morphology of NPs is crucial to their toxicity. The effects of NPs on organisms might be regulated by morphological structure (Hamilton et al., 2009; Oosthuizen et al., 2012; Silva et al., 2013; Tong et al., 2013; Medina-Reyes et al., 2015). However, how different shapes (mainly including nanobelts, nanorods, nanotubes, and nanospheres) of TiO_2 NPs modulate their transportation to the brain, how they get excreted from the brain, and how they have toxic effects on neurons or glia cells are still largely unknown. Moreover, the surface coating can regulate the physicochemical properties of TiO_2 NPs, which might influence their toxicity (Rossi et al., 2010; Dabrunz et al., 2011). However, whether TiO_2 NPs coated with inorganic or organic materials could alleviate or exacerbate the harmful impacts on the brain is unclear as well. In the study by Liu et al. (2010b), mouse neural stem cells C17.2 were treated with SiO_2-coated rutile TiO_2 NPs. This exposure induced elevation of the β-tublin-positive cells, which indicated that the TiO_2 NPs can induce the C17.2 differentiating into neurons. Zhang et al. (2011) investigated how different size, morphology, and surface modification of TiO_2 NPs in rutile form regulated their toxic effects on the brain after female mice were exposed by intranasal instillation every other day for 30 days. The experimental groups included group A (microsized, hydrophobic, rod like, and no coating), group B (nanosized, hydrophobic, needle-like, and no coating), group C (nanosized, hydrophilic, needle-like, and coated with silica), and group D (nanosized, hydrophilic, rod-like, and coated with silica). Results collected from the study demonstrated that (1) microsized TiO_2 could not be detected in the brain regions. The Ti contents in cerebellum region demonstrated no significant difference in the five groups (four experimental groups and one control group). Significantly increased Ti contents in the cerebral cortex were detected in groups C and D. In the striatum region, the Ti contents were markedly elevated in groups B, C, and D; (2) the neuronal loss in cerebral cortex and hippocampus CA1 region was significant in groups B, C, and D; (3) the levels of norepinephrine in hippocampus, cerebral cortex, cerebellum, and striatum were significantly reduced in groups C and D; (4) the levels of dopamine, dihydroxy-phenylacetic acid, herpes virus ateles, 5-hydroxytryptamine, 5-hydroxyindoleacetic acid in the four sub-brain regions were significantly affected in groups C and D. These findings suggested that the size, shape, and surface modification could modulate the toxic effects of TiO_2 NPs on the brain. However, studies about effects of shape and surface modification on neurotoxicity of TiO_2 NPs are insufficient, which need further investigations.

6.4 Administration Route

As for in vivo studies, the administration routes play an important part in neurotoxicity of NPs (Espinosa-Cristobal et al., 2013; Choi et al., 2015). The Ti contents in the brain regions of mice/rats could not be detected when the TiO_2 NPs were administered via intravenous injection (Fabian et al., 2008; Sugibayashi et al., 2008; Patri et al., 2009; Shinohara et al., 2014a). But when the pregnant mice were treated with TiO_2 NPs, their uterine weights were lowered with smaller fetuses, which indirectly suggested that the TiO_2 NPs could induce fetal resorption and retard fetal growth via intravenous injection (Yamashita et al., 2011). Ti could not be detected in the brain when mice were exposed to TiO_2 NPs by inhalation (Yin et al., 2014). Based on current studies, intranasal instillation or nasal administration was the most effective pathway for TiO_2 NPs to be translocated into brain (Wang et al., 2008b). This might be due to the retrograde axonal translocation of NPs directly from nose to brain. Therefore, more researches are needed to investigate how and why the different administration routes lead to different biodistribution of NPs.

7. SUMMARY

With the rapid development of nanotechnology, numerous nanomaterial-based products are widely used at present, such as consumer products, food additives, cosmetics, and drug carriers. Meanwhile, concerns on health risks about unexpected exposure to TiO_2 NPs are arising. In in vivo studies, once animals were exposed to TiO_2 NPs, the NPs could be translocated into the brain mainly through the BBB and nose–brain pathway. Besides, TiO_2 NPs may affect brain development of embryo by crossing the placental barrier. The TiO_2 NPs contents accumulated in the brain regions are tiny at one exposure, but its elimination from the brain was limited. Therefore, long-term or chronic exposure to TiO_2 NPs could potentially lead to the gradually increased TiO_2 NPs contents in the brain, which may eventually induce impairments on the neurons and glial cells and lead to brain dysfunction as a consequence.

Several in vivo and in vitro studies have demonstrated that TiO_2 NPs, for their nanoscale, are toxic to brain. However, as the experimental parameters used in all of the current studies were not standardized (such as different administration routes, experimental animals, crystal forms, and different shape and size), the conclusions from those studies are not comparable and some of them might even be conflicting. Therefore, it is urgent to standardize experiments on assessing the neurotoxicity of TiO_2 NPs. In addition, all the research objects in those experiments are only consisted of animals (such as mice and rats) or cells (such as PC12, U373, and C6). In consequence, neurotoxic data of TiO_2 NPs collected from those studies might be inappropriate to determine their neurotoxic effects on humans. For another, the mechanisms underlying the neurotoxicity of TiO_2 NPs are still not clear. For example, oxidative stress or ROS might be the main mechanism underlying neurotoxicity of TiO_2 NPs. Therefore, the mechanisms of nanoneurotoxicity are still an important topic for future research. On the other hand, the toxic effects of different physicochemical characteristics of TiO_2 NPs on the brain are still unclear and should be investigated intensively, which includes the crystal forms, shape, size, surface modifications, and so on. To decrease translocation rate of TiO_2 NPs into the brain and neurotoxicity, it is urgent to seek out the optimum parameters of physicochemical properties of TiO_2 NPs to improve the biosafety of TiO_2 NPs-based products. Moreover, because TiO_2 NPs could induce neuron or glial cell death and disturb the homeostasis in brain, the possible relationship between the TiO_2 NPs exposure and neurodegenerative diseases or psychiatric disorders needs further investigation.

REFERENCES

Auttachoat, W., McLoughlin, C.E., White Jr., K.L., Smith, M.J., 2014. Route-dependent systemic and local immune effects following exposure to solutions prepared from titanium dioxide nanoparticles. J. Immunotoxicol. 11, 273–282.

Baello, S., Iqbal, M., Bloise, E., Javam, M., Gibb, W., Matthews, S.G., 2014. TGF-beta 1 regulation of multidrug resistance P-glycoprotein in the developing male blood–brain barrier. Endocrinology 155, 475–484.

Ballabh, P., Braun, A., Nedergaard, M., 2004. The blood–brain barrier: an overview – structure, regulation, and clinical implications. Neurobiol. Dis. 16, 1–13.

Barbu, E., Molnar, E., Tsibouklis, J., Gorecki, D.C., 2009. The potential for nanoparticle-based drug delivery to the brain: overcoming the blood–brain barrier. Expert Opin. Drug Deliv. 6, 553–565.

Ben Younes, N.R., Amara, S., Mrad, I., Ben-Slama, I., Jeljeli, M., Omri, K., El Ghoul, J., El Mir, L., Ben Rhouma, K., Abdelmelek, H., Sakly, M., 2015. Subacute toxicity of titanium dioxide (TiO_2) nanoparticles in male rats: emotional behavior and pathophysiological examination. Environ. Sci. Pollut. Res. Int. 22, 8728–8737.

Bird, C.M., Burgess, N., 2008. The hippocampus and memory: insights from spatial processing. Nat. Rev. Neurosci. 9, 182–194.

Block, M.L., Elder, A., Auten, R.L., Bilbo, S.D., Chen, H., Chen, J.-C., Cory-Slechta, D.A., Costa, D., Diaz-Sanchez, D., Dorman, D.C., Gold, D.R., Gray, K., Jeng, H.A., Kaufman, J.D., Kleinman, M.T., Kirshner, A., Lawler, C., Miller, D.S., Nadadur, S.S., Ritz, B., Semmens, E.O., Tonelli, L.H., Veronesi, B., Wright, R.O., Wright, R.J., 2012. The outdoor air pollution and brain health workshop. Neurotoxicology 33, 972–984.

Brieger, K., Schiavone, S., Miller Jr., F.J., Krause, K.-H., 2012. Reactive oxygen species: from health to disease. Swiss Med. Wkly. 142.

Brun, E., Carriere, M., Mabondzo, A., 2012. In vitro evidence of dysregulation of blood–brain barrier function after acute and repeated/long-term exposure to TiO_2 nanoparticles. Biomaterials 33, 886–896.

Candelario-Jalil, E., Yang, Y., Rosenberg, G.A., 2009. Diverse roles of matrix metalloproteinases and tissue inhibitors of metalloproteinases in neuroinflammation and cerebral ischemia. Neuroscience 158, 983–994.

Catauro, M., Bollino, F., Papale, F., Marciano, S., Pacifico, S., 2015. TiO_2/PCL hybrid materials synthesized via sol-gel technique for biomedical applications. Mater. Sci. Eng. C Mater. Biol. Appl. 47, 135–141.

Chen, X., Mao, S.S., 2007. Titanium dioxide nanomaterials: synthesis, properties, modifications, and applications. Chem. Rev. 107, 2891–2959.

Chen, J., Dong, X., Zhao, J., Tang, G., 2009. In vivo acute toxicity of titanium dioxide nanoparticles to mice after intraperitioneal injection. J. Appl. Toxicol. 29, 330–337.

Chen, L.Q., Fang, L., Ling, J., Ding, C.Z., Kang, B., Huang, C.Z., 2015. Nanotoxicity of silver nanoparticles to red blood cells: size dependent adsorption, uptake, and hemolytic activity. Chem. Res. Toxicol. 28, 501–509.

Cheng, F., Sajedin, S.M., Kelly, S.M., Lee, A.F., Kornherr, A., 2014. UV-stable paper coated with APTES-modified P25 TiO_2 nanoparticles. Carbohydr. Polym. 114, 246–252.

Cheung, Z.H., Ip, N.Y., 2011. Autophagy deregulation in neurodegenerative diseases – recent advances and future perspectives. J. Neurochem. 118, 317–325.

Cho, W.-S., Kang, B.-C., Lee, J.K., Jeong, J., Che, J.-H., Seok, S.H., 2013. Comparative absorption, distribution, and excretion of titanium dioxide and zinc oxide nanoparticles after repeated oral administration. Part. Fibre Toxicol. 10.

Choi, J., Kim, H., Kim, P., Jo, E., Kim, H.-M., Lee, M.-Y., Jin, S.M., Park, K., 2015. Toxicity of zinc oxide nanoparticles in rats treated by two different routes: single intravenous injection and single oral administration. J. Toxicol. Environ. Health A 78, 226–243.

Chu, M., Wu, Q., Yang, H., Yuan, R., Hou, S., Yang, Y., Zou, Y., Xu, S., Xu, K., Ji, A., Sheng, L., 2010. Transfer of quantum dots from pregnant mice to pups across the placental barrier. Small 6, 670–678.

Clemente, Z., Castro, V.L., Feitosa, L.O., Lima, R., Jonsson, C.M., Maia, A.H.N., Fraceto, L.F., 2015. Biomarker evaluation in fish after prolonged exposure to nano-TiO_2: influence of illumination conditions and crystal phase. J. Nanosci. Nanotechnol. 15, 5424–5433.

Cui, Y., Chen, X., Zhou, Z., Lei, Y., Ma, M., Cao, R., Sun, T., Xu, J., Huo, M., Cao, R., Wen, C., Che, Y., 2014. Prenatal exposure to nanoparticulate titanium dioxide enhances depressive-like behaviors in adult rats. Chemosphere 96, 99–104.

Dabrunz, A., Duester, L., Prasse, C., Seitz, F., Rosenfeldt, R., Schilde, C., Schaumann, G.E., Schulz, R., 2011. Biological surface coating and molting inhibition as mechanisms of TiO_2 nanoparticle toxicity in *Daphnia magna*. PLoS One 6.

D'Aversa, T.G., Eugenin, E.A., Lopez, L., Berman, J.W., 2013. Myelin basic protein induces inflammatory mediators from primary human endothelial cells and blood–brain barrier disruption: implications for the pathogenesis of multiple sclerosis. Neuropathol. Appl. Neurobiol. 39, 270–283.

De Vos, K.J., Grierson, A.J., Ackerley, S., Miller, C.C.J., 2008. Role of axonal transport in neurodegenerative diseases. Annu. Rev. Neurosci. 151–173.

Di Bona, K.R., Xu, Y., Ramirez, P.A., DeLaine, J., Parker, C., Bao, Y., Rasco, J.F., 2014. Surface charge and dosage dependent potential developmental toxicity and biodistribution of iron oxide nanoparticles in pregnant CD-1 mice. Reprod. Toxicol. 50, 36–42.

Dominguez, A., Suarez-Merino, B., Goni-de-Cerio, F., 2014. Nanoparticles and blood–brain barrier: the key to central nervous system diseases. J. Nanosci. Nanotechnol. 14, 766–779.

Ekblad, M., Gissler, M., Lehtonen, L., Korkeila, J., 2010. Prenatal smoking exposure and the risk of psychiatric morbidity into young adulthood. Arch. Gen. Psychiatry 67, 841–849.

Espinosa-Cristobal, L.F., Martinez-Castanon, G.A., Loyola-Rodriguez, J.P., Patino-Marin, N., Reyes-Macias, J.F., Vargas-Morales, J.M., Ruiz, F., 2013. Toxicity, distribution, and accumulation of silver nanoparticles in Wistar rats. J. Nanopart. Res. 15.

Fabian, E., Landsiedel, R., Ma-Hock, L., Wiench, K., Wohlleben, W., van Ravenzwaay, B., 2008. Tissue distribution and toxicity of intravenously administered titanium dioxide nanoparticles in rats. Arch. Toxicol. 82, 151–157.

Fang, W., Lv, P., Geng, X., Shang, E., Yang, Q., Sha, L., Li, Y., 2013. Penetration of verapamil across blood–brain barrier following cerebral ischemia depending on both paracellular pathway and P-glycoprotein transportation. Neurochem. Int. 62, 23–30.

Fujioka, K., Hanada, S., Inoue, Y., Sato, K., Hirakuri, K., Shiraishi, K., Kanaya, F., Ikeda, K., Usui, R., Yamamoto, K., Kim, S.U., Manome, Y., 2014. Effects of silica and titanium oxide particles on a human neural stem cell line: morphology, mitochondrial activity, and gene expression of differentiation markers. Int. J. Mol. Sci. 15, 11742–11759.

Geraets, L., Oomen, A.G., Krystek, P., Jacobsen, N.R., Wallin, H., Laurentie, M., Verharen, H.W., Brandon, E.F.A., de Jong, W.H., 2014. Tissue distribution and elimination after oral and intravenous administration of different titanium dioxide nanoparticles in rats. Part. Fibre Toxicol. 11.

Gitrowski, C., Al-Jubory, A.R., Handy, R.D., 2014. Uptake of different crystal structures of TiO_2 nanoparticles by Caco-2 intestinal cells. Toxicol. Lett. 226, 264–276.

Gustafsson, A., Jonasson, S., Sandstrom, T., Lorentzen, J.C., Bucht, A., 2014. Genetic variation influences immune responses in sensitive rats following exposure to TiO_2 nanoparticles. Toxicology 326, 74–85.

Hamilton Jr., R.F., Wu, N., Porter, D., Buford, M., Wolfarth, M., Holian, A., 2009. Particle length-dependent titanium dioxide nanomaterials toxicity and bioactivity. Part. Fibre Toxicol. 6.

Hara, T., Nakamura, K., Matsui, M., Yamamoto, A., Nakahara, Y., Suzuki-Migishima, R., Yokoyama, M., Mishima, K., Saito, I., Okano, H., Mizushima, N., 2006. Suppression of basal autophagy in neural cells causes neurodegenerative disease in mice. Nature 441, 885–889.

Hong, F., Hong, J., Wang, L., Zhou, Y., Liu, D., Xu, B., Yu, X., Sheng, L., 2015. Chronic exposure to nanoparticulate TiO_2 causes renal fibrosis involving activation of the Wnt pathway in mouse kidney. J. Agric. Food Chem. 63, 1639–1647.

Hosking, M.P., Liu, L., Ransohoff, R.M., Lane, T.E., 2009. A protective role for ELR plus chemokines during acute viral encephalomyelitis. PLoS Pathog. 5.

Hougaard, K.S., Jackson, P., Jensen, K.A., Sloth, J.J., Loeschner, K., Larsen, E.H., Birkedal, R.K., Vibenholt, A., Boisen, A.-M.Z., Wallin, H., Vogel, U., 2010. Effects of prenatal exposure to surface-coated nanosized titanium dioxide (UV-Titan). A study in mice. Part. Fibre Toxicol. 7.

Howland, J.G., Wang, Y.T., 2008. Synaptic plasticity in learning and memory: stress effects in the hippocampus. Prog. Brain Res. 169, 145–158.

Hu, R., Gong, X., Duan, Y., Li, N., Che, Y., Cui, Y., et al., 2010. Neurotoxicological effects and the impairment of spatial recognition memory in mice caused by exposure to TiO_2 nanoparticles. Biomaterials 31(31), 8043–8050. 10.1016/j.biomaterials.2010.07.011.

Hu, R., Zheng, L., Zhang, T., Gao, G., Cui, Y., Cheng, Z., Cheng, J., Hong, M., Tang, M., Hong, F., 2011. Molecular mechanism of hippocampal apoptosis of mice following exposure to titanium dioxide nanoparticles. J. Hazard. Mater. 191, 32–40.

Hua, Z.K., Tang, F., Bai, Y.H., Wei, B., Xu, Y.M., Ye, D.M., Wang, X.J., Chen, T., Wang, G., 2015. A titanium dioxide nano film encapsulation for metal orthopedic implants in diathermy rehabilitation therapy. Nanosci. Nanotechnol. Lett. 7, 268–271.

Huerta-Garcia, E., Perez-Arizti, J.A., Marquez-Ramirez, S.G., Delgado-Buenrostro, N.L., Chirino, Y.I., Iglesias, G.G., Lopez-Marure, R., 2014. Titanium dioxide nanoparticles induce strong oxidative stress and mitochondrial damage in glial cells. Free Radic. Biol. Med. 73, 84–94.

Kenzaoui, B.H., Bernasconi, C.C., Guney-Ayra, S., Juillerat-Jeanneret, L., 2012. Induction of oxidative stress, lysosome activation and autophagy by nanoparticles in human brain-derived endothelial cells. Biochem. J. 441, 813–821.

Khan, M.Z., Ashraf, M., Hussain, T., Rehman, A., Malik, M.M., Raza, Z.A., Nawab, Y., Zia, Q., 2015. In situ deposition of TiO_2 nanoparticles on polyester fabric and study of its functional properties. Fibers Polym. 16, 1092–1097.

Levine, B., Kroemer, G., 2008. Autophagy in the pathogenesis of disease. Cell 132, 27–42.

Li, Y., Li, J., Yin, J., Li, W., Kang, C., Huang, Q., Li, Q., 2010. Systematic influence induced by 3 nm titanium dioxide following intratracheal instillation of mice. J. Nanosci. Nanotechnol. 10, 8544–8549.

Li, J., O, W., Li, W., Jiang, Z.-G., Ghanbari, H.A., 2013. Oxidative stress and neurodegenerative disorders. Int. J. Mol. Sci. 14, 24438–24475.

Lin, X., Li, J., Ma, S., Liu, G., Yang, K., Tong, M., Lin, D., 2014. Toxicity of TiO_2 nanoparticles to *Escherichia coli*: effects of particle size, crystal phase and water chemistry. PLoS One 9.

Liu, S., Xu, L., Zhang, T., Ren, G., Yang, Z., 2010a. Oxidative stress and apoptosis induced by nanosized titanium dioxide in PC12 cells. Toxicology 267, 172–177.

Liu, X., Ren, X., Deng, X., Huo, Y., Xie, J., Huang, H., Jiao, Z., Wu, M., Liu, Y., Wen, T., 2010b. A protein interaction network for the analysis of the neuronal differentiation of neural stem cells in response to titanium dioxide nanoparticles. Biomaterials 31, 3063–3070.

Lockman, P.R., Koziara, J.M., Mumper, R.J., Allen, D.D., 2004. Nanoparticle surface charges alter blood–brain barrier integrity and permeability. J. Drug Target. 12, 635–641.

Logan, N., Sherif, A., Cross, A.J., Collins, S.N., Traynor, A., Bozec, L., Parkin, I.P., Brett, P., 2015. TiO_2-coated CoCrMo: improving the osteogenic differentiation and adhesion of mesenchymal stem cells in vitro. J. Biomed. Mater. Res. A 103, 1208–1217.

Louboutin, J.-P., Strayer, D.S., 2013. Relationship between the chemokine receptor CCR5 and microglia in neurological disorders: consequences of targeting CCR5 on neuroinflammation, neuronal death and regeneration in a model of epilepsy. CNS Neurol. Disord. Drug Targets 12, 815–829.

Ma, L., Liu, J., Li, N., Wang, J., Duan, Y., Yan, J., Liu, H., Wang, H., Hong, F., 2010. Oxidative stress in the brain of mice caused by translocated nanoparticulate TiO_2 delivered to the abdominal cavity. Biomaterials 31, 99–105.

Ma, Y., Wang, X., Jia, Y., Chen, X., Han, H., Li, C., 2014. Titanium dioxide-based nanomaterials for photocatalytic fuel generations. Chem. Rev. 114, 9987–10043.

Marquez-Ramirez, S.G., Delgado-Buenrostro, N.L., Chirino, Y.I., Iglesias, G.G., Lopez-Marure, R., 2012. Titanium dioxide nanoparticles inhibit proliferation and induce morphological changes and apoptosis in glial cells. Toxicology 302, 146–156.

Martirosyan, A., Schneider, Y.-J., 2014. Engineered nanomaterials in food: implications for food safety and consumer health. Int. J. Environ. Res. Public Health 11, 5720–5750.

Massaad, C.A., Klann, E., 2011. Reactive oxygen species in the regulation of synaptic plasticity and memory. Antioxid. Redox Signal. 14, 2013–2054.

Medina-Reyes, E.I., Deciga-Alcaraz, A., Freyre-Fonseca, V., Delgado-Buenrostro, N.L., Flores-Flores, J.O., Gutierrez-Lopez, G.F., Sanchez-Perez, Y., Garcia-Cuellar, C.M., Pedraza-Chaverri, J., Chirino, Y.I., 2015. Titanium dioxide nanoparticles induce an adaptive inflammatory response and invasion and proliferation of lung epithelial cells in chorioallantoic membrane. Environ. Res. 136, 424–434.

Meena, R., Kumar, S., Paulraj, R., 2015. Titanium oxide (TiO_2) nanoparticles in induction of apoptosis and inflammatory response in brain. J. Nanopart. Res. 17.

Mestre, L., Inigo, P.M., Mecha, M., Correa, F.G., Hernangomez-Herrero, M., Loria, F., Docagne, F., Borrell, J., Guaza, C., 2011. Anandamide inhibits Theiler's virus induced VCAM-1 in brain endothelial cells and reduces leukocyte transmigration in a model of blood–brain barrier by activation of CB1 receptors. J. Neuroinflammation 8.

Mizushima, N., Komatsu, M., 2011. Autophagy: renovation of cells and tissues. Cell 147, 728–741.

Mizushima, N., Levine, B., Cuervo, A.M., Klionsky, D.J., 2008. Autophagy fights disease through cellular self-digestion. Nature 451, 1069–1075.

Mohammadipour, A., Fazel, A., Haghir, H., Motejaded, F., Rafatpanah, H., Zabihi, H., Hosseini, M., Bideskan, A.E., 2014. Maternal exposure to titanium dioxide nanoparticles during pregnancy; impaired memory and decreased hippocampal cell proliferation in rat offspring. Environ. Toxicol. Pharmacol. 37, 617–625.

Montazer, M., Pakdel, E., 2011. Functionality of nano titanium dioxide on textiles with future aspects: focus on wool. J. Photochem. Photobiol. C 12, 293–303.

Nel, A., Xia, T., Madler, L., Li, N., 2006. Toxic potential of materials at the nanolevel. Science 311, 622–627.

Newman, M.D., Stotland, M., Ellis, J.I., 2009. The safety of nanosized particles in titanium dioxide- and zinc oxide-based sunscreens. J. Am. Acad. Dermatol. 61, 685–692.

Nixon, R.A., 2013. The role of autophagy in neurodegenerative disease. Nat. Med. 19, 983–997.

Numano, T., Xu, J., Futakuchi, M., Fukamachi, K., Alexander, D.B., Furukawa, F., Kanno, J., Hirose, A., Tsuda, H., Suzui, M., 2014. Comparative study of toxic effects of anatase and rutile type nanosized titanium dioxide particles in vivo and in vitro. Asian Pac. J. Cancer Prev. 15, 929–935.

O'Connor, M.J., Paley, B., 2009. Psychiatric conditions associated with prenatal alcohol exposure. Dev. Disabil. Res. Rev. 15, 225–234.

Oosthuizen, M.A., Oberholzer, H.M., Scriba, M.R., van der Spuy, W.J., Pretorius, E., 2012. Evaluation of the morphological changes in the lungs of BALB/c mice after inhalation of spherical and rod-shaped titanium nanoparticles. Micron 43, 863–869.

Patri, A., Umbreit, T., Zheng, J., Nagashima, K., Goering, P., Francke-Carroll, S., Gordon, E., Weaver, J., Miller, T., Sadrieh, N., McNeil, S., Stratmeyer, M., 2009. Energy dispersive X-ray analysis of titanium dioxide nanoparticle distribution after intravenous and subcutaneous injection in mice. J. Appl. Toxicol. 29, 662–672.

Perlson, E., Maday, S., Fu, M.-M., Moughamian, A.J., Holzbaur, E.L.F., 2010. Retrograde axonal transport: pathways to cell death? Trends Neurosci. 33, 335–344.

Podila, R., Brown, J.M., 2013. Toxicity of engineered nanomaterials: a physicochemical perspective. J. Biochem. Mol. Toxicol. 27, 50–55.

Prodana, M., Duta, M., Ionita, D., Bojin, D., Stan, M.S., Dinischiotu, A., Demetrescu, I., 2015. A new complex ceramic coating with carbon nano-tubes, hydroxyapatite and TiO_2 nanotubes on Ti surface for biomedical applications. Ceram. Int. 41, 6318–6325.

Rang, F.J., Boonstra, J., 2014. Causes and consequences of age-related changes in DNA methylation: a role for ROS? Biology 3, 403–425.

Roney, C., Kulkarni, P., Arora, V., Antich, P., Bonte, F., Wu, A.M., Mallikarjuana, N.N., Manohar, S., Liang, H.F., Kulkarni, A.R., Sung, H.W., Sairam, M., Aminabhavi, T.M., 2005. Targeted nanoparticles for drug delivery through the blood–brain barrier for Alzheimer's disease. J. Control. Release 108, 193–214.

Rossi, E.M., Pylkkanen, L., Koivisto, A.J., Vippola, M., Jensen, K.A., Miettinen, M., Sirola, K., Nykasenoja, H., Karisola, P., Stjernvall, T., Vanhala, E., Kiilunen, M., Pasanen, P., Makinen, M., Hameri, K., Joutsensaari, J., Tuomi, T., Jokiniemi, J., Wolff, H., Savolainen, K., Matikainen, S., Alenius, H., 2010. Airway exposure to silica-coated TiO_2 nanoparticles induces pulmonary neutrophilia in mice. Toxicol. Sci. 113, 422–433.

Schaeppi, M.G., Jaquet, V., Belli, D.C., Krause, K.-H., 2008. Hyperinflammation in chronic granulomatous disease and anti-inflammatory role of the phagocyte NADPH oxidase. Semin. Immunopathol. 30, 255–271.

Sekar, D., Falcioni, M.L., Barucca, G., Falcioni, G., 2014. DNA damage and repair following in vitro exposure to two different forms of titanium dioxide nanoparticles on trout erythrocyte. Environ. Toxicol. 29, 117–127.

Semmler-Behnke, M., Lipka, J., Wenk, A., Hirn, S., Schaeffler, M., Tian, F., Schmid, G., Oberdoerster, G., Kreyling, W.G., 2014. Size dependent translocation and fetal accumulation of gold nanoparticles from maternal blood in the rat. Part. Fibre Toxicol. 11.

Sheng, L., Ze, Y., Wang, L., Yu, X., Hong, J., Zhao, X., Ze, X., Liu, D., Xu, B., Zhu, Y., Long, Y., Lin, A., Zhang, C., Zhao, Y., Hong, F., 2015. Mechanisms of TiO_2 nanoparticle-induced neuronal apoptosis in rat primary cultured hippocampal neurons. J. Biomed. Mater. Res. A 103, 1141–1149.

Shi, H.B., Magaye, R., Castranova, V., Zhao, J.S., 2013. Titanium dioxide nanoparticles: a review of current toxicological data. Part. Fibre Toxicol. 10, 33.

Shimizu, M., Tainaka, H., Oba, T., Mizuo, K., Umezawa, M., Takeda, K., 2009. Maternal exposure to nanoparticulate titanium dioxide during the prenatal period alters gene expression related to brain development in the mouse. Part. Fibre Toxicol. 6.

Shin, J.A., Lee, E.J., Seo, S.M., Kim, H.S., Kang, J.L., Park, E.M., 2010. Nanosized titanium dioxide enhanced inflammatory responses in the septic brain of mouse. Neuroscience 165, 445–454.

Shinohara, N., Danno, N., Ichinose, T., Sasaki, T., Fukui, H., Honda, K., Gamo, M., 2014a. Tissue distribution and clearance of intravenously administered titanium dioxide (TiO_2) nanoparticles. Nanotoxicology 8, 132–141.

Shinohara, N., Oshima, Y., Kobayashi, T., Imatanaka, N., Nakai, M., Ichinose, T., Sasaki, T., Zhang, G., Fukui, H., Gamo, M., 2014b. Dose-dependent clearance kinetics of intratracheally administered titanium dioxide nanoparticles in rat lung. Toxicology 325, 1–11.

Silva, R.M., Teesy, C., Franzi, L., Weir, A., Westerhoff, P., Evans, J.E., Pinkerton, K.E., 2013. Biological response to nano-scale titanium dioxide (TiO_2): role of particle dose, shape, and retention. J. Toxicol. Environ. Health A 76, 953–972.

Smith, J.A., Park, S., Krause, J.S., Banik, N.L., 2013. Oxidative stress, DNA damage, and the telomeric complex as therapeutic targets in acute neurodegeneration. Neurochem. Int. 62, 764–775.

Song, B., Liu, J., Feng, X., Wei, L., Shao, L., 2015. A review on potential neurotoxicity of titanium dioxide nanoparticles. Nanoscale Res. Lett. 10, 1042.

Sorce, S., Krause, K.-H., 2009. NOX enzymes in the central nervous system: from signaling to disease. Antioxid. Redox Signal. 11, 2481–2504.

Sugibayashi, K., Todo, H., Kimura, E., 2008. Safety evaluation of titanium dioxide nanoparticles by their absorption and elimination profiles. J. Toxicol. Sci. 33, 293–298.

Takahashi, Y., Mizuo, K., Shinkai, Y., Oshio, S., Takeda, K., 2010. Prenatal exposure to titanium dioxide nanoparticles increases dopamine levels in the prefrontal cortex and neostriatum of mice. J. Toxicol. Sci. 35, 749–756.

Thomas, M.P., Chartrand, K., Reynolds, A., Vitvitsky, V., Banerjee, R., Gendelman, H.E., 2007. Ion channel blockade attenuates aggregated alpha synuclein induction of microglial reactive oxygen species: relevance for the pathogenesis of Parkinson's disease. J. Neurochem. 100, 503–519.

Tong, T., Shereef, A., Wu, J., Binh, C.T., Kelly, J.J., Gaillard, J.-F., Gray, K.A., 2013. Effects of material morphology on the phototoxicity of nano-TiO_2 to bacteria. Environ. Sci. Technol. 47, 12486–12495.

Tsyganova, N.A., Khairullin, R.M., Terentyuk, G.S., Khlebtsov, B.N., Bogatyrev, V.A., Dykman, L.A., Erykov, S.N., Khlebtsov, N.G., 2014. Penetration of pegylated gold nanoparticles through rat placental barrier. Bull. Exp. Biol. Med. 157, 383–385.

Umezawa, M., Tainaka, H., Kawashima, N., Shimizu, M., Takeda, K., 2012. Effect of fetal exposure to titanium dioxide nanoparticle on brain development – brain region information. J. Toxicol. Sci. 37, 1247–1252.

Valdiglesias, V., Costa, C., Sharma, V., Kilic, G., Pasaro, E., Teixeira, J.P., Dhawan, A., Laffon, B., 2013. Comparative study on effects of two different types of titanium dioxide nanoparticles on human neuronal cells. Food Chem. Toxicol. 57, 352–361.

Wang, J., Chen, C., Liu, Y., Jiao, F., Li, W., Lao, F., Li, Y., Li, B., Ge, C., Zhou, G., Gao, Y., Zhao, Y., Chai, Z., 2008a. Potential neurological lesion after nasal instillation of TiO_2 nanoparticles in the anatase and rutile crystal phases. Toxicol. Lett. 183, 72–80.

Wang, J., Li, Y., Li, W., Chen, C., Li, B., Zhao, Y., 2008b. Biological effect of intranasally instilled titanium dioxide nanoparticles on female mice. Nano 3, 279–285.

Wang, J., Liu, Y., Jiao, F., Lao, F., Li, W., Gu, Y., Li, Y., Ge, C., Zhou, G., Li, B., Zhao, Y., Chai, Z., Chen, C., 2008c. Time-dependent translocation and potential impairment on central nervous system by intranasally instilled TiO_2 nanoparticles. Toxicology 254, 82–90.

Wu, J., Liu, W., Xue, C., Zhou, S., Lan, F., Bi, L., Xu, H., Yang, X., Zeng, F.-D., 2009. Toxicity and penetration of TiO_2 nanoparticles in hairless mice and porcine skin after subchronic dermal exposure. Toxicol. Lett. 191, 1–8.

Wu, J., Sun, J., Xue, Y., 2010. Involvement of JNK and P53 activation in G2/M cell cycle arrest and apoptosis induced by titanium dioxide nanoparticles in neuron cells. Toxicol. Lett. 199, 269–276.

Wu, Q., Li, J., Zhang, W., Qian, H., She, W., Pan, H., Wen, J., Zhang, X., Liu, X., Jiang, X., 2014a. Antibacterial property, angiogenic and osteogenic activity of Cu-incorporated TiO_2 coating. J. Mater. Chem. B 2, 6738–6748.

Wu, S., Weng, Z., Liu, X., Yeung, K.W.K., Chu, P.K., 2014b. Functionalized TiO_2-based nanomaterials for biomedical applications. Adv. Funct. Mater. 24, 5464–5481.

Xue, Y., Wu, J., Sun, J., 2012. Four types of inorganic nanoparticles stimulate the inflammatory reaction in brain microglia and damage neurons in vitro. Toxicol. Lett. 214, 91–98.

Yamashita, K., Yoshioka, Y., Higashisaka, K., Mimura, K., Morishita, Y., Nozaki, M., Yoshida, T., Ogura, T., Nabeshi, H., Nagano, K., Abe, Y., Kamada, H., Monobe, Y., Imazawa, T., Aoshima, H., Shishido, K., Kawai, Y., Mayumi, T., Tsunoda, S., Itoh, N., Yoshikawa, T., Yanagihara, I., Saito, S., Tsutsumi, Y., 2011. Silica and titanium dioxide nanoparticles cause pregnancy complications in mice. Nat. Nanotechnol. 6, 321–328.

Yin, J., Kang, C., Li, Y., Li, Q., Zhang, X., Li, W., 2014. Aerosol inhalation exposure study of respiratory toxicity induced by 20 nm anatase titanium dioxide nanoparticles. Toxicol. Res. 3, 367–374.

Yu, J., Yang, Y.L., Fan, R.Q., Li, L., Wei, L.G., 2015. Mechanism of performance enhancement via fluorine doped titanium dioxide nanoparticles in dye sensitized solar cells. J. Fluor. Chem. 176, 71–77.

Ze, Y., Zheng, L., Zhao, X., Gui, S., Sang, X., Su, J., Guan, N., Zhu, L., Sheng, L., Hu, R., Cheng, J., Cheng, Z., Sun, Q., Wang, L., Hong, F., 2013. Molecular mechanism of titanium dioxide nanoparticles-induced oxidative injury in the brain of mice. Chemosphere 92, 1183–1189.

Ze, Y., Hu, R., Wang, X., Sang, X., Ze, X., Li, B., Su, J., Wang, Y., Guan, N., Zhao, X., Gui, S., Zhu, L., Cheng, Z., Cheng, J., Sheng, L., Sun, Q., Wang, L., Hong, F., 2014a. Neurotoxicity and gene-expressed profile in brain-injured mice caused by exposure to titanium dioxide nanoparticles. J. Biomed. Mater. Res. A 102, 470–478.

Ze, Y., Sheng, L., Zhao, X., Hong, J., Ze, X., Yu, X., Pan, X., Lin, A., Zhao, Y., Zhang, C., Zhou, Q., Wang, L., Hong, F., 2014b. TiO_2 nanoparticles induced hippocampal neuroinflammation in mice. PLoS One 9.

Ze, Y., Sheng, L., Zhao, X., Ze, X., Wang, X., Zhou, Q., et al., 2014c. Neurotoxic characteristics of spatial recognition damage of the hippocampus in mice following subchronic peroral exposure to TiO_2 nanoparticles. J. Hazard. Mater. 264, 219–229. 10.1016/j.jhazmat.2013.10.072.

Zhang, Y., Yu, C., Huang, G., Wang, C., Wen, L., 2010. Nano rare-earth oxides induced size-dependent vacuolization: an independent pathway from autophagy. Int. J. Nanomedicine 5, 601–609.

Zhang, L., Bai, R., Li, B., Ge, C., Du, J., Liu, Y., Le Guyader, L., Zhao, Y., Wu, Y., He, S., Ma, Y., Chen, C., 2011. Rutile TiO_2 particles exert size and surface coating dependent retention and lesions on the murine brain. Toxicol. Lett. 207, 73–81.

Zhang, R.H., Zhang, W.W., Bai, X.Y., Song, X.T., Wang, C.Y., Gao, X.X., Tian, X.B., Liu, F.Z., 2015. Discussion on the development of nano Ag/TiO_2 coating bracket and its antibacterial property and biocompatibility in orthodontic treatment. Pak. J. Pharm. Sci. 28, 807–810.

Chapter 6

The Application, Neurotoxicity, and Related Mechanism of Iron Oxide Nanoparticles

G. Kiliç[1,*], N. Fernández-Bertólez[2,*], C. Costa[3,4], F. Brandão[3,4], J.P. Teixeira[3,4], E. Pásaro[2], B. Laffon[2], V. Valdiglesias[2]
[1]*Karolinska Institutet, Stockholm, Sweden;* [2]*Universidade da Coruña, A Coruña, Spain;* [3]*Portuguese National Institute of Health, Porto, Portugal;* [4]*University of Porto, Porto, Portugal*

Chapter Outline

1. IRON OXIDE NANOPARTICLES

Among the different types of magnetic nanoparticles, iron oxide nanoparticles (IONs) are of particular interest because of their unique intrinsic magnetic properties, so-called superparamagnetism. This feature, together with their high colloidal stability, makes them very attractive for a broad range of uses. Thus they are by far the most frequently used nanoconstituents for biomedical applications including magnetic separation of biological materials (He et al., 2014), magnetic resonance imaging (MRI) (Iv et al., 2015), targeted drug delivery (Tse et al., 2014; Vasir and Labhasetwar, 2005), anticancer hyperthermia therapy

* These authors contributed equally to this chapter

Neurotoxicity of Nanomaterials and Nanomedicine. http://dx.doi.org/10.1016/B978-0-12-804598-5.00006-4

(Dan et al., 2015; Moroz et al., 2002), magnetotransfections (Adams et al., 2013), gene therapy (Li et al., 2014), molecular/cellular tracking (Sharifi et al., 2015), and tissue repair (Ziv-Polat et al., 2010). In the literature, if the individual particles are larger than 50 nm (size of core/shell), IONs are classified as superparamagnetic iron oxides (SPIOs), or ultrasmall superparamagnetic iron oxides (USPIOs) for particles smaller than 50 nm (Estelrich et al., 2015). In this chapter, to avoid confusion, the term IONs will refer to both types of magnetic IONs (SPIOs and USPIOs).

Iron oxide can exist in different chemical compositions, such as magnetite (Fe_3O_4) or maghemite (γ-Fe_2O_3). Because the crystalline structures of magnetite and maghemite are nearly identical, the two oxides have very similar physical properties (Estelrich et al., 2015), whereas their magnetic properties are entirely different (Sharma et al., 2014). Thus magnetization is higher for magnetite than for maghemite (Turcheniuk et al., 2013). However, most IONs employed in biomedical applications have not been categorized as composition of their core. It has been demonstrated that naked IONs have high chemical activity, and are easily oxidized in air (especially magnetite), resulting in loss of magnetism and dispersibility. Therefore providing proper surface coating and developing some effective protection strategies are very important to prevent agglomeration of the particles, increase circulation time, and provide biocompatibility (Mahdavi et al., 2013; Mahmoudi et al., 2009; Wu et al., 2008). Although the iron oxide core is responsible for the magnetic properties of IONs, the ligand coat is essential to stabilize IONs in physiological media. Thus biocompatibility and bioavailability of IONs can be enhanced by covering their surfaces with various different materials and numerous biological molecules. The choice of coating is mainly determined by the desired application concerning functionalization, stability, or size, because every material has advantages and drawbacks (Petters et al., 2014a). Once surface coatings degrade, the iron oxide core can be metabolized easily and release iron ions, which can be transported by proteins like ferritin, transferritin, and hemosiderin (Santhosh and Ulrih, 2013), and incorporated into the body's iron pool (Almeida et al., 2011; Riviere, 2014).

Besides stability and biocompatibility, coating of IONs may also give functionalization to which further drugs, genes, or imaging agents can be conjugated (Mahmoudi et al., 2011). Thus the coating of IONs has frequently been modified with fluorescent dyes for imaging (Bertorelle et al., 2006; Ge et al., 2009; Petters et al., 2014b; Yan et al., 2013), and targeting molecules (Agemy et al., 2011; Kumar et al., 2012), drugs (Hassan and Gallo, 1993; Jenkins et al., 2011), or nucleic acids (Choi et al., 2015; Krötz et al., 2003; Kumar et al., 2010). Although IONs are recognized as one of the most prominent nanomaterials because of their unique physicochemical properties and their great biocompatibility (Mahmoudi et al., 2011; Pisanic et al., 2009), multiple studies have called these claims into question. Researches indicated that these nanomaterials not only reach the brain (Liu et al., 2013), but also can cause a certain degree of neurotoxicity (Migliore et al., 2015).

The purpose of this chapter is to review the most promising applications of IONs, induced toxicity, particularly in neural systems, as well as the potential mechanisms of this toxicity, and to determine how these mechanisms should be considered in the nanomedicine field. For this purpose, in vitro and in vivo studies on ION neurotoxicity published so far were compiled, to comprehensively analyze available data describing the ION toxicological profile, as well as to highlight the more imperative necessities of further research in this field.

2. APPLICATIONS OF IRON OXIDE NANOPARTICLES

2.1 Magnetic Resonance Imaging Contrast Agent

The diagnoses of tumors, tissue damage, and neurological disorders have been more accurately made with the aid of various imaging techniques (Ittrich et al., 2013). Among those, MRI has become an important and useful technique in biomedical applications. In MRI technique a special chemical medium called contrast agent is used, which moves in the direction of the magnetic field and provides an improved image of the desired anatomical site for accurate diagnoses (Felton et al., 2014).

IONs are proved to be good contrast agents because of their superparamagnetic properties and high relaxivity, which provide significant advantages over traditional contrast agents (Lin et al., 2008; Rosen et al., 2012). Based on their ability to oxidize/reduce or to form complex compounds with different biochemical agents, and their biocompatible and biodegradable nature, the US Food and Drug Administration approved the clinical use of IONs for imaging (Lu et al., 2010; Reimer and Balzer, 2003; Sharma et al., 2014).

Therefore IONs are currently used in a wide number of MRI-based clinical applications. They have been mostly used for tumor imaging and imaging of the gastrointestinal tract and central nervous system (CNS) for neurovascular, neurooncological, or neuroinflammatory processes (reviewed by Ittrich et al., 2013).

IONs offer the possibility to differentiate cancerous tissue from surrounding healthy tissue through targeting mechanisms: (1) passive targeting, which relies on the anatomical and physiological differences between cancerous and normal tissues or (2) active targeting, which relies on ligands conjugated to the ION surface to recognize specific markers on cancerous tissue surface (Jin et al., 2014; Rosen et al., 2012).

As previously mentioned, once synthesized, the iron oxide core is coated with a biocompatible coating; the choice of the coating material to be used makes a substantial difference in its ability to serve as contrast agent and cellular response material (Felton et al., 2014). Yet, the targeting efficiency of IONs can be further improved by either employing tumor pathological characters (low pH environment), or conjugation biomarkers on their surface such as peptides, antibodies, or small molecules (Figuerola et al., 2010; Jin et al., 2014).

Regardless of the existence of the blood–brain barrier (BBB), ION-based brain-targeting delivery systems have been used for a long time for imaging and treatment of various brain diseases (Enochs et al., 1999; Saleh et al., 2007).

Magnetic resonance tracking of ION-labeled cells is a noninvasive technique to track transplanted cells in longitudinal brain tumor detection and characterization studies in living animals (Neri et al., 2008). One of the most recently designed MRI contrast agent is an epidermal growth factor receptor (EGFR) monoclonal antibody-conjugated ION designed for the EGFR-specific detection of glioma in Wistar rat model bearing C6 glioma (Mu et al., 2015).

2.2 Hyperthermia

Hyperthermia has been used as a medical treatment for cancer in both traditional and modern medicine because cancerous cells are more sensitive to temperature than normal cells (Lin et al., 2008). Magnetic nanoparticle-mediated hyperthermia is a promising therapy for the selective apoptosis of tumor cells through controlled heating of the damaged tissue, which involves targeting IONs to tumor tissue and then applying an external alternating magnetic field to generate heat (Laurent et al., 2011; Wang et al., 2013a). If the temperature can be maintained above the therapeutic threshold of 42°C for 30min or more, tumors might be destroyed (Figuerola et al., 2010). The delivery of heat by IONs is mainly the result of energy loss processes during the reorientation of the magnetization in the magnetic field due to rotation of nanoparticles in the surrounding medium (Hilger, and Kaiser, 2012).

With consideration of the known biological effects of temperatures higher than 42°C on phospholipids, proteins, and nucleic acids, hyperthermia treatment might ultimately lead to the deterioration of integrity of cellular structures, macromolecules, and the impairment of enzyme activity, particularly DNA repair systems (Hilger, 2013). Yet, compared with previous hyperthermia approaches, ION-mediated hyperthermia is more attractive as it offers a way to ensure that only specific target tissue is heated and eventually destroyed. It has been revealed that magnetic nanoparticles, either directly injected into tumors or delivered intravenously, can circulate through the blood stream and reach target organs by active targeting through surface ligands (Torres-Lugo and Rinaldi, 2013).

One of the first clinical approaches for treatments of tumors via ION-based magnetic heating was already done by intratumoral application of aminosilane-coated IONs to treat glioblastoma multiforme (Maier-Hauff et al., 2006). After this treatment, the investigation of brain autopsies showed the presence of dispersed or aggregated nanoparticles within the tumor tissue. Likewise, using IONs and alternating magnetic fields, Zhao et al. (2012) found that head and neck cancerous tumor cells in mice can be destroyed without harming normal cells.

In this context, magnetic hyperthermia has the advantage that the heat source is directly in contact with the target cells (Grüttner et al., 2013). Intensive research activities have shown that the therapeutic outcome is distinctly influenced not only by the structural features of the magnetic nanoparticle formulations but also by distribution patterns of tumor tissue (Hilger, 2013).

2.3 Drug Delivery

Drug delivery refers to the pharmaceutical approach of transporting drugs to a desired place in the body with reduced toxicity for nontarget tissues and increased therapeutic effect of the drug in the desirable locations (Hola et al., 2015). It is well known that if IONs are synthesized for binding to biocompatible polymers, eg, dextran, and then coated with other biomaterials or drugs, their targeting and trafficking inside the body will be much easier (Sharma et al., 2014).

The functional properties of IONs as an imaging agent and their ability to be manipulated under a magnetic field make them attractive delivery carriers in vitro and in vivo (Kim et al., 2012). Because of these properties, magnetically driven ION-based delivery systems have already been used in clinical applications for extracellular, intracellular, and site-specific targeted delivery of biotherapeutics to tumorous, inflammatory, and infectious sites under the influence of an external magnetic field (Mok and Zhang, 2013). Several biotherapeutic agents including chemotherapeutics, radiotherapeutics, antiinflammatory agents, antibodies, peptides, oligonucleotides as well as genes can be integrated on IONs to deliver those agents to specific locations (Kim et al., 2012; Wahajuddin and Arora, 2012).

The method of magnetically guided drug targeting involves (1) the immobilization of a drug on ION, (2) the injection of the drug/carrier complex into the subject, and (3) the use of high-gradient external magnetic fields to guide the complex and concentrate it at the desired locations (Estelrich et al., 2015). These magnetic drug-bearing nanostructures rely on external magnetic field guidance to reach their target tissue; they acquire the capability to drag the encapsulated or attached drug to the target site in the body and are inactivated on the removal of the magnetic field (Laurent et al., 2014).

3. MECHANISMS OF ION TOXICITY

Given the variety of medical applications of IONs, understanding the potential risks related to their exposure is crucial. Thus much attention has been increasingly paid to their biocompatibility in the human body.

Many studies have demonstrated that most of the toxicity of ION arises from the production of an excessive amount of reactive oxygen species (ROS), including free radicals such as the superoxide anion, hydroxyl radicals, and hydrogen peroxide (Nel et al., 2006; Unfried et al., 2007). There are four known primary sources of oxidative stress in response to IONs: (1) direct generation of ROS from the surface of IONs, (2) production of ROS via leaching of iron molecules, (3) alteration of mitochondrial and other organelle functions, and (4) induction of cell signaling pathways (Liu et al., 2013; Pisanic et al., 2009).

Although the mammalian cellular machinery includes defense mechanisms against oxidative stress, such as antioxidant enzymes and glutathione, these defenses can be overwhelmed by the oxidizing capacity of the nanoparticles leading to high ROS levels that can damage cells by peroxidizing lipids, disrupting DNA, modulating gene transcription, altering proteins and resulting in decrease in physiological function, and even cell apoptosis/death (Lei et al., 2013).

As IONs are required to be magnetically targeted to a particular tissue/organ for the biomedical applications, iron content may increase in the cell, either by the iron present on the surface of the particles themselves or molecular iron that has leached from the nanoparticles (Pisanic et al., 2009). It is well documented that transition metals including iron can generate ROS via mechanisms such as Fenton-type reactions (Manke et al., 2013). Following internalization via several mechanisms, IONs are likely to be degraded into iron ions by hydrolyzing enzymes inside lysosomes (Sabella et al., 2014). This free iron can react with hydrogen peroxide to produce hydroxyl radicals. These highly active free radicals generated by Fenton-type reactions are then able to react with any biomolecule (eg, lipids, proteins, DNA) within the cell (Rahman, 2007), which can lead to an imbalance in its homeostasis and cause abnormal cellular responses, including cytotoxicity, oxidative stress, epigenetic events, and DNA damage, which may eventually initiate carcinogenesis (Ankamwar et al., 2010; Bulte et al., 2001; Häfeli et al., 2009; Veranth et al., 2007).

In addition to the Fenton reaction, IONs can cause structural damage to ATP-generating mitochondria, and significantly affect electron flow and act to alter mitochondrial functionality producing altered membrane potential (Khan et al., 2012), cytochrome c release (Mukhopadhyay et al., 2012), and uncoupling of oxidative phosphorylation (Sarkar et al., 2014). IONs can also activate and upregulate plasma membrane proteins such as nicotinamide adenine dinucleotide phosphate (NADPH) oxidase (Könczöl et al., 2013), which may also contribute to the underlying mechanisms associated with toxicity.

Furthermore, exposure of IONs might cause cellular stress and alterations in cell morphology and cell motility (Cromer Berman et al., 2013; Wu and Zhang, 2010). Direct and indirect contact with IONs can also affect the structure of DNA as well as its replication process (Patil et al., 2015), and there is enough evidence that IONs upregulate genes that are associated with endothelial layer integrity (Kalive et al., 2012) and lysosomal function (Kedziorek et al., 2010), caspase (Alarifi et al., 2013) and cytokine (Couto et al., 2015) activation, and changes in iron metabolism–related genes (Liu and Wang, 2013).

4. NEUROTOXICITY

Owing to the important functions of iron in the cellular metabolism, brain iron metabolism and the transport of this metal into the brain have been extensively studied and are subject of numerous reviews (Benarroch, 2009; Crichton et al., 2011; Hohnholt et al., 2013; Mc Carthy et al., 2015; Moos et al., 2007). Although iron plays an important role in the human body, its metabolism in human brain has been shown to be different than in other human organs (Galazka-Friedman, 2009). The brain is especially sensitive to the effects of oxidative stress, as it highly depends on iron for its high respiratory activity, as well as for the production of numerous neurotransmitters and myelinogenesis (Pisanic et al., 2009). This high oxidative metabolism and resulting generation of ROS make the brain particularly susceptible to ROS imbalance compared with other organs, which is thought to be related to neurodegeneration and associated disorders (Skaper et al., 1999). It has also been well established that iron dysregulation is involved in the pathogenesis of several neurodegenerative diseases including Parkinson and Alzheimer diseases (Myhre et al., 2013).

There has been increasing incidence of such neurodegenerative diseases and primary brain tumors (Win-Shwe and Fujimaki, 2011). Although the exact etiology of these diseases is unknown, exposure to iron nanoparticles may contribute to increase the risk as they may enter the human body, intentionally or not, by various routes, including inhalation, dermal penetration, ingestion, and injection, and may then be distributed by means of systemic circulation to various tissues.

Furthermore, several studies have shown the ION migration to the brain of exposed in vivo models (reviewed by Petters et al., 2014a). Besides, considering the fact that nanoparticles have also been shown to be somehow toxic to neuronal type cells (Valdiglesias et al., 2013a,b), this gives rise to concerns regarding the potential neurotoxic effects of exposure to IONs, due to their increasing usage in the treatment of CNS malignancies. Their capacity to be conjugated to antibodies and peptides allow for specific targeting of tumors and disruption of associated active signaling pathways, delivery of drugs to target tissues across the BBB, monitoring and tracking neural stem cells used for therapy, and hyperthermic ablation of tumors in thermotherapy (Iv et al., 2015).

4.1 Mechanisms and Pathways of ION Entry to the Brain

Similar to other types of nanoparticles, IONs can enter the human body through several routes, where they eventually interact with numerous biomolecules and cell surfaces or components (Shang et al., 2014). IONs can reach the brain via various mechanisms and pathways, which can simply be listed as: (1) after peripheral injection, from blood by passing the BBB; (2) after inhalation, through the olfactory neuronal pathway via uptake by sensory nerve endings and translocation through the axon into the brain; and (3) by direct injection into the brain (reviewed by Petters et al., 2014a).

One mechanism for the transport of IONs across the BBB appears to be receptor-mediated endocytosis followed by transcytosis through the capillary endothelium (Kreuter, 2015). The receptors expressed at the BBB endothelial cells, including lactoferrin (Lf) receptor (Wohlfart et al., 2012) and transferrin receptor (Mills et al., 2010), can mediate uptake of the nanoparticles. Several studies with Lf-conjugated ION (magnetite) showed that they could effectively penetrate the BBB in healthy rodents by transcytosis (Hsiao and Huang, 2011; Huang et al., 2007; Xie et al., 2011).

Therefore when a brain delivery probe was prepared by covalently conjugating Lf to a poly-ethylene glycol (PEG)-coated ION, as a platform for brain-targeted theranostics, the efficacy of the ION-Lf conjugate was proven to facilitate the transport of the nanoparticles across the BBB by receptor-mediated transcytosis, in Sprague–Dawley rats as well as in in vitro BBB model (Qiao et al., 2012). Likewise, Yan et al. (2013) reported evidence of penetration of PEGylated and transferrin-coated fluorescent ION from the blood into the brain, and it was also verified in their study that not only were these IONs distributed to cytoplasm, dendrites, axons, and synapses of neurons, but also that they maintained intrinsic functions of bare magnetite. Similarly, Dan et al. (2013) demonstrated that IONs functionalized with platelet-endothelial cell adhesion molecule (PECAM-1) antibody significantly increased ION flux across the in vitro BBB model and enhanced accumulation of IONs in brain of male rats by cell adhesion molecule–mediated endocytosis. Hence, modifications of the ION surface with groups that target proteins/receptors specifically expressed in brain capillary endothelial cells are likely to improve the permeation of IONs into the brain (Petters et al., 2014a).

In the case of in vitro BBB model systems, Thomsen et al. (2013) demonstrated that the ability of IONs to pass through this barrier was significantly stronger in the presence of an external magnetic force, and Hoff et al. (2013) showed that the rate of passing through the BBB strongly depends on the physicochemical properties of the ION.

Yet, there are not enough studies in the literature to verify that BBB stays intact after the crossing of ION-based nanocarriers. For instance, Veiseh et al. (2009) designed a nanoprobe composed of IONs coated with a PEGylated chitosan, conjugated with a targeting ligand, chlorotoxin, which selectively binds to a variety of cancers including glioma and medulloblastoma. When IONs were injected intravenously to transgenic mice models of brain tumors, they were found to be able to traverse the BBB, without compromising the integrity of it, which was verified by using the Evans blue exclusion assay. Imam et al. (2015) speculated that IONs may adversely affect the BBB barrier by crossing it. Their study with an in vitro model of the BBB of rat brain and in vivo experiments with the dopaminergic system of male adult rats contained evidence of compromised integrity of the BBB after a single exposure to 10 nm IONs, evaluated by trans-endothelial electrical resistance permeability measurements. These studies that have investigated the effect of IONs on the BBB used different methodological approaches and IONs with different surface properties. Therefore the question remains as to whether the passage of the specifically designed IONs compromises the integrity of the BBB after they cross it.

Evidence suggests that the transit through the gastrointestinal tract may strongly influence the subsequent access of IONs into the brain. Several groups studied the gastrointestinal uptake and tissue/organ distribution of IONs in rodents; significant iron accumulation in the brain after oral administration was found (Kumari et al., 2012, 2013; Singh et al., 2013). These data would be interpreted as evidence for transfer of IONs through the BBB after oral administration.

IONs may also enter the body and eventually gain access to the brain by inhalation. Intranasally administered IONs can follow the olfactory neuronal pathway to reach brain; thus nanoparticles in the mucosa of the nasal region may be transported along the olfactory nerve into the olfactory bulb (Wang et al., 2009, 2011). Furthermore, fluorescent IONs applied through the nose-only exposure chamber system have been reported to be distributed into the mice brain, documented by MRI (Kwon et al., 2008). In addition, after intranasal instillation of IONs (magnetite), the particles were found to be deposited particularly at high concentrations in the rat striata and hippocampus. Indeed, most IONs were retained in the striata for a minimum of 14 days where they induced oxidative damage (Wu et al., 2013a). Similarly, Wang et al. (2011) also showed that intranasal exposure of IONs (maghemite) can lead to pathological alterations in olfactory bulb, hippocampus, and striatum, and cause microglial proliferation, activation, and recruitment in these areas, especially in the olfactory bulb.

The use of IONs in biomedical research on CNS is progressively gaining importance; therefore, as a consequence, the number of toxicological reports, looking at the possible neurotoxicity of these nanoparticles is growing. It is well established that IONs can enter the brain and can be localized in neurons, astrocytes, and macrophages, where depending on the character of the nanoparticle, eg, surface coatings and particle size, they may induce a wide range of toxic effects. Therefore in the following section, published studies on effects of IONs on CNS, both in vitro and in vivo, are reviewed and compiled.

4.2 In Vitro Studies on ION Neurotoxicity

In vitro studies are essential to investigate the effects of IONs on the different types of cells of the nervous system, from neurons to glial cells. This section describes and discusses the current knowledge on toxicity of IONs in cultured cell models of neurons, astrocytes, microglia, and oligodendrocytes.

4.2.1 Neurons

Most studies reported so far on the consequences of exposure to different IONs on cultured neuronal cells have been performed on the PC12 cell line, regularly used as a model of neuronal differentiation (Guroff, 1985). However, after treatment of these cells with different concentrations of IONs, results of such studies are contradictory: they reported either a decrease (Deng et al., 2014; Pisanic et al., 2007) or an increase (Marcus et al., 2015) in neurite number, and an enhancing effect on neurite length (Deng et al., 2014; Kim et al., 2011; Pisanic et al., 2007; Soenen et al., 2011). From their detailed study,

Wu et al. (2013a) concluded that ION (magnetite) exposure results in a dose-dependent decrease in viability and in cellular oxidative stress in PC12 cells. Cell cycle arrest in G_2/M phase, decrease in cellular viability, and increase in mRNA levels of p53 were also observed when PC12 cells were incubated with different concentrations of ION (Wu and Sun, 2011). Deng et al. (2014) observed that, after exposure to different concentrations of IONs (6.25–100 μg/mL), ROS were generated in PC12 cells in a concentration-dependent manner, and the levels of ROS in the cells treated with silk fibroin–coated IONs were lower than those obtained with uncoated IONs. Besides, Soenen et al. (2011) tested four different ION types (coated with dextran, carboxy-dextran, lipid, and citrate) in PC12 cells as well as in c17.2 neural progenitor cells, and their results indicated that the citrate-coated IONs were the most toxic nanoparticles and the lipid-coated ones were the least toxic, under their experimental conditions. Similarly, Rivet et al. (2012) reported different ION toxicological profiles according to their coating chemical composition in primary cortical neurons derived from embryonic chick forebrains. When toxicity experiments were performed in human neuroblastoma SH-SY5Y cell line with poly-(L-lysine)-coated IONs, authors observed dose-dependent cytotoxicity and significant ROS production only at high concentrations (>20 μg/mL) (Riggio et al., 2012). In a more detailed study after a 24-h exposure of SH-SY5Y cells to IONs, Imam et al. (2015) observed inhibition of cell proliferation, significant reductions in the number of active mitochondria, and a dose-dependent increase in ROS. Likewise, Costa et al. (2015) observed that silica-coated IONs significantly decreased SH-SY5Y viability only after 24 h of exposure. Besides, despite being effectively internalized by the cells, these IONs produced low cytotoxicity (cell cycle alterations, apoptosis, and membrane integrity assessment) but dose- and time-dependent increase in DNA damage not related to the production of double-strand breaks or chromosome loss (Kiliç et al., 2015)

Opposite to these studies, Petters and Dringen (2015) reported that exposure of primary rat cerebellar granule neurons with dimercaptosuccinic acid (DMSA)-coated IONs did not induce any acute toxicity, but caused ION accumulation as demonstrated by the increase in the specific iron content of the cells. Furthermore, results of Gramowski et al. (2010) showed that exposure to IONs (maghemite) did not induce changes in the ROS level in neuronal networks of mice-derived primary frontal cortex cell culture.

4.2.2 Astrocytes

Astrocytes are glial cells that form much of the cell population in the CNS (Amthor, 2011; Tsai et al., 2012). Concurrently, most of the studies available in the literature on interactions of glial cells with IONs are carried out in cultured astrocytes. Although these studies have been performed by using different types of IONs and various incubation conditions, they generally demonstrate that cultured astrocytes are not acutely damaged by the presence of IONs (reviewed by Hohnholt et al., 2013).

Exposure of primary astrocyte cultures to citrate-coated IONs did not cause any acute loss in cell viability, but resulted in an increase in the cellular iron content (Geppert et al., 2009). Similarly, no ION-related cytotoxic effects were observed in rat primary astrocytes treated with polyethyleneimine-coated, rhodamine B isothiocyanate-tagged IONs (Yiu et al., 2012). Besides, DMSA-coated IONs in the presence of an external static magnetic field did not compromise cell viability of astrocyte primary cells that were cultured from the brains of newborn Wistar rats (Lamkowsky et al., 2012). On the contrary, when Geppert et al. (2012) exposed cultured astrocytes with the same properties to the same type of ION, the authors observed a time- and concentration-dependent accumulation of cellular iron, and ROS formation induction. Furthermore, when Au et al. (2007) studied the effects of ION chemical formulation on rat astrocytes at a specified concentration (175 μM), they found significant effects on mitochondrial function as well as decreased cell viability.

Although astrocytes are frequently used as primary or secondary culture from rodent brains (Souza et al., 2013), astroglia-rich primary cultures are mostly contaminated by other glial cells. Therefore the heterogeneity of astrocytes within the living brain constitutes a challenge for astrocyte culture and may modify the experimental outcome (Saura, 2007). In this regard, Petters and Dringen (2014) compared primary (APC) and secondary (ASC) rat astrocyte cultures derived from brains of neonatal Wistar rats, for accumulation of IONs. They observed no difference in ION accumulation between the microglia-containing APC and the ASC that hardly contained microglial cells. Accordingly, other independent studies showed no dextran-coated ION effects either on rat ASC (Ding et al., 2010) or on APC (Neubert et al., 2015).

Apart from these studies with primary or secondary astrocyte cultures from rodents, there are also other cell lines used as astrocyte models, for instance, C6 rat glioma cells and human glioblastoma-astrocytoma cells (Barth and Kaur, 2009). In vitro studies of IONs conjugated with recombinant human epidermal growth factor showed high intracellular incorporation and absence of influence on C6 cell viability and proliferation (Shevtsov et al., 2014). Ankamwar et al. (2010) performed cytotoxicity tests using various human glia cell lines (D54MG, G9T, SF126, U87, U251, U373) treated with IONs (magnetite) coated with bipolar surfactant tetramethylammonium 11-aminoundecanoate, which exhibited cytotoxicity at the highest concentration (100 μg/mL) only. However, Chen et al. (2012) showed concentration-dependent cytotoxicity of IONs (magnetite and maghemite) on human glioma U251 cells, and an increase of H_2O_2-induced cell damage. On the contrary, Wang et al. (2012) observed that cell viability and proliferation capacity of U251 cells and their differentiated progenies were not affected by poly(L-lysine)-modified IONs; yet IONs could reduce intracellular H_2O_2 levels and accelerate the progression of cell cycle (Wang et al., 2013b).

4.2.3 Microglia

Among the different types of brain cells, microglia are particularly relevant in the accumulation of large amounts of IONs (Jenkins et al., 2013), because of their phagocytic activity. Furthermore, their presence in mixed cultures appears to reduce the internalization efficiency of IONs by other types of cultured brain cells (Pickard and Chari, 2010). Although excess iron can directly produce toxic effects on neurons according to the classical oxidative stress pattern, it is believed microglia are one of the prime mediators and transgressors of neurotoxicity induced by ROS (Block et al., 2007; Pisanic et al., 2009). Thus, to investigate the mechanisms involved in ION-induced microglial toxicity, Petters et al. (2015) applied fluorescent DMSA-coated IONs to primary cultures of microglial cells. Exposure to IONs for 6h caused a strong concentration-dependent increase in the microglial iron content, which was accompanied by a substantial generation of ROS and cell toxicity. Furthermore, treatment of cultured rat microglial cells with IONs caused a time- and concentration-dependent uptake of the particles, and exposure to higher concentrations of IONs (Luther et al., 2013) or for longer incubation periods (Pickard and Chari, 2010) severely compromised cell viability. Yet, short-term exposure of mice microglial culture to carboxydextran-coated IONs (1–50 μg iron/mL for 30 min) did not cause toxicity (Wu et al., 2013b). Besides, toxicity induced by IONs was reported to be different in microglial cell lines than in microglial primary cultures (Petters et al., 2014a), because contradictory results were observed (Wang et al., 2011; Xue et al., 2012).

Microglia serve many of the same functions in the CNS that macrophages do for the rest of the body, including phagocytosis, antigen presentation, and cytokine release (Pisanic et al., 2009). However, the reports on the consequences of ION exposure on microglial functions are controversial (reviewed by Petters et al., 2014a). IONs have been reported to increase the secretion of interleukin (IL)-1b, IL-6, and tumor necrosis factor (TNF)-α in rat primary microglia (Xue et al., 2012); to decrease the secretion of IL-1b but not of TNF-α in murine primary microglia (Wu et al., 2013b); or to not affect their secretion in BV2 microglial cells (Wang et al., 2011).

4.2.4 Oligodendrocytes

There is little information available on the uptake and metabolism of IONs in oligodendrocytes. Nevertheless, a lower ION uptake in cultured oligodendrocytes has been reported when compared with microglial and astrocyte cultures (Jenkins et al., 2013). The knowledge on the consequences of ION treatment in oligodendrocytes has been obtained from experiments on oligodendroglial origin OLN-93 cells. These cells accumulate efficiently IONs coated with citrate (Hohnholt et al., 2010), DMSA (Hohnholt and Dringen, 2011; Hohnholt et al., 2011), or fluorescently labeled DMSA (Petters et al., 2014a). Moreover, despite the large amounts of iron accumulated, OLN-93 cells did not show substantial ROS production or alteration in the cellular thiol reduction potential after incubation with DMSA-coated IONs for up to 48h (Hohnholt et al., 2011); losses in cell viability and changes in cellular morphology after incubation with citrate-coated IONs for up to 72h were not detected (Hohnholt et al., 2010).

Considering all these in vitro studies on neural cells, it seems that ION exposure has little or no effect on astrocytes and oligodendrocytes, although the latter have been poorly studied so far; so toxicity in this case cannot be completely discarded. However, diverse neurotoxic effects were described after treatment of microglial and neuronal cells with IONs, mostly related to oxidative stress and cell cycle alterations. Still, results of these studies are often contradictory and the toxicity reported is in general hardly consistent. Inconsistencies observed in the effects of ION exposure on nervous system cells might be due to differences in culture conditions (Luther et al., 2013), concentrations employed (Ankamwar et al., 2010), and the type of ION tested (Soenen et al., 2011), apart from the use of primary cultures or cell lines (Petters et al., 2014a).

4.3 In Vivo Studies on ION Neurotoxicity

The CNS is composed of the brain and the spinal cord. Both are delicate organs in the human body, which must be protected from injury (Hu and Gao, 2010). As previously mentioned, IONs have the capacity to reach the brain via various mechanisms and pathways; thus they may subsequently influence brain physiology and cause severe side effects. Therefore in vivo experiments are necessary to assess sensitivity of organ system to nanoparticles, as well as to consider pharmacokinetic factors, the time course of brain development, and cell-to-cell interactions that cannot be modeled in vitro (Win-Shwe and Fujimaki, 2011). Yet, the number of in vivo studies on potential neurotoxicity of IONs is quite restricted.

As already mentioned, there is evidence in the literature that IONs can be distributed to brain after intraperitoneal, intravenous, or oral administrations, which was interpreted as evidence for the transfer of IONs through the BBB. In the study by Kim et al. (2006), after intraperitoneal administration of silica-coated ION-containing rhodamine B isothiocyanate for 4 weeks into mice, the nanoparticles were detected in the brain, without producing apparent toxicity. This observation was confirmed in other ION studies by Jain et al. (2008) with intravenously administered oleic acid–coated IONs, and by Muldoon et al. (2005) with intracerebrally inoculated or intraarterially administered clinical iron oxide nanoagents (ferumoxtran-10, ferumoxides, and ferumoxytol) to normal brains of rats.

Although these studies note that there are no adverse effects of IONs, there are also notable exceptions. One was the study by Bourrinet et al. (2006) where ferumoxtran-10 was intravenously injected to mice, rats, rabbits, dogs, and monkeys, and induction of neurobehavioral and neurovegetative effects at moderate doses was shown. More obvious toxicity was observed when IONs coated with cross-linked aminated dextran was intraperitoneally injected to zebrafish (de Oliveira et al., 2014). They not only caused gradual increase in brain iron contents, but also induced toxicity by the inhibition of acetylcholinesterase (AChE) and triggered apoptosis (de Oliveira et al., 2014). In their experiments, Kumari et al. (2013) found that IONs caused significant inhibition of a neurotoxicity biomarker (ie, AChE), and significant dose-dependent inhibition of total, Na^+-K^+, Mg^{2+}, and Ca^{2+}-ATPases of brains of Wistar rats treated for 14 days via the oral route. This observation suggested that IONs may affect the various steps in the metabolic pathways of the neurotransmitters and inhibit the AChE enzyme, which is considered as a sign of neurotoxicity. Similarly, in an earlier study of the same research group, when the same type of nanoparticles were applied to Wistar rats via oral route for a longer time (28 days in repeated doses of 30, 300, and 1000 mg/kg/day), results showed that brain activities of total, Na^+-K^+, Mg^{2+}, and Ca^{2+}-ATPases were significantly inhibited in dose-dependent order, indicating that nerve conduction might have been affected by IONs (Kumari et al., 2012). The authors suggested that the observed decreases in the enzyme activities may be due to the interaction of these nanoparticles with the enzymes, causing alterations in the cell membrane permeability and inducing enzyme inhibition (Kumari et al., 2012).

Wu et al. (2013a) observed regional distribution of IONs (magnetite) in brains of Sprague-Dawley rats after intranasal instillation for 7 days. The particles were found to be deposited at particularly high concentrations in the rat striatum and hippocampus. Over half of the IONs were retained in the striata for a minimum of 14 days, and induced oxidative damage in the striatum but not in the hippocampus. Similarly, ION exposure in mice caused neuronal fatty degeneration in the hippocampus 14 days after intranasal instillation (Wang et al., 2007). In another study, when IONs were intranasally instilled every other day into both nostrils of mice for a period of 40 days, it led to pathological alterations in olfactory bulb, hippocampus, and striatum, and also induced microglial activation in brain, especially in the olfactory bulb (Wang et al., 2011).

In a very thorough study, Kim et al. (2013) exposed Sprague–Dawley rats by intraneural injection into an intact sciatic nerve to four types of IONs with different surface and core chemistries, namely DMSA-coated IONs (both maghemite and magnetite), PEG-coated magnetite nanoparticles, and PEG-Au-coated magnetite nanoparticles. In this study, they found out that 48 h after injection of IONs, levels of mitogen-activated protein kinase/extracellular signal-regulated kinase and caspase 3 (as indicators of inflammation and apoptosis, respectively), were significantly increased in nerves of all ION-injected animals. Although DMSA-maghemite stimulated higher levels of caspase 3 activation compared to DMSA-magnetite, the authors were unable to clarify the specific influence of the iron core oxidation state on neurotoxicity. Moreover, macrophages were the main cells to internalize DMSA-maghemite nanoparticles in nerve, although no apoptosis was observed in these cells, whereas endothelial cells, Schwann cells, neurons, and T cells were vulnerable to cell death.

Therefore results of all these studies suggest the possibility of an adverse impact of IONs on the nervous system. Nevertheless, different factors related to the nanoparticle—eg, biocompatibility, iron release capacity, presence and chemical features of coating—or to the experimental setting—eg, exposure conditions/pathway, type of tissue/cell affected, and ION concentration—may influence the potential neurotoxicity induced by ION exposure (Ankamwar et al., 2010; Petters et al., 2014a; Soenen et al., 2011; Wang et al., 2011). The potential long-term toxicity of IONs on CNS is especially important concerning the use of these nanomaterials for neurobiological applications and the sensitivity of this tissue to their effects, but further investigations are needed to deeply define and characterize this impact.

5. CONCLUSIONS

IONs display many unique properties that provide them with numerous advantages and opportunities for their usage in biomedical applications. Although initial studies on the toxicity of IONs resulted in an early assumption that they are highly biocompatible on the basis of cell viability assessments, new data are coming up on other toxicity end points suggesting that caution should be taken, especially when these nanomaterials are designed to be introduced in the organism. Multiple studies have revealed that IONs can induce toxic effects in both in vitro cell cultures and in vivo systems. Yet, those studies have also demonstrated that nanoparticle toxicity is extremely complex, and the biological activity of IONs depends on a variety of physicochemical properties such as particle size, magnetic particle core chemical composition, presence of coating, and surface properties.

IONs were demonstrated to easily reach the brain after oral, inhalation, intraperitoneal, or intravenous administration. Besides, many ION applications are directly related to the nervous system. Thus special interest has been focused on potential neurotoxicity of these nanoparticles. A number of studies have addressed the potential neurotoxic effects of IONs on different neural cell types, ie, neurons, astrocytes, microglia, and oligodendrocytes, and in animal models. Results of these studies pointed out that, under specific conditions, mainly depending on surface properties and concentrations tested, IONs can induce several cytotoxic effects including oxidative damage, cell cycle alterations, and apoptosis. Nevertheless, results are often contradictory and the exact mechanism of ION human neurotoxicity remains unclear. Besides, there is an important gap in knowledge regarding the specific effects on genetic material and cellular repair systems, which are crucial to be addressed because alterations at these levels could lead to the onset and development of severe pathologies, including cancer. Therefore to advance in the knowledge on ION toxicity mechanism and fully describe the potential risks associated with ION exposure, further investigation is encouraged by carrying out in vitro studies dealing with the potential genotoxic effects of different types of IONs, and also by means of well-designed in vivo toxicity studies using appropriate dosing methods and supported by systemic and quantitative pharmacokinetic studies. Information obtained from such studies would help to define margins of safety and conditions of use of these nanomaterials to avoid or minimize risks, especially those related to their exponentially growing applications in the biomedical field.

REFERENCES

Adams, C.F., Pickard, M.R., Chari, D.M., 2013. Magnetic nanoparticle mediated transfection of neural stem cell suspension cultures is enhanced by applied oscillating magnetic fields. Nanomed. Nanotechnol. Biol. Med. 9, 737–741.

Agemy, L., Friedmann-Morvinski, D., Kotamraju, V.R., Roth, L., Sugahara, K.N., Girard, O.M., et al., 2011. Targeted nanoparticle enhanced proapoptotic peptide as potential therapy for glioblastoma. Proc. Natl. Acad. Sci. USA 108, 17450–17455.

Alarifi, S., Ali, D., Alkahtani, S., Verma, A., Ahamed, M., Ahmed, M., et al., 2013. Induction of oxidative stress, DNA damage, and apoptosis in a malignant human skin melanoma cell line after exposure to zinc oxide nanoparticles. Int. J. Nanomed. 8, 983–993.

Almeida, J.P.M., Chen, A.L., Foster, A., Drezek, R., 2011. In vivo biodistribution of nanoparticles. Nanomedicine 6, 815–835.

Amthor, F., 2011. Neuroscience for Dummies, first ed. John Wiley & Sons.

Ankamwar, B., Lai, T.C., Huang, J.H., Liu, R.S., Hsiao, M., Chen, C.H., et al., 2010. Biocompatibility of Fe(3)O(4) nanoparticles evaluated by in vitro cytotoxicity assays using normal, glia and breast cancer cells. Nanotechnology 21, 75102.

Au, C., Mutkus, L., Dobson, A., Riffle, J., Lalli, J., Aschner, M., 2007. Effects of nanoparticles on the adhesion and cell viability on astrocytes. Biol. Trace Elem. Res. 120, 248–256.

Barth, R.F., Kaur, B., 2009. Rat brain tumor models in experimental neuro-oncology: the C6, 9L, T9, RG2, F98, BT4C, RT-2 and CNS-1 gliomas. J. Neurooncol. 94, 299–312.

Benarroch, E.E., 2009. Brain iron homeostasis and neurodegenerative disease. Neurology 72, 1436–1440.

Bertorelle, F., Wilhelm, C., Roger, J., Gazeau, F., Ménager, C., Cabuil, V., 2006. Fluorescence-modified superparamagnetic nanoparticles: intracellular uptake and use in cellular imaging. Langmuir 22, 5385–5391.

Block, M.L., Zecca, L., Hong, J.-S., 2007. Microglia-mediated neurotoxicity: uncovering the molecular mechanisms. Nat. Rev. Neurosci. 8, 57–69.

Bourrinet, P., Bengele, H.H., Bonnemain, B., Dencausse, A., Idee, J.-M., Jacobs, P.M., et al., 2006. Preclinical safety and pharmacokinetic profile of ferumoxtran-10, an ultrasmall superparamagnetic iron oxide magnetic resonance contrast agent. Invest. Radiol. 41, 313–324.

Bulte, J.W.M., Douglas, T., Witwer, B., Zhang, S.-C., Strable, E., Lewis, B.K., et al., 2001. Magnetodendrimers allow endosomal magnetic labeling and in vivo tracking of stem cells. Nat. Biotechnol. 19, 1141–1147.

Chen, Z., Yin, J.-J., Zhou, Y.-T., Zhang, Y., Song, L., Song, M., et al., 2012. Dual enzyme-like activities of iron oxide nanoparticles and their implication for diminishing cytotoxicity. ACS Nano 6, 4001–4012.

Choi, J.-W., Park, J.W., Na, Y., Jung, S.-J., Hwang, J.K., Choi, D., et al., 2015. Using a magnetic field to redirect an oncolytic adenovirus complexed with iron oxide augments gene therapy efficacy. Biomaterials 65, 163–174.

Costa, C., Brandão, F., Bessa, M.J., Costa, S., Valdiglesias, V., Kiliç, G., et al., 2015. In vitro cytotoxicity of superparamagnetic iron oxide nanoparticles on neuronal and glial cells. Evaluation of nanoparticle interference with viability tests. J. Appl. Toxicol. http://dx.doi.org/10.1002/jat.3213.

Couto, D., Sousa, R., Andrade, L., Leander, M., Lopez-Quintela, M.A., Rivas, J., et al., 2015. Polyacrylic acid coated and non-coated iron oxide nanoparticles are not genotoxic to human T lymphocytes. Toxicol. Lett. 234, 67–73.

Crichton, R.R., Dexter, D.T., Ward, R.J., 2011. Brain iron metabolism and its perturbation in neurological diseases. J. Neural Transm. 1996 (118), 301–314.

Cromer Berman, S.M., Kshitiz null, W.C.J., Orukari, I., Levchenko, A., Bulte, J.W.M., et al., 2013. Cell motility of neural stem cells is reduced after SPIO-labeling, which is mitigated after exocytosis. Magn. Reson. Med. 69, 255–262.

Dan, M., Scott, D.F., Hardy, P.A., Wydra, R.J., Hilt, J.Z., Yokel, R.A., et al., 2013. Block copolymer cross-linked nanoassemblies improve particle stability and biocompatibility of superparamagnetic iron oxide nanoparticles. Pharm. Res. 30, 552–561.

Dan, M., Bae, Y., Pittman, T.A., Yokel, R.A., 2015. Alternating magnetic field-induced hyperthermia increases iron oxide nanoparticle cell association/uptake and flux in blood–brain barrier models. Pharm. Res. 32, 1615–1625.

de Oliveira, G.M.T., Kist, L.W., Pereira, T.C.B., Bortolotto, J.W., Paquete, F.L., de Oliveira, E.M.N., et al., 2014. Transient modulation of acetylcholinesterase activity caused by exposure to dextran-coated iron oxide nanoparticles in brain of adult zebrafish. Comp. Biochem. Physiol. Toxicol. Pharmacol. 162, 77–84.

Deng, M., Huang, Z., Zou, Y., Yin, G., Liu, J., Gu, J., 2014. Fabrication and neuron cytocompatibility of iron oxide nanoparticles coated with silk-fibroin peptides. Colloids Surf. B Biointerfaces 116, 465–471.

Ding, J., Tao, K., Li, J., Song, S., Sun, K., 2010. Cell-specific cytotoxicity of dextran-stabilized magnetite nanoparticles. Colloids Surf. B Biointerfaces 79, 184–190.

Enochs, W.S., Harsh, G., Hochberg, F., Weissleder, R., 1999. Improved delineation of human brain tumors on MR images using a long-circulating, superparamagnetic iron oxide agent. J. Magn. Reson. Imaging 9, 228–232.

Estelrich, J., Escribano, E., Queralt, J., Busquets, M.A., 2015. Iron oxide nanoparticles for magnetically-guided and magnetically-responsive drug delivery. Int. J. Mol. Sci. 16, 8070–8101.

Felton, C., Karmakar, A., Gartia, Y., Ramidi, P., Biris, A.S., Ghosh, A., 2014. Magnetic nanoparticles as contrast agents in biomedical imaging: recent advances in iron- and manganese-based magnetic nanoparticles. Drug Metab. Rev. 46, 142–154.

Figuerola, A., Di Corato, R., Manna, L., Pellegrino, T., 2010. From iron oxide nanoparticles towards advanced iron-based inorganic materials designed for biomedical applications. Pharmacol. Res. 62, 126–143.

Galazka-Friedman, J., 2009. Iron as a risk factor in neurological diseases. In: Gajbhiye, N.S., Date, S.K. (Eds.), ICAME 2007. Springer Berlin Heidelberg, pp. 31–44.

Ge, Y., Zhang, Y., He, S., Nie, F., Teng, G., Gu, N., 2009. Fluorescence modified chitosan-coated magnetic nanoparticles for high-efficient cellular imaging. Nanoscale Res. Lett. 4, 287–295.

Geppert, M., Hohnholt, M., Gaetjen, L., Grunwald, I., Bäumer, M., Dringen, R., 2009. Accumulation of iron oxide nanoparticles by cultured brain astrocytes. J. Biomed. Nanotechnol. 5, 285–293.

Geppert, M., Hohnholt, M.C., Nürnberger, S., Dringen, R., 2012. Ferritin up-regulation and transient ROS production in cultured brain astrocytes after loading with iron oxide nanoparticles. Acta Biomater. 8, 3832–3839.

Gramowski, A., Flossdorf, J., Bhattacharya, K., Jonas, L., Lantow, M., Rahman, Q., et al., 2010. Nanoparticles induce changes of the electrical activity of neuronal networks on microelectrode array neurochips. Environ. Health Perspect. 118, 1363–1369.

Grüttner, C., Müller, K., Teller, J., Westphal, F., 2013. Synthesis and functionalisation of magnetic nanoparticles for hyperthermia applications. Int. J. Hyperther. 29, 777–789.

Guroff, G., 1985. PC12 cells as a model of neuronal differentiation (cited September 10, 2015). In: Bottenstein, J.E., Sato, G. (Eds.), Cell Culture in the Neurosciences. Springer US, pp. 245–272.

Häfeli, U.O., Riffle, J.S., Harris-Shekhawat, L., Carmichael-Baranauskas, A., Mark, F., Dailey, J.P., et al., 2009. Cell uptake and in vitro toxicity of magnetic nanoparticles suitable for drug delivery. Mol. Pharm. 6, 1417–1428.

Hassan, E.E., Gallo, J.M., 1993. Targeting anticancer drugs to the brain. I: enhanced brain delivery of oxantrazole following administration in magnetic cationic microspheres. J. Drug Target. 1, 7–14.

He, J., Huang, M., Wang, D., Zhang, Z., Li, G., 2014. Magnetic separation techniques in sample preparation for biological analysis: a review. J. Pharm. Biomed. Anal. 101, 84–101.

Hilger, I., Kaiser, W.A., 2012. Iron oxide-based nanostructures for MRI and magnetic hyperthermia. Nanomedicine 7, 1443–1459.

Hilger, I., 2013. In vivo applications of magnetic nanoparticle hyperthermia. Int. J. Hyperther. 29, 828–834.

Hoff, D., Sheikh, L., Bhattacharya, S., Nayar, S., Webster, T.J., 2013. Comparison study of ferrofluid and powder iron oxide nanoparticle permeability across the blood–brain barrier. Int. J. Nanomed. 8, 703–710.

Hohnholt, M.C., Geppert, M., Dringen, R., 2010. Effects of iron chelators, iron salts, and iron oxide nanoparticles on the proliferation and the iron content of oligodendroglial OLN-93 cells. Neurochem. Res. 35, 1259–1268.

Hohnholt, M.C., Geppert, M., Dringen, R., 2011. Treatment with iron oxide nanoparticles induces ferritin synthesis but not oxidative stress in oligodendroglial cells. Acta Biomater. 7, 3946–3954.

Hohnholt, M.C., Geppert, M., Luther, E.M., Petters, C., Bulcke, F., Dringen, R., 2013. Handling of iron oxide and silver nanoparticles by astrocytes. Neurochem. Res. 38, 227–239.

Hohnholt, M.C., Dringen, R., 2011. Iron-dependent formation of reactive oxygen species and glutathione depletion after accumulation of magnetic iron oxide nanoparticles by oligodendroglial cells. J. Nanopart. Res. 13, 6761–6774.

Hola, K., Markova, Z., Zoppellaro, G., Tucek, J., Zboril, R., 2015. Tailored functionalization of iron oxide nanoparticles for MRI, drug delivery, magnetic separation and immobilization of biosubstances. Biotechnol. Adv. 33, 1162–1176.

Hsiao, I.-L., Huang, Y.-J., 2011. Titanium oxide shell coatings decrease the cytotoxicity of ZnO nanoparticles. Chem. Res. Toxicol. 24, 303–313.

Hu, Y.-L., Gao, J.-Q., 2010. Potential neurotoxicity of nanoparticles. Int. J. Pharm. 394, 115–121.

Huang, R., Ke, W., Qu, Y., Zhu, J., Pei, Y., Jiang, C., 2007. Characterization of lactoferrin receptor in brain endothelial capillary cells and mouse brain. J. Biomed. Sci. 14, 121–128.

Imam, S.Z., Lantz-McPeak, S.M., Cuevas, E., Rosas-Hernandez, H., Liachenko, S., Zhang, Y., et al., 2015. Iron oxide nanoparticles induce dopaminergic damage: in vitro pathways and in vivo imaging reveals mechanism of neuronal damage. Mol. Neurobiol. 1–14.

Ittrich, H., Peldschus, K., Raabe, N., Kaul, M., Adam, G., 2013. Superparamagnetic iron oxide nanoparticles in biomedicine: applications and developments in diagnostics and therapy. RöFo 185, 1149–1166.

Iv, M., Telischak, N., Feng, D., Holdsworth, S.J., Yeom, K.W., Daldrup-Link, H.E., 2015. Clinical applications of iron oxide nanoparticles for magnetic resonance imaging of brain tumors. Nanomedicine 10, 993–1018.

Jain, T.K., Reddy, M.K., Morales, M.A., Leslie-Pelecky, D.L., Labhasetwar, V., 2008. Biodistribution, clearance, and biocompatibility of iron oxide magnetic nanoparticles in rats. Mol. Pharm. 5, 316–327.

Jenkins, S.I., Pickard, M.R., Granger, N., Chari, D.M., 2011. Magnetic nanoparticle-mediated gene transfer to oligodendrocyte precursor cell transplant populations is enhanced by magnetofection strategies. ACS Nano 5, 6527–6538.

Jenkins, S.I., Pickard, M.R., Furness, D.N., Yiu, H.H.P., Chari, D.M., 2013. Differences in magnetic particle uptake by CNS neuroglial subclasses: implications for neural tissue engineering. Nanomedicine 8, 951–968.

Jin, R., Lin, B., Li, D., Ai, H., 2014. Superparamagnetic iron oxide nanoparticles for MR imaging and therapy: design considerations and clinical applications. Curr. Opin. Pharmacol. 18C, 18–27.

Kalive, M., Zhang, W., Chen, Y., Capco, D.G., 2012. Human intestinal epithelial cells exhibit a cellular response indicating a potential toxicity upon exposure to hematite nanoparticles. Cell Biol. Toxicol. 28, 343–368.

Kedziorek, D.A., Muja, N., Walczak, P., Ruiz-Cabello, J., Gilad, A.A., Jie, C.C., et al., 2010. Gene expression profiling reveals early cellular responses to intracellular magnetic labeling with superparamagnetic iron oxide nanoparticles. Magn. Reson. Med. 63, 1031–1043.

Khan, M.I., Mohammad, A., Patil, G., Naqvi, S.A.H., Chauhan, L.K.S., Ahmad, I., 2012. Induction of ROS, mitochondrial damage and autophagy in lung epithelial cancer cells by iron oxide nanoparticles. Biomaterials 33, 1477–1488.

Kiliç, G., Costa, C., Fernández-Bertólez, N., Pásaro, E., Teixeira, J.P., Laffon, B., et al., 2015. In vitro toxicity evaluation of silica-coated iron oxide nanoparticles in human SHSY5Y neuronal cells. Toxicol. Res. 5, 235–247.

Kim, J.S., Yoon, T.-J., Yu, K.N., Kim, B.G., Park, S.J., Kim, H.W., et al., 2006. Toxicity and tissue distribution of magnetic nanoparticles in mice. Toxicol. Sci. 89, 338–347.

Kim, J.A., Lee, N., Kim, B.H., Rhee, W.J., Yoon, S., Hyeon, T., et al., 2011. Enhancement of neurite outgrowth in PC12 cells by iron oxide nanoparticles. Biomaterials 32, 2871–2877.

Kim, J.-E., Shin, J.-Y., Cho, M.-H., 2012. Magnetic nanoparticles: an update of application for drug delivery and possible toxic effects. Arch. Toxicol. 86, 685–700.

Kim, Y., Kong, S.D., Chen, L.-H., Pisanic II, T.R., Jin, S., Shubayev, V.I., 2013. In vivo nanoneurotoxicity screening using oxidative stress and neuroinflammation paradigms. Nanomed. Nanotechnol. Biol. Med. 9, 1057–1066.

Könczöl, M., Weiss, A., Stangenberg, E., Gminski, R., Garcia-Käufer, M., Gieré, R., et al., 2013. Cell-cycle changes and oxidative stress response to magnetite in A549 human lung cells. Chem. Res. Toxicol. 26, 693–702.

Kreuter, J., 2015. Influence of chronobiology on the nanoparticle-mediated drug uptake into the brain. Pharmaceutics 7, 3–9.

Krötz, F., de Wit, C., Sohn, H.Y., Zahler, S., Gloe, T., Pohl, U., et al., 2003. Magnetofection–a highly efficient tool for antisense oligonucleotide delivery in vitro and in vivo. Mol. Ther. 7, 700–710.

Kumar, M., Yigit, M., Dai, G., Moore, A., Medarova, Z., 2010. Image-guided breast tumor therapy using a small interfering RNA nanodrug. Cancer Res. 70, 7553–7561.

Kumar, M., Singh, G., Arora, V., Mewar, S., Sharma, U., Jagannathan, N.R., et al., 2012. Cellular interaction of folic acid conjugated superparamagnetic iron oxide nanoparticles and its use as contrast agent for targeted magnetic imaging of tumor cells. Int. J. Nanomed. 7, 3503–3516.

Kumari, M., Rajak, S., Singh, S.P., Kumari, S.I., Kumar, P.U., Murty, U.S.N., et al., 2012. Repeated oral dose toxicity of iron oxide nanoparticles: biochemical and histopathological alterations in different tissues of rats. J. Nanosci. Nanotechnol. 12, 2149–2159.

Kumari, M., Rajak, S., Singh, S.P., Murty, U.S.N., Mahboob, M., Grover, P., et al., 2013. Biochemical alterations induced by acute oral doses of iron oxide nanoparticles in Wistar rats. Drug Chem. Toxicol. 36, 296–305.

Kwon, J.-T., Hwang, S.-K., Jin, H., Kim, D.-S., Minai-Tehrani, A., Yoon, H.-J., et al., 2008. Body distribution of inhaled fluorescent magnetic nanoparticles in the mice. J. Occup. Health 50, 1–6.

Lamkowsky, M.-C., Geppert, M., Schmidt, M.M., Dringen, R., 2012. Magnetic field-induced acceleration of the accumulation of magnetic iron oxide nanoparticles by cultured brain astrocytes. J. Biomed. Mater. Res. A 100, 323–334.

Laurent, S., Dutz, S., Häfeli, U.O., Mahmoudi, M., 2011. Magnetic fluid hyperthermia: focus on superparamagnetic iron oxide nanoparticles. Adv. Colloid Interface Sci. 166, 8–23.

Laurent, S., Saei, A.A., Behzadi, S., Panahifar, A., Mahmoudi, M., 2014. Superparamagnetic iron oxide nanoparticles for delivery of therapeutic agents: opportunities and challenges. Expert Opin. Drug Deliv. 11, 1449–1470.

Lei, L., Ling-Ling, J., Yun, Z., Gang, L., 2013. Toxicity of superparamagnetic iron oxide nanoparticles: research strategies and implications for nanomedicine. Chin. Phys. B 22, 127503.

Li, D., Tang, X., Pulli, B., Lin, C., Zhao, P., Cheng, J., et al., 2014. Theranostic nanoparticles based on bioreducible polyethylenimine-coated iron oxide for reduction-responsive gene delivery and magnetic resonance imaging. Int. J. Nanomed. 9, 3347–3361.

Lin, M.M., Kim, D.K., El Haj, A.J., Dobson, J., 2008. Development of superparamagnetic iron oxide nanoparticles (SPIONS) for translation to clinical applications. IEEE Trans. Nanobiosci. 7, 298–305.

Liu, Y., Wang, J., 2013. Effects of DMSA-coated Fe_3O_4 nanoparticles on the transcription of genes related to iron and osmosis homeostasis. Toxicol. Sci. 131, 521–536.

Liu, G., Gao, J., Ai, H., Chen, X., 2013. Applications and potential toxicity of magnetic iron oxide nanoparticles. Small 9, 1533–1545.

Lu, M., Cohen, M.H., Rieves, D., Pazdur, R., 2010. FDA report: ferumoxytol for intravenous iron therapy in adult patients with chronic kidney disease. Am. J. Hematol. 85, 315–319.

Luther, E.M., Petters, C., Bulcke, F., Kaltz, A., Thiel, K., Bickmeyer, U., et al., 2013. Endocytotic uptake of iron oxide nanoparticles by cultured brain microglial cells. Acta Biomater. 9, 8454–8465.

Mahdavi, M., Ahmad, M.B., Haron, M.J., Namvar, F., Nadi, B., Rahman, M.Z.A., et al., 2013. Synthesis, surface modification and characterisation of biocompatible magnetic iron oxide nanoparticles for biomedical applications. Molecules 18, 7533–7548.

Mahmoudi, M., Simchi, A., Milani, A.S., Stroeve, P., 2009. Cell toxicity of superparamagnetic iron oxide nanoparticles. J. Colloid Interface Sci. 336, 510–518.

Mahmoudi, M., Sahraian, M.A., Shokrgozar, M.A., Laurent, S., 2011. Superparamagnetic iron oxide nanoparticles: promises for diagnosis and treatment of multiple sclerosis. ACS Chem. Neurosci. 2, 118–140.

Maier-Hauff, K., Rothe, R., Scholz, R., Gneveckow, U., Wust, P., Thiesen, B., et al., 2006. Intracranial thermotherapy using magnetic nanoparticles combined with external beam radiotherapy: results of a feasibility study on patients with glioblastoma multiforme. J. Neurooncol. 81, 53–60.

Manke, A., Wang, L., Rojanasakul, Y., 2013. Mechanisms of nanoparticle-induced oxidative stress and toxicity, mechanisms of nanoparticle-induced oxidative stress and toxicity. BioMed. Res. Int. 2013, e942916.

Marcus, M., Skaat, H., Alon, N., Margel, S., Shefi, O., 2015. NGF-conjugated iron oxide nanoparticles promote differentiation and outgrowth of PC12 cells. Nanoscale 7, 1058–1066.

Mc Carthy, D.J., Malhotra, M., O'Mahony, A.M., Cryan, J.F., O'Driscoll, C.M., 2015. Nanoparticles and the blood–brain barrier: advancing from in-vitro models towards therapeutic significance. Pharm. Res. 32, 1161–1185.

Migliore, L., Uboldi, C., Di Bucchianico, S., Coppedè, F., 2015. Nanomaterials and neurodegeneration. Environ. Mol. Mutagen. 56, 149–170.

Mills, E., Dong, X., Wang, F., Xu, H., 2010. Mechanisms of brain iron transport: insight into neurodegeneration and CNS disorders. Future Med. Chem. 2, 51.

Mok, H., Zhang, M., 2013. Superparamagnetic iron oxide nanoparticle-based delivery systems for biotherapeutics. Expert Opin. Drug Deliv. 10, 73–87.

Moos, T., Rosengren Nielsen, T., Skjørringe, T., Morgan, E.H., 2007. Iron trafficking inside the brain. J. Neurochem. 103, 1730–1740.

Moroz, P., Jones, S.K., Gray, B.N., 2002. Magnetically mediated hyperthermia: current status and future directions. Int. J. Hyperther. 18, 267–284.

Mu, K., Zhang, S., Ai, T., Jiang, J., Yao, Y., Jiang, L., et al., 2015. Monoclonal antibody-conjugated superparamagnetic iron oxide nanoparticles for imaging of epidermal growth factor receptor-targeted cells and gliomas. Mol. Imaging 14.

Mukhopadhyay, A., Joshi, N., Chattopadhyay, K., De, G., 2012. A facile synthesis of PEG-coated magnetite (Fe_3O_4) nanoparticles and their prevention of the reduction of cytochrome c. ACS Appl. Mater. Interfaces 4, 142–149.

Muldoon, L.L., Sandor, M., Pinkston, K.E., Neuwelt, E.A., 2005. Imaging, distribution, and toxicity of superparamagnetic iron oxide magnetic resonance nanoparticles in the rat brain and intracerebral tumor. Neurosurgery 57, 785–796 discussion 785–796.

Myhre, O., Utkilen, H., Duale, N., Brunborg, G., Hofer, T., 2013. Metal dyshomeostasis and inflammation in Alzheimer's and Parkinson's diseases: possible impact of environmental exposures. Oxid. Med. Cell. Longev. 2013, e726954.

Nel, A., Xia, T., Mädler, L., Li, N., 2006. Toxic potential of materials at the nanolevel. Science 311, 622–627.

Neri, M., Maderna, C., Cavazzin, C., Deidda-Vigoriti, V., Politi, L.S., Scotti, G., et al., 2008. Efficient in vitro labeling of human neural precursor cells with superparamagnetic iron oxide particles: relevance for in vivo cell tracking. Stem Cells 26, 505–516.

Neubert, J., Wagner, S., Kiwit, J., Bräuer, A.U., Glumm, J., 2015. New findings about iron oxide nanoparticles and their different effects on murine primary brain cells. Int. J. Nanomed. 10, 2033–2049.

Patil, U.S., Adireddy, S., Jaiswal, A., Mandava, S., Lee, B.R., Chrisey, D.B., 2015. In vitro/in vivo toxicity evaluation and quantification of iron oxide nanoparticles. Int. J. Mol. Sci. 16, 24417–24450.

Petters, C., Dringen, R., 2014. Comparison of primary and secondary rat astrocyte cultures regarding glucose and glutathione metabolism and the accumulation of iron oxide nanoparticles. Neurochem. Res. 39, 46–58.

Petters, C., Dringen, R., 2015. Accumulation of iron oxide nanoparticles by cultured primary neurons. Neurochem. Int. 81, 1–9.

Petters, C., Irrsack, E., Koch, M., Dringen, R., 2014a. Uptake and metabolism of iron oxide nanoparticles in brain cells. Neurochem. Res. 39, 1648–1660.

Petters, C., Bulcke, F., Thiel, K., Bickmeyer, U., Dringen, R., 2014b. Uptake of fluorescent iron oxide nanoparticles by oligodendroglial OLN-93 cells. Neurochem. Res. 39, 372–383.

Petters, C., Thiel, K., Dringen, R., 2015. Lysosomal iron liberation is responsible for the vulnerability of brain microglial cells to iron oxide nanoparticles: comparison with neurons and astrocytes. Nanotoxicology 1–11.

Pickard, M.R., Chari, D.M., 2010. Robust uptake of magnetic nanoparticles (MNPs) by central nervous system (CNS) microglia: implications for particle uptake in mixed neural cell populations. Int. J. Mol. Sci. 11, 967–981.

Pisanic 2nd, T.R., Blackwell, J.D., Shubayev, V.I., Fiñones, R.R., Jin, S., 2007. Nanotoxicity of iron oxide nanoparticle internalization in growing neurons. Biomaterials 28, 2572–2581.

Pisanic, T.R., Jin, S., Shubayev, V.I., 2009. Iron oxide magnetic nanoparticle nanotoxicity: incidence and mechanisms. In: Sahu, S.C., Casciano, D.A. (Eds.), Nanotoxicity. John Wiley & Sons, Ltd, pp. 397–425.

Qiao, R., Jia, Q., Hüwel, S., Xia, R., Liu, T., Gao, F., et al., 2012. Receptor-mediated delivery of magnetic nanoparticles across the blood–brain barrier. ACS Nano 6, 3304–3310.

Rahman, K., 2007. Studies on free radicals, antioxidants, and co-factors. Clin. Interv. Aging 2, 219–236.

Reimer, P., Balzer, T., 2003. Ferucarbotran (Resovist): a new clinically approved RES-specific contrast agent for contrast-enhanced MRI of the liver: properties, clinical development, and applications. Eur. Radiol. 13, 1266–1276.

Riggio, C., Calatayud, M.P., Hoskins, C., Pinkernelle, J., Sanz, B., Torres, T.E., et al., 2012. Poly-l-lysine-coated magnetic nanoparticles as intracellular actuators for neural guidance. Int. J. Nanomed. 7, 3155–3166.

Rivet, C.J., Yuan, Y., Borca-Tasciuc, D.-A., Gilbert, R.J., 2012. Altering iron oxide nanoparticle surface properties induce cortical neuron cytotoxicity. Chem. Res. Toxicol. 25, 153–161.

Riviere, J., 2014. Pharmacokinetics and biodistribution of nanomaterials. In: Nanotoxicology Progress Toward Nanomedicine, second ed. CRC Press, pp. 135–152.

Rosen, J.E., Chan, L., Shieh, D.-B., Gu, F.X., 2012. Iron oxide nanoparticles for targeted cancer imaging and diagnostics. Nanomed. Nanotechnol. Biol. Med. 8, 275–290.

Sabella, S., Carney, R.P., Brunetti, V., Malvindi, M.A., Al-Juffali, N., Vecchio, G., et al., 2014. A general mechanism for intracellular toxicity of metal-containing nanoparticles. Nanoscale 6, 7052–7061.

Saleh, A., Schroeter, M., Ringelstein, A., Hartung, H.-P., Siebler, M., Mödder, U., et al., 2007. Iron oxide particle-enhanced MRI suggests variability of brain inflammation at early stages after ischemic stroke. Stroke 38, 2733–2737.

Santhosh, P.B., Ulrih, N.P., 2013. Multifunctional superparamagnetic iron oxide nanoparticles: promising tools in cancer theranostics. Cancer Lett. 336, 8–17.

Sarkar, A., Ghosh, M., Sil, P.C., 2014. Nanotoxicity: oxidative stress mediated toxicity of metal and metal oxide nanoparticles. J. Nanosci. Nanotechnol. 14, 730–743.

Saura, J., 2007. Microglial cells in astroglial cultures: a cautionary note. J. Neuroinflammation 4, 26.

Shang, L., Nienhaus, K., Nienhaus, G.U., 2014. Engineered nanoparticles interacting with cells: size matters. J. Nanobiotechnol. 12, 5.

Sharifi, S., Seyednejad, H., Laurent, S., Atyabi, F., Saei, A.A., Mahmoudi, M., 2015. Superparamagnetic iron oxide nanoparticles for in vivo molecular and cellular imaging. Contrast Media Mol. Imaging 10, 329–355.

Sharma, H.S., Menon, P.K., Lafuente, J.V., Aguilar, Z.P., Wang, Y.A., Muresanu, D.F., et al., 2014. The role of functionalized magnetic iron oxide nanoparticles in the central nervous system injury and repair: new potentials for neuroprotection with cerebrolysin therapy. J. Nanosci. Nanotechnol. 14, 577–595.

Shevtsov, M.A., Nikolaev, B.P., Yakovleva, L.Y., Marchenko, Y.Y., Dobrodumov, A.V., Mikhrina, A.L., et al., 2014. Superparamagnetic iron oxide nanoparticles conjugated with epidermal growth factor (SPION–EGF) for targeting brain tumors. Int. J. Nanomed. 9, 273–287.

Singh, S.P., Rahman, M.F., Murty, U.S.N., Mahboob, M., Grover, P., 2013. Comparative study of genotoxicity and tissue distribution of nano and micron sized iron oxide in rats after acute oral treatment. Toxicol. Appl. Pharmacol. 266, 56–66.

Skaper, S.D., Floreani, M., Ceccon, M., Facci, L., Giusti, P., 1999. Excitotoxicity, oxidative stress, and the neuroprotective potential of melatonin. Ann. NY Acad. Sci. 890, 107–118.

Soenen, S.J.H., Himmelreich, U., Nuytten, N., De Cuyper, M., 2011. Cytotoxic effects of iron oxide nanoparticles and implications for safety in cell labelling. Biomaterials 32, 195–205.

Souza, D.G., Bellaver, B., Souza, D.O., Quincozes-Santos, A., 2013. Characterization of adult rat astrocyte cultures. PLoS One 8, e60282.

Thomsen, L.B., Linemann, T., Pondman, K.M., Lichota, J., Kim, K.S., Pieters, R.J., et al., 2013. Uptake and transport of superparamagnetic iron oxide nanoparticles through human brain capillary endothelial cells. ACS Chem. Neurosci. 4, 1352–1360.

Torres-Lugo, M., Rinaldi, C., 2013. Thermal potentiation of chemotherapy by magnetic nanoparticles. Nanomedicine 8, 1689–1707.

Tsai, H.-H., Li, H., Fuentealba, L.C., Molofsky, A.V., Taveira-Marques, R., Zhuang, H., et al., 2012. Regional astrocyte allocation regulates CNS synaptogenesis and repair. Science 337, 358–362.

Tse, B.W.-C., Cowin, G.J., Soekmadji, C., Jovanovic, L., Vasireddy, R.S., Ling, M.-T., et al., 2014. PSMA-targeting iron oxide magnetic nanoparticles enhance MRI of preclinical prostate cancer. Nanomedicine 1–12.

Turcheniuk, K., Tarasevych, A.V., Kukhar, V.P., Boukherroub, R., Szunerits, S., 2013. Recent advances in surface chemistry strategies for the fabrication of functional iron oxide based magnetic nanoparticles. Nanoscale 5, 10729–10752.

Unfried, K., Albrecht, C., Klotz, L.-O., Von Mikecz, A., Grether-Beck, S., Schins, R.P.F., 2007. Cellular responses to nanoparticles: target structures and mechanisms. Nanotoxicology 1, 52–71.

Valdiglesias, V., Costa, C., Sharma, V., Kiliç, G., Pásaro, E., Teixeira, J.P., et al., 2013a. Comparative study on effects of two different types of titanium dioxide nanoparticles on human neuronal cells. Food Chem. Toxicol. 57, 352–361.

Valdiglesias, V., Costa, C., Kiliç, G., Costa, S., Pásaro, E., Laffon, B., et al., 2013b. Neuronal cytotoxicity and genotoxicity induced by zinc oxide nanoparticles. Environ. Int. 55, 92–100.

Vasir, J.K., Labhasetwar, V., 2005. Targeted drug delivery in cancer therapy. Technol. Cancer Res. Treat. 4, 363–374.

Veiseh, O., Sun, C., Fang, C., Bhattarai, N., Gunn, J., Kievit, F., et al., 2009. Specific targeting of brain tumors with an optical/magnetic resonance imaging nanoprobe across the blood–brain barrier. Cancer Res. 69, 6200–6207.

Veranth, J.M., Kaser, E.G., Veranth, M.M., Koch, M., Yost, G.S., 2007. Cytokine responses of human lung cells (BEAS-2B) treated with micron-sized and nanoparticles of metal oxides compared to soil dusts. Part Fibre Toxicol. 4, 2.

Wahajuddin, Arora, S., 2012. Superparamagnetic iron oxide nanoparticles: magnetic nanoplatforms as drug carriers. Int. J. Nanomed. 7, 3445–3471.

Wang, B., Feng, W.Y., Wang, M., Shi, J.W., Zhang, F., Ouyang, H., et al., 2007. Transport of intranasally instilled fine Fe_2O_3 particles into the brain: micro-distribution, chemical states, and histopathological observation. Biol. Trace Elem. Res. 118, 233–243.

Wang, B., Feng, W., Zhu, M., Wang, Y., Wang, M., Gu, Y., et al., 2009. Neurotoxicity of low-dose repeatedly intranasal instillation of nano- and submicron-sized ferric oxide particles in mice. J. Nanopart. Res. 11, 41–53.

Wang, Y., Wang, B., Zhu, M.-T., Li, M., Wang, H.-J., Wang, M., et al., 2011. Microglial activation, recruitment and phagocytosis as linked phenomena in ferric oxide nanoparticle exposure. Toxicol. Lett. 205, 26–37.

Wang, X., Wei, F., Liu, A., Wang, L., Wang, J.-C., Ren, L., et al., 2012. Cancer stem cell labeling using poly(L-lysine)-modified iron oxide nanoparticles. Biomaterials 33, 3719–3732.

Wang, Y.-X.J., Xuan, S., Port, M., Idee, J.-M., 2013a. Recent advances in superparamagnetic iron oxide nanoparticles for cellular imaging and targeted therapy research. Curr. Pharm. Des. 19, 6575–6593.

Wang, X., Tu, Q., Zhao, B., An, Y., Wang, J.-C., Liu, W., et al., 2013b. Effects of poly(L-lysine)-modified Fe_3O_4 nanoparticles on endogenous reactive oxygen species in cancer stem cells. Biomaterials 34, 1155–1169.

Win-Shwe, T.-T., Fujimaki, H., 2011. Nanoparticles and neurotoxicity. Int. J. Mol. Sci. 12, 6267–6280.

Wohlfart, S., Gelperina, S., Kreuter, J., 2012. Transport of drugs across the blood–brain barrier by nanoparticles. J. Control. Release 161, 264–273.

Wu, J., Sun, J., 2011. Investigation on mechanism of growth arrest induced by iron oxide nanoparticles in PC12 cells. J. Nanosci. Nanotechnol. 11, 11079–11083.

Wu, X., Zhang, M., 2010. Toxic effects of iron oxide nanoparticles on human umbilical vein endothelial cells. Int. J. Nanomed. 5, 385–399.

Wu, W., He, Q., Jiang, C., 2008. Magnetic iron oxide nanoparticles: synthesis and surface functionalization strategies. Nanoscale Res. Lett. 3, 397–415.

Wu, J., Ding, T., Sun, J., 2013a. Neurotoxic potential of iron oxide nanoparticles in the rat brain striatum and hippocampus. Neurotoxicology 34, 243–253.

Wu, H.-Y., Chung, M.-C., Wang, C.-C., Huang, C.-H., Liang, H.-J., Jan, T.-R., 2013b. Iron oxide nanoparticles suppress the production of IL-1beta via the secretory lysosomal pathway in murine microglial cells. Part Fibre Toxicol. 10, 46.

Xie, H., Zhu, Y., Jiang, W., Zhou, Q., Yang, H., Gu, N., et al., 2011. Lactoferrin-conjugated superparamagnetic iron oxide nanoparticles as a specific MRI contrast agent for detection of brain glioma in vivo. Biomaterials 32, 495–502.

Xue, Y., Wu, J., Sun, J., 2012. Four types of inorganic nanoparticles stimulate the inflammatory reaction in brain microglia and damage neurons in vitro. Toxicol. Lett. 214, 91–98.

Yan, F., Wang, Y., He, S., Ku, S., Gu, W., Ye, L., 2013. Transferrin-conjugated, fluorescein-loaded magnetic nanoparticles for targeted delivery across the blood–brain barrier. J. Mater. Sci. Mater. Med. 24, 2371–2379.

Yiu, H.H.P., Pickard, M.R., Olariu, C.I., Williams, S.R., Chari, D.M., Rosseinsky, M.J., 2012. Fe_3O_4-PEI-RITC magnetic nanoparticles with imaging and gene transfer capability: development of a tool for neural cell transplantation therapies. Pharm. Res. 29, 1328–1343.

Zhao, Q., Wang, L., Cheng, R., Mao, L., Arnold, R.D., Howerth, E.W., et al., 2012. Magnetic nanoparticle-based hyperthermia for head & neck cancer in mouse models. Theranostics 2, 113–121.

Ziv-Polat, O., Topaz, M., Brosh, T., Margel, S., 2010. Enhancement of incisional wound healing by thrombin conjugated iron oxide nanoparticles. Biomaterials 31, 741–747.

Chapter 7

The Application, Neurotoxicity, and Related Mechanisms of Silver Nanoparticles

L. Strużyńska
Mossakowski Medical Research Centre, Polish Academy of Sciences, Warsaw, Poland

Chapter Outline

1. INTRODUCTION

Silver in its ionized form is known to have strong antibacterial properties and has been used in food preservation and water disinfection since ancient times (Barillo and Marx, 2014). In medicinal applications, silver compounds have been used for wound care and cure of infections such as syphilis and gonorrhea, and also for treatment of mental diseases and epilepsy (Marshall and Schneider, 1977). Silver is currently in widespread use as a component of optics, electrical conductors, paints, soldering materials, and mirror production.

Neurotoxicity of Nanomaterials and Nanomedicine. http://dx.doi.org/10.1016/B978-0-12-804598-5.00007-6

The last few decades have witnessed an increasing interest in developing nanoengineering methods to construct nanomaterials for use in a wide range of applications. The development of nanotechnologies has induced the rediscovery of biocidal activity of silver for medical applications. The unique physicochemical properties of the nanosized form of silver tend to be more effective or different from silver in the bulk form and this has led to extensive use of silver nanoparticles (AgNPs). Moreover, these properties may be enhanced or modified by coating the surface of AgNPs with different substances and chemical groups (Mody et al., 2010) (Fig. 7.1).

Owing to the strong antibacterial and antifungal activity of AgNPs, the market for commercially available AgNP-containing goods has increased significantly. Currently, about 30% of all nanoparticle-based products contain AgNPs (Rejeski, 2011). The long list includes home appliances such as refrigerators, dishwashers, and washing machines containing an AgNP-based core,

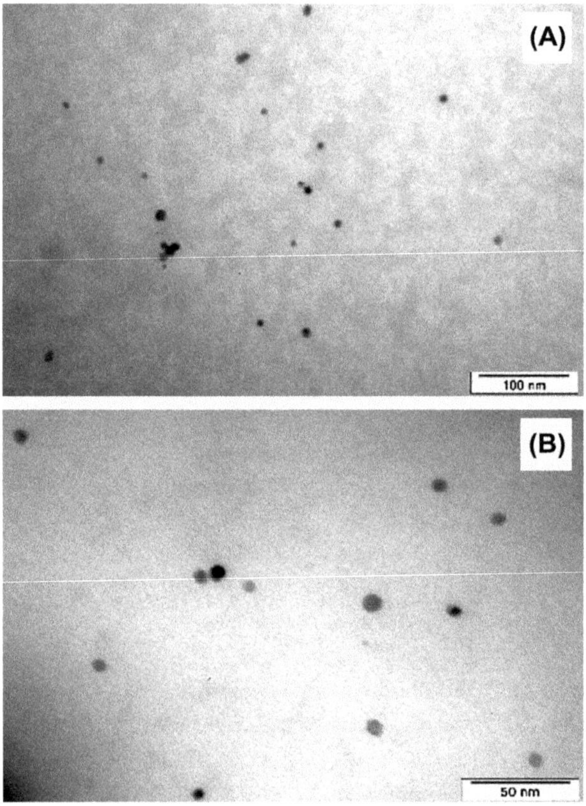

FIGURE 7.1 Transmission electron microscopy image of commercially available (A) citrate-stabilized AgNPs of a homogeneous size (mean diameter of 10 ± 4 nm; Sigma–Aldrich) and (B) PVP-coated AgNPs of different sizes (mean diameter <100 nm; Sigma–Aldrich).

which generates silver ions during work cycles (Chen and Schluesener, 2008; Marambio-Jones and Hoek, 2010). The antibacterial properties of AgNPs have led to development of AgNP-coated textiles for use in sports and medicine as well as development of cosmetic and hygienic products (personal cleansers, deodorants, soaps, toothbrushes, and pastes). Cleansers, detergents, disinfectants, and paints including AgNPs are used at locations requiring elevated hygienic standards, such as hospitals and restaurants (Fewtrell, 2014; Gaillet and Rouanet, 2015). Global production of AgNPs is estimated to be 320 tons/year (Gottschalk et al., 2010) and future scenarios have led to predictions that applications of AgNPs will increase. The widespread use of engineered AgNPs may be considered a substantial source of silver exposure. Common usage of reactive AgNPs with high cytotoxic potency toward microbes has raised substantial questions about their human safety profile.

2. CURRENT AND FUTURE APPLICATIONS OF AGNPs IN MEDICINE

Apart from the extended use of AgNPs in many consumer products, the last few decades have witnessed growing prominence of AgNPs in medicine. The ability of AgNPs to generate reactive oxygen species (ROS), which inactivate microbial enzymes (Matsumura et al., 2003), leading to the inhibition of growth of microorganisms, is considered to be the main feature provided by AgNPs. Strong antimicrobial activity (reviewed by Dos Santos et al., 2014) makes AgNPs attractive for use as a coating material for various medical devices used in surgical procedures such as central venous and bladder catheters, stents or cardiovascular implants, prostheses, or endotracheal tubes. Such coatings minimize the risk of infections (Ge et al., 2014) and reduce the inflammatory response (Grunkemeier et al., 2006). AgNPs not only exhibit a direct bacteriotoxic effect but also prevent formation of biofilm, which is formed by masses of bacterial cells conglomerating on the inhabited surface. Biofilms are responsible for enhanced virulence and increased resistance to drugs and disinfectants. Wound healing is another application that takes advantage of the biocidal properties of AgNPs. Dressings and gels containing AgNPs significantly reduce the period required for healing of ulcers, burns, or cases of epidermal necrolysis (Chaloupka et al., 2010; Huang et al., 2007).

Potent activity against gram-positive and gram-negative bacteria also makes AgNPs very attractive for use as an alternative treatment of antibiotic-resistant bacterial diseases (Ge et al., 2014). Apart from antibacterial activity, AgNPs also exert antifungal and antiviral effects. Many species of highly pathogenic fungi involved in various severe infections such as *Aspergillus* sp. and *Candida* sp. have been reported to be sensitive to treatment with AgNPs (for a review see Ge et al., 2014). A synergistic effect has been observed when treatment with AgNPs is combined with commercially available antifungal drugs (Gajbhiye et al., 2009). Moreover, AgNPs exhibit very effective antiviral properties against herpes simplex virus (Baram-Pinto et al., 2009), hepatitis B virus (Lu et al., 2008), and human immunodeficiency virus (HIV) 1 virus. With regard to HIV-1, it has been demonstrated that AgNPs inhibit both the initial and postentry stages of the viral life cycle (Lara et al., 2010).

It has been proved that AgNPs combine antimicrobial properties with antiinflammatory activity in an animal model of allergic rhinitis (Shin and Ye, 2012) and in a human clinical study where AgNP-coated wound dressings were found to accelerate healing of ulcers by inhibiting both infection and the inflammatory response (Sibbald et al., 2007).

Although widely recognized and applied, the antimicrobial properties of AgNPs represent only one of their characteristics. Nanoparticles of noble metals, including silver, also have many novel and unique biomedical properties arising from confinement of photons within particle bodies and this leads to significant scattering and absorption of electromagnetic radiation. In addition, exogenous molecules are attracted to the exterior of the particles and this effect is useful in the development of drug delivery strategies (Dreaden and El-Sayed, 2012).

The rapid development of nanotechnology gives us important opportunities to investigate AgNPs in the context of conventional medicine as well as in diagnosis strategies that employ biomarkers and imaging systems and in treatments of many disorders including cancer and central nervous system (CNS) pathologies. Theragnosis, an innovative concept in next-generation medicine combining diagnosis with therapy, is expected to benefit from the unique features of AgNPs (Leite et al., 2015).

Early diagnosis is extremely important in the case of cancer. AgNPs have already been reported as a promising tool in noninvasive methods of cancer detection (Lin et al., 2011). In addition, AgNPs have been found to act as potent sensitizers in the treatment of radioresistant glioblastoma (Xu et al., 2009). A radiosensitizer is a compound that selectively and locally absorbs radiation during cancer radiotherapy. Such a tactic can minimize damage of normal tissue around tumors during targeted destruction of cancer cells (Dos Santos et al., 2014).

Apart from enhancing radiosensitivity of cancers, AgNPs demonstrate direct cytotoxic effects on cancer cells. Death of glioma C6 cells via an apoptotic pathway (Wang et al., 2013) and oxidative stress–induced damage of DNA and subsequent death of acute myeloid leukemia cells (Guo et al., 2013) have been reported.

AgNPs are also investigated as drug carriers or elements of nanoparticulate drug-carrier systems in targeted drug delivery to the site of the disease. Anatomical and physiological differences between normal and diseased tissue, such as leaky tumor vasculature, favor enhanced accumulation of NPs inside pathological lesions in passive targeting as opposed to active targeting, in which the carrier becomes conjugated to the cell- or tissue-specific ligand (Parveen et al., 2012). Moreover, modification with cell-penetrating peptide (TAT), which enhances cell penetration of AgNPs, has been proposed as a strategy for overcoming the multidrug resistance (MDR) problem in cancer treatment (Liu et al., 2012b). The mechanism of drug efflux from cancer cells, mediated by ATP-binding cassette transporters, underlies the MDR phenomenon and is responsible for ineffective chemotherapy.

Furthermore, searching for effective but safe strategies for treatment of brain pathologies is a challenge. Nanoparticles, including AgNPs, carrying chemical compounds and possessing proper characteristics may cross the blood–brain barrier (BBB) to modulate brain cells (Leite et al., 2015). Applications of AgNPs are also being developed in regenerative medicine as components of implants and scaffolds (Ge et al., 2014) and as inserts for total joint replacement (Morley et al., 2007).

The use of AgNPs for controlling growth of microorganisms carries a potential risk of toxic effects in humans. The fate of nanoparticles in human cells and mechanisms of toxicity are poorly known. The high reactivity of AgNPs, which allows them to freely cross cell membranes, is of great concern in neuroscience. These interactions must be solved if the promise of AgNP-based nanotools is to be realized. Therefore, investigations using animal models are necessary to develop biosafe nonthrombogenic and nonneurotoxic nanoparticles.

3. BIODISTRIBUTION OF AGNPs IN MAMMALIAN ORGANISMS

Organisms may be exposed to AgNPs via different routes: through the gastrointestinal tract, dermal contact, or inhalation. Oral and inhalatory absorption dominates over dermal contact. Once absorbed into an organism, AgNPs are further biodistributed in tissue compartments.

After oral administration, silver is quickly absorbed from the gastrointestinal tract into the blood, where its concentration remains stable over several hours and is significantly deposited in the serum compartment after a 2-week exposure period. The rate of silver absorption is reported to be equal regardless of the silver formulation (ionic/nano) (Skalska et al., 2015) or higher for ionic silver (Loeschner et al., 2011). The bioavailability is lower for AgNPs than for ionic form of silver as a result of fecal excretion (68% vs. 49%) and lower absorption in organs (Loeschner et al., 2011; van der Zande et al., 2012). Studies on biodistribution of AgNPs administered to rats by oral gavage or intravenous injection indicate that the highest absorption occurs in liver, spleen, kidney, and gonads, with the lowest observed in lungs and brain (Loeschner et al., 2011; Kim et al., 2010; Lee et al., 2013). When administered via intranasal or inhalatory routes, enhanced silver concentrations were identified in blood, liver, lungs, kidney, olfactory bulb, and brain (Sung et al., 2009; Genter et al., 2012).

Transmission electron microscopy (TEM) is a useful method to demonstrate the presence of AgNPs in brain tissue. Dark round nanosized granules are observed in a diameter range similar to that of nanoparticles administered to animals. A modified TEM method for tissue processing in which the phase of staining with lead and uranium is omitted gives better results because the only visible dark matter (being potentially of the metallic form) is silver. While analyzing noncontrasted brain specimens, nanoparticle-like structures have been reported to be located between lamellae of myelin sheaths and in the lysosomal compartment of neurons and endothelial cells of microvessels (De Matteis et al., 2015; Skalska et al., 2015; see also Fig. 7.2). Using the same method of analysis, nanosized granules were detected previously in the lysosomal cell compartment of astrocytic and neuronal cultures treated with AgNPs (Haase et al., 2012).

Based on the previous studies, it can be assumed that AgNPs enter the vertebrate brain via the upper respiratory tract and become mobilized along nerves in the olfactory bulb when administered by inhalation, or via the BBB when

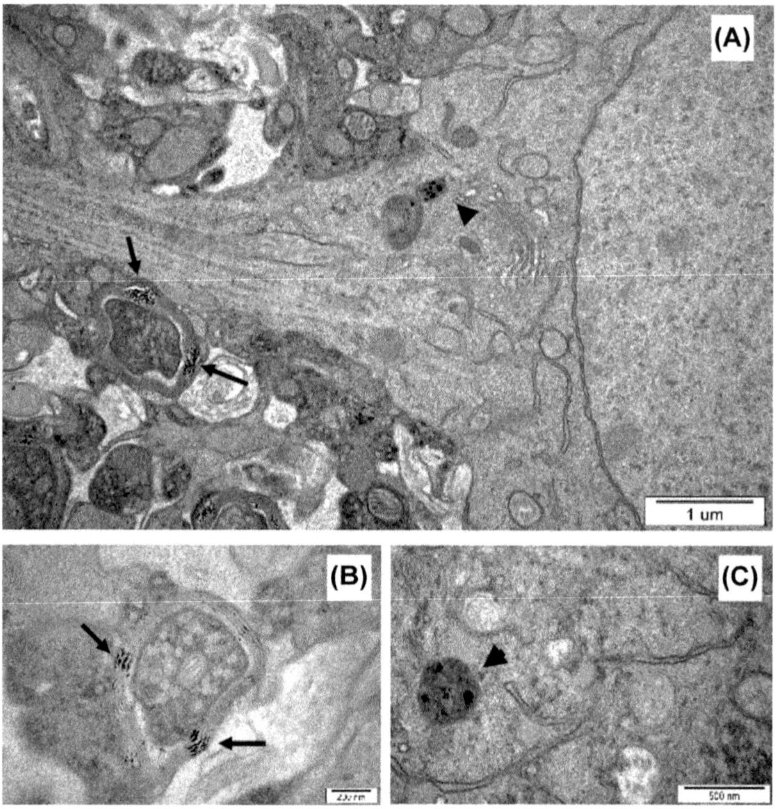

FIGURE 7.2 Transmission electron microscopy images demonstrating the presence of nanoparticles in rat brain after 2-week exposure to 10-nm AgNPs. AgNPs are visible as round dark granules located between lamellae of myelin sheaths (A, B; *arrows*) and inside lysosomes in neurons (A, C; *arrowheads*) in ultrathin sections that were not contrasted with heavy metals during tissue processing.

administered systemically or orally (Hoet et al., 2004; Sharma et al., 2009a). In addition, it was found that the brain is not cleared of accumulated AgNPs, even during a prolonged recovery period (Lee et al., 2013). AgNPs of small diameter have a long half-life in the brain, indicating that the BBB impedes their transfer back to the blood (van der Zande et al., 2012). The tendency of AgNPs to accumulate in the brain creates a risk of significant long-lasting pathological effects.

4. AGNP-INDUCED NEUROTOXICITY

Experimental studies revealed that engineered AgNPs may be efficiently absorbed while administered by different routes and accumulated in brain. Hence it is highly probable that AgNPs liberated from commercial nanoproducts may be considered as a substantial source of exposure. In addition, advances in nanoparticle-based approaches in the field of theragnosis of neurological disorders open wide possibilities for diagnosis and treatment of brain tumors, brain infections (including HIV), and affective or neurodegenerative disorders. Development of novel brain-targeted drugs holds the promise of providing efficient therapies in specific brain areas that were previously restricted by specialized CNS barriers including the BBB and the cerebrospinal fluid barrier. AgNPs are candidates for such targeted drugs and therefore there is significant interest in investigating their potential neurotoxic effects (Yang et al., 2010). Such knowledge will be useful during the course of development of AgNPs that are free of side effects. Currently available information about AgNP-associated hazards is limited and requires verification.

4.1 Adverse Effects of AgNPs in Brain Cells

Based on in vitro experiments, there is no doubt that AgNPs have toxic effects on neuronal cells. Exposure to AgNPs was shown to decrease viability of cultured neurons (Haase et al., 2012; Xu et al., 2013; Yin et al., 2013; Ziemińska et al., 2014) and astroglia-rich primary cultures (Haase et al., 2012), although it appears that astrocytes are generally less sensitive than neurons. During exposure of astrocytes to AgNPs, metallothioneins were shown to be upregulated (Luther et al., 2012). These metal-binding proteins represent a protective mechanism induced in the cells under conditions of metal toxicity. In addition, inhibition of axonal outgrowth and induction of degeneration of neurites of cultured cortical neurons were shown to be connected with cytoskeletal components (β-tubulin and F-actin) and a deficiency of synaptic proteins (Xu et al., 2013).

It is still being considered whether AgNPs exert neurotoxic properties at the time of administration to an organism. The previous experimental data concerning neurotoxicity of ionic silver demonstrated preferential accumulation of granular silver-containing deposits in large motor neurons and protoplasmic astrocytes in areas of the CNS responsible for motor control. The deposits were found to be predominately located in hippocampus and pons of rodents chronically exposed by oral gavage (Rungby and Danscher, 1983). However, there is a contradictory study claiming that pathological and behavioral changes induced by this metal are rarely observed (Lansdown, 2007).

There is little evidence of neurotoxic action of silver in its "nano" form in vivo. It appears, however, that AgNPs penetrating the brain cause swelling of astrocytes located outside the BBB, and appearance of pyknotic and necrotic neurons (Tang et al., 2009). Histopathological analysis of the cerebellum of neonatal rats (Yin et al., 2015) and hippocampus of adult rats (Liu et al., 2012a) exposed intranasally to AgNPs showed edema, cell degeneration, and necrotic areas.

Prolonged oral exposure of rats to commercially available small (10 nm) AgNPs revealed that the AgNPs are efficiently absorbed into the blood followed by changes in synapses at both the ultrastructural and biochemical levels (Skalska et al., 2015). This experimental design permits us to assess the profile of toxic events in animals exposed to relatively low doses of AgNPs (0.2 mg/kg body weight), which better reflects potential exposure of humans to AgNPs. It should be stressed that it is very helpful to conduct studies that investigate the induction of pathological processes in the brain by very low concentrations of nanoparticles, especially because the vast majority of studies on neurotoxicity have been done using in vitro systems with high doses of AgNPs.

4.2 AgNPs Influence Blood–Brain Barrier Function

Because it is apparent that AgNPs reach the brain parenchyma, it is important to characterize the route of passage of AgNPs to the brain and their related, potentially toxic, mechanisms. It is known that the BBB is a protective gateway to the brain, which strictly regulates transport of different substances between blood and brain at both anatomical and metabolic levels. Thus, the BBB maintains brain homeostasis for proper functioning of the CNS. This highly selective and specialized structure is composed of a basement membrane and endothelial cells connected with tight junctions, which interact with pericytes, perivascular astrocytes, and neurons, creating a functional neurovascular unit (Hawkins and Davis, 2005). At the anatomical level, the function of barrier, which restricts transport of molecules to the brain, is maintained by microvascular endothelial cells. Pericytes are coating endothelial cells from the abluminal side which are ensheathed by the basal lamina of collagen, fibronectin, and laminin. Pericytes interact with endothelial cells which are involved in regulating barrier properties. Similarly, perivascular astrocytic end feet maintain barrier integrity via dynamic interactions with the endothelium.

The tight junctions form a physical barrier consisting transmembrane and cytoplasmic proteins predominately including junctional adhesion molecules, occludins, claudins, zonula occludens (ZO), and cingulin, which are connected to cytoskeletal proteins. Another feature distinguishing cerebral capillaries from other types of microvessels is a lack of pinocytosis and endothelial cell fenestration. As a result, there is an increase in the number of mitochondria relative to peripheral endothelial cells (Hawkins and Davis, 2005; Saunders et al., 2008).

At the biochemical level, compounds infiltrate the brain predominately via carrier-mediated transport and receptor-mediated transcytosis and less frequently by passive diffusion. The ability of a substance to cross the BBB depends mainly on its size and lipophilicity (Bhaskar et al., 2010). Using an in vitro model, it was revealed that AgNPs pass the BBB mainly by transcytosis but accumulate inside endothelial cells as well (Tang et al., 2010). Rats treated chronically with 10 nm AgNPs were found to accumulate AgNPs in lysosomes of endothelial cells (Skalska et al., 2015). On this basis, it is presumed likely that AgNPs influence the function of endothelial cells by increasing permeability of microvessels. Indeed, existing in vitro data have identified perforations in the monolayer of rat brain microvessel endothelial cells exposed to AgNPs (Trickler et al., 2010). Moreover, AgNP-induced discontinuous tight junctions and decreased expression of ZO-1 protein were observed in parallel with increased permeability of endothelial cells in a triple coculture BBB model consisting of endothelial cells, pericytes, and astrocytes (Xu et al., 2015). A marked BBB disintegration in AgNP-treated rodents receiving AgNPs via intravenous, intraperitoneal, or intracerebroventricular administration has also been reported (Sharma et al., 2009a, 2010). Our own unpublished data support these findings as well. Perivascular edema with swollen end feet of astrocytes surrounding capillaries indicates enhanced permeability of microvessels in brain of adult rats exposed chronically to low levels of AgNPs (Fig. 7.3).

Regression analysis has indicated a strong inverse correlation between the size of AgNPs and microvessel leakage in the brain of exposed rats (Sharma et al., 2013). This study suggests that breakdown of the BBB function could play a crucial role in AgNP-induced neurotoxicity. Vasogenic changes lead to brain edema, which is one of the main causes of additional brain tissue injuries caused by mechanical compression of cells and tissue. Morphological analysis of neuronal, glial, and myelin injuries showed a close correlation between edematous neuropil and cellular damage. Swollen areas contain many neurons, astrocytes, and axons with extensive damage, suggesting that edema formation and cellular injury induced by AgNP exposure are interlinked processes.

Furthermore, AgNP-induced release of proinflammatory cytokines such as interleukin (IL)-1β and tumor necrosis factor α in rat brain microvessels may additionally enhance the breakdown of the BBB (Trickler et al., 2010). However, it is also likely that very small 4–6 nm AgNPs can pass through the endothelial cell membrane without damaging them (Tang et al., 2008).

BBB integrity is often changed during the course of brain diseases, and during attack of pathogenic factors or inflammatory conditions. In dysfunctional microvessels transport pathways such as transcellular transport and transcytosis are affected. Paradoxically, an impaired BBB may provide the opportunity

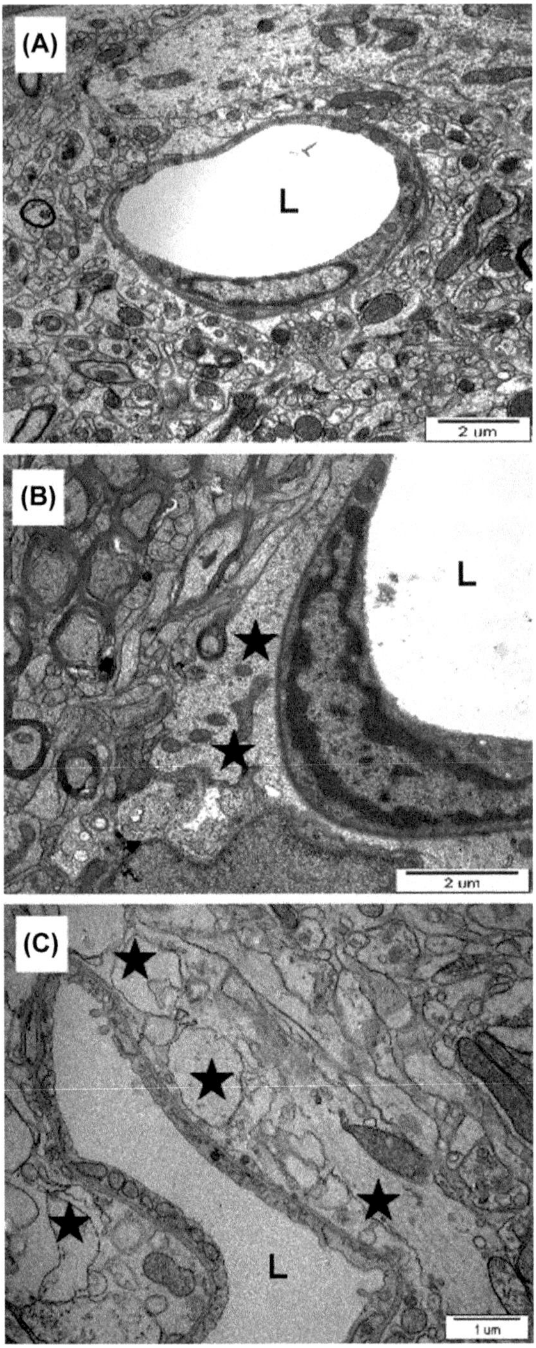

FIGURE 7.3 Representative transmission electron microscopy images of microvessels in cortical sections of control (A) and AgNP-treated (B, C) rat brain. Enhanced permeability of microvessels produces local swelling (B) or massive swelling (C) of adjacent astrocytic end feet (*asterisks*); *L*, lumen of vessel.

for drugs to reach the diseased brain. Modulated opening of tight junctions to enhance passage of substance through the BBB has been suggested as a strategy for transporting drugs into the CNS (Chen and Liu, 2012).

From the perspective of the application of AgNPs as drug carriers for treatment of brain disorders, it is very important to characterize their impact on neuronal tissue under nonphysiological conditions. Under pathological conditions connected with enhanced permeability of cerebral microvessels, accumulation of AgNPs in brain parenchyma was found to be enhanced and aggravated by existing brain pathology. In experimental rats subjected to heat stress that were previously treated with AgNPs, exacerbation of BBB dysfunction was observed, which was found to lead to edema formation, impairment of cognitive and motor functions, and brain damage (Sharma et al., 2009b). Similarly, during sleep deprivation, which represents severe stress that leads to alteration of brain metabolism and BBB function, AgNPs intensified the disruption of the BBB by 1.5- to 4-fold in a manner depending on the period of wakefulness (Sharma et al., 2015).

Interestingly, the negative characteristics of AgNPs can be harnessed for positive purposes. AgNPs have been confirmed to enhance expression of antiangiogenic activity resulting from inhibition of vascular endothelial growth factor–induced proliferation. There is evidence that AgNPs induce apoptosis in bovine retinal endothelial cells by enhancing caspase-3 activity with parallel inhibition of a cell survival pathway connected with PI3 kinase/Akt (Kalishwaralal et al., 2009). Abnormal angiogenesis is pivotal for acceleration of tumor growth and metastasis. Thus, it was postulated that the ability to inhibit angiogenesis might be significant for developing AgNP-based methods for mitigating the progression of malignant cancers (Gurunathan et al., 2009).

5. CELLULAR AND MOLECULAR MECHANISMS OF AGNPs NEUROTOXICITY

Mechanisms of AgNPs toxicity in general are currently under intensive investigation. However, specific mechanisms that account for their neurotoxicity are still unclear. Studies on neurotoxicity of AgNPs should be intensified and focused on animal models because of significant accumulation of silver in the brain relative to other organs, and the high susceptibility of the brain to oxidative stress which underlies many neuropathological states.

5.1 Mitochondria, Oxidative Stress, Inflammation, and Cell Death

The brain is particularly vulnerable to oxidative stress because it has a high metabolic rate that leads to production of large amounts of ROS generated during metabolic activity of mitochondria and peroxisomes. ROS are also by-products of metabolism of certain neurotransmitters induced during inactivation of the monoaminergic neurotransmitters such as serotonin, dopamine, noradrenaline, and adrenaline by monoamine oxidase activity. As such, release of ROS might be triggered by overactivation of the N-methyl-D-aspartate (NMDA) class of glutamatergic receptors. Moreover, the brain contains high levels of polyunsaturated fatty acids, which are sensitive to oxidative damage. Excessive ROS production may lead to the oxidative stress, especially when accompanied by a deficiency of cellular antioxidant defense systems, among which the glutathione (GSH/GSSG) system is particularly important.

ROS are highly reactive molecules that can readily modify the functions of cellular biomolecules. Thus, any imbalance between ROS and antioxidants will lead to increased lipid peroxidation and oxidation of DNA, RNA, and proteins.

The vast majority of data from in vitro and in vivo studies strongly support the concept that oxidative imbalance and subsequent oxidative stress are among the most important pathological events in different cells and organs during exposure to AgNPs (for a review see Kim and Ryu, 2013, Skalska and Struzynska, 2015).

Existing data indicate that AgNPs accumulate in the cellular lysosomal compartment and there is a high probability that silver ions are released from the surface of AgNPs at low pH (about 4.8). The acidic environment of lysosomes provides the most effective conditions for liberating ions from endocytosed AgNPs (Setyawati et al., 2014). Silver in its ionic form exhibits a high affinity for protein thiol groups and promotes their oxidation. Reduced glutathione (GSH) has also been shown to be a target of ionic silver (Baldi et al., 1988), which significantly depletes the pool of this major cellular antioxidant defense system. The influence of AgNPs on the level of SH groups has been confirmed in a myelin fraction isolated from brains of exposed adult rats (Dąbrowska-Bouta et al., 2015). Furthermore, in the same model, exposure to low doses of small-diameter AgNPs (10 nm) significantly decreases the ratio of the reduced to the oxidized form of GSH in brain homogenates (Skalska et al., 2014).

As stated earlier, diminishing of the pool of reduced GSH may cause a cellular redox imbalance leading to further oxidative stress. Under conditions of oxidative stress, proteins become oxidized and undergo changes in their three-dimensional structure and/or degradation, which leads to loss of their physiological functions. Decreased levels of myelin-related proteins: myelin-associated glycoprotein (MAG), myelin oligodendrocyte protein (MOG) and 2′,3′-cyclic nucleotide 3′-phosphodiesterase (CNP) (Dabrowska-Bouta et al., 2015) and synaptically bound proteins [synapsin, synaptophysin I, postsynaptic density protein 95 (PSD-95)] (Skalska et al., 2015) observed in rats exposed to 10 nm AgNPs may indicate the existence of such a mechanism.

Neuronal mitochondria are targets of AgNPs. Interactions of AgNPs with these structures may perturb their function (Xu et al., 2013; Ziemińska et al., 2014). Mitochondrial dysfunction further leads to increased production of free radicals, thereby exacerbating the cycle of oxidative stress. Indeed, such a mechanism was confirmed by several studies using different experimental animal models or in vitro systems. Generation of ROS was reported in mixed cultures of primary neural cells (Haase et al., 2012) and in primary cultures of cerebellar granule cells (CGCs) (Yin et al., 2013; Ziemińska et al., 2014). It appears that neurons are more susceptible/vulnerable to AgNPs because loading of primary astrocytes with AgNPs was not found to alter significantly either the level of total/reduced GSH or ROS production (Luther et al., 2012). Our electron microscopy studies clearly show ultrastructural changes in neuronal mitochondria of rat brains after exposure to polyvinylpyrrolidone (PVP)-coated AgNPs (mixture <100 nm) (Fig. 7.4). Swelling of these intracellular structures

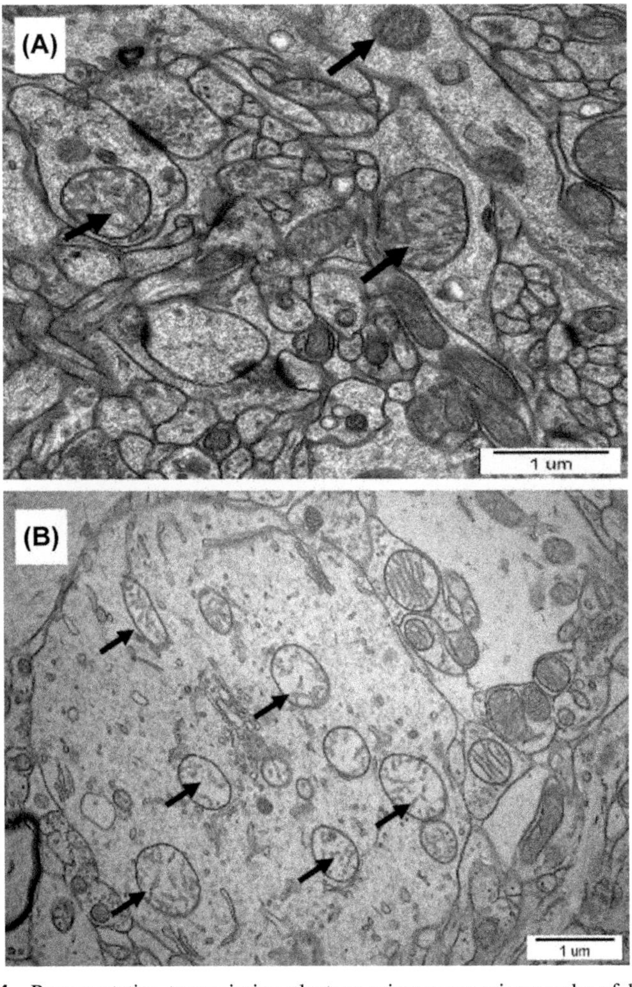

FIGURE 7.4 Representative transmission electron microscopy micrographs of hippocampal sections obtained from rats: (A) image exhibiting normal structure of mitochondria in controls; (B) ultrastructurally changed mitochondria in AgNPs-treated animals (*arrows*).

and damage of cristae are suggestive of dysfunctionality. In accordance with these observations are studies demonstrating increased ROS production in hippocampus of rats after intranasal AgNPs administration (Liu et al., 2012a). Consequently, upregulation of oxidative stress–related genes in different regions of mouse brain was observed after a single intraperitoneal injection of AgNPs (Rahman et al., 2009).

Nonenzymatic thiol systems including GSH, thioredoxins, and cysteine are not only efficient antioxidants but also function in cell signaling and physiological regulation. Redox-sensitive thiols are critical for redox-related signaling pathways mediated mainly by transcription factors Nrf-2 and nuclear factor κB. Moreover, regulatory pathways of ROS production and inflammation are cross-linked. ROS are necessary for the activation of the inflammasome, a multiprotein inflammatory complex.

Although the connection between AgNP-induced ROS generation and inflammatory response has been proved in cultured macrophages (Nishanth et al., 2011), and in liver (Ebabe et al., 2013) and lungs (Sung et al., 2008) of exposed rats, it is less evident in the case of brain. As stated earlier, AgNPs may induce the release of proinflammatory cytokines in rat brain microvessels leading to the subsequent increase in BBB permeability (Trickler et al., 2010). Secretion of IL-1β cytokine and induction of gene expression of CXCL13 chemokine has been observed in cultured murine microglial BV-2 cells exposed to 3–5 nm AgNPs (Huang et al., 2015). These observations indicate proinflammatory action of AgNPs in animal cells in contradiction to reports on their antiinflammatory effect toward bacterial cells (see Section 2).

It is well known that sustained oxidative stress may lead to cell injury and death. Thus, a contribution of oxidative stress to AgNP-mediated neurotoxicity has also been considered. In vitro studies have indicated that there are linkages between decreased viability of CGCs and the drop of mitochondrial membrane potential, enhanced ROS production, and subsequent oxidative stress (Rahman et al., 2009; Ziemińska et al., 2014). Limited in vivo studies confirm the connection between AgNP-induced oxidative stress and cell death in the CNS. Increased ROS production was found to be correlated with histopathological changes in pyramidal neurons indicating cellular death after intranasal administration of AgNPs (Liu et al., 2012a). Likewise, it should be emphasized that AgNP-induced mitochondrial dysfunction is a significant factor involved in activation of caspase-3 and parallel histopathological alterations indicating apoptotic-like cell death in cerebellum of exposed neonatal rats (Yin et al., 2013, 2015).

5.2 Interactions With Cellular Calcium and NMDA Glutamate Receptors

Induction of oxidative stress–mediated cell death is a well-documented mechanism of AgNPs toxicity. However, it seems that the mechanisms of neurotoxic action may be more complex. New lines of studies have indicated that AgNPs may affect calcium homeostasis. Cultured neurons have been shown to respond to administration of AgNPs with a rapid and strong increase in intracellular calcium levels (Haase et al., 2012; Ziemińska et al., 2014). Similarly, PVP-coated AgNPs in a concentration range of 25–75 μg/mL were found to evoke a dose-dependent increase in extracellular calcium uptake in cultured CGCs (Ziemińska et al., 2014). Results of this study revealed that destabilization of calcium homeostasis is a significant mechanism of AgNPs neurotoxicity connected with overactivation of glutamatergic NMDA receptors (NMDARs), because this effect was found to be abolished by the addition of MK-801, a noncompetitive inhibitor of NMDAR. Moreover, calcium imbalance was followed by decreased mitochondrial membrane potential, increased ROS production, and cell death, which also were partially or fully reversed by MK-801. It is known that NMDAR-mediated overloading of neurons with extracellular calcium and deregulation of intracellular calcium signaling pathways are the basis of the excitotoxicity phenomenon (Orrenius, 2007; Zhivotovsky and Orrenius, 2011). Excitotoxicity links overactivation of glutamate receptors, mitochondrial dysfunction, production of ROS, and the induction of oxidative/nitrosative stress, with all these mechanisms leading to degeneration of glutamatergic neurons (Bonfoco et al., 1995). These interrelationships observed under conditions of AgNPs exposure indicate an important contribution of the excitotoxicity-like mechanism to AgNP-induced neurotoxicity.

One study has revealed that, concurrently with induction of calcium imbalance, AgNPs influence zinc homeostasis in CGCs, increasing both extracellular and intracellular zinc levels (Ziemińska and Strużyńska, 2016). It is assumed that destabilization of cellular zinc equilibrium is partially a result of its excessive release from glutamatergic synapses as a consequence of NMDAR overactivation, because zinc is largely colocalized with the excitatory neurotransmitter glutamate in glutamatergic neurons (Frederickson, 1989) and coreleased on cell activation (Assaf and Chung, 1984). Reversible effect of MK-801 on Zn levels indicates that it selectively concerns NMDARs. Other non-NMDA ionotropic glutamate receptors, like α-amino-3-hydroxy-5-methyl-4-isoxazolepropionic acid (AMPA) and kainate (KA) receptors, do not contribute to AgNP-induced cell death, because administration of 6-cyano-7-nitroquinoxaline-2,3-dione (CNQX) (an antagonist of AMPA/KA glutamate receptors) does not have a positive effect on cell viability.

There has been no direct in vivo evidence that NMDARs are involved in the mechanisms of AgNP-induced neurotoxicity in exposed animals. However, decreased levels of the protein PSD-95, which is a member of membrane-associated guanylate kinase (MAGUK) family, has been reported (Skalska et al., 2015) and is discussed in Section 5.3. This protein interacts at the postsynaptic membrane of neurons with other MAGUK proteins to form a scaffold for the clustering of NMDA receptors, AMPA receptors, potassium ion channels, and associated signaling proteins, thereby playing an important role in synaptic plasticity and the phenomenon of long-term potentiation (LTP) (Meyer et al., 2014). Moreover, AgNPs were found to induce profound upregulation of neuronal nitric oxide synthase (NOS) in the cerebral cortex of exposed animals (Sharma et al., 2013). This could represent an indirect marker characterizing events downstream of NMDA receptor activation. The key events are massive influx of Ca^{2+} to neurons via NMDA receptor channels followed by NOS activation, generation of nitric oxide, activation of guanylate cyclase, and production of cGMP (Izumi et al., 2008).

5.3 AgNP-Induced Neurodegeneration

Oxidative stress, neuroinflammation, and cell death are hallmarks of many pathological states of the CNS, including neurodegenerative disorders. Based on in vitro studies, it has been hypothesized that AgNPs may take part in progression of neurodegenerative states by accelerating amyloid beta (Aβ) generation and deposition of Aβ1-40 and Aβ1-42 fragments. Furthermore, expression of amyloid β precursor protein (APP) was found to increase with exposure to AgNPs. Concomitantly, expression of low-density lipoprotein receptor and neprilysin, both of which are involved in enhanced uptake and degradation of Aβ, were found to be downregulated by AgNPs (Huang et al., 2015). Upregulation of APP and deposition of Aβ resulting in amyloid plaque formation underlie the pathogenesis of Alzheimer's disease (Dong et al., 2012).

Investigations of animal models have revealed features indicative of AgNP-induced synaptic degeneration. In ultrastructural images of brain specimens obtained from rats subjected to low doses of small (10 nm) AgNPs (Skalska et al., 2015), swollen neuropil was observed and the vast majority of synapses were found to be altered in both hippocampus and cerebral cortex of exposed rats. Pathological changes in the synapses were found to include enhanced packing density of synaptic vesicles as well as thickened and barely discernible postsynaptic density (Fig. 7.5B). These changes were observed irrespective of the administered form of silver (ionic silver or AgNPs). However, certain features, observed mainly in hippocampus, were found to be specific for AgNPs. Examples include disturbed presynaptic membranes with synaptic vesicle clusters located extrasynaptically in the neuropil (Fig. 7.5C) and the presence of myelin-like bodies. The formation of these structures, which are remnants of fragmented membranes and organelles, suggests that significant synaptic degeneration accompanies AgNPs exposure. Similar observations were made earlier in an experimentally evoked pathological state of the optic nerve (excitotoxicity-induced axonopathy) (Saggu et al., 2010) and in a case of retinal neurodegeneration (Schuettauf et al., 2004). It was also suggested that myelin-like bodies develop during the initial stage of autophagy in the process of lysosomal degradation of cellular organelles and proteins impaired under stress conditions (Kelekar, 2005). These observations are in line with data on AgNP-induced oxidative stress leading to oxidative damage of molecules (see Section 5.1).

Changes in the ultrastructure of the presynaptic part of the synapse were found to be accompanied by significant reduction of expression of presynaptic

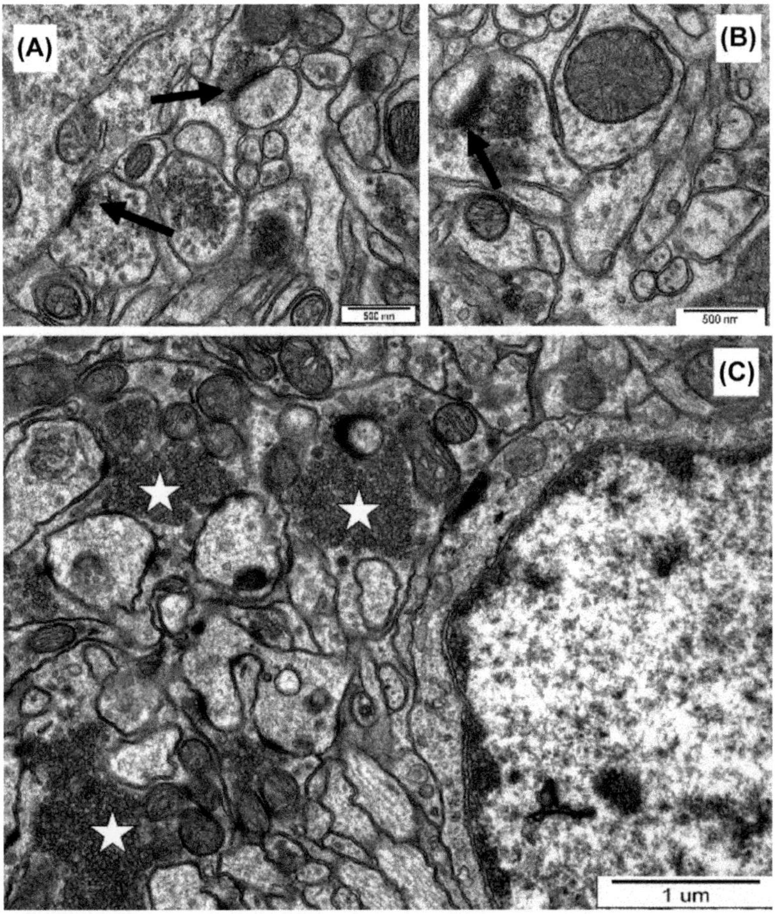

FIGURE 7.5 Representative transmission electron microscopy micrographs of hippocampal sections obtained from control (A) and AgNP-treated (B, C) rats. (A) Normal structure of synapse with proper distribution of synaptic vesicles in the presynaptic part and normal appearance of synaptic cleft (*arrows*). (B) Structural changes in synapse in the form of a blurred and thickened synaptic cleft without clearly marked pre- and postsynaptic membranes (*arrow*). (C) Changes in neuropil after exposure to AgNPs, including blurred synaptic clefts (*arrows*) and disturbed presynaptic membranes with aggregations of free synaptic vesicles loosely located in neuropil (*asterisks*).

proteins, synapsin I and synaptophysin, both of which play roles in the exocytosis machinery involved in the movements of synaptic vesicles along actin filaments, including fusion, exocytosis, and recycling (De Camilli and Greengard, 1986; Greengard et al., 1993; Li et al., 1995). Synapsin I deficiency was shown to result in a reduction of the releasable pool of synaptic vesicles docking to the presynaptic membrane of the synaptic cleft with simultaneous sequestration in the reserve pool (Yamagata, 2003). This may explain the observation of formation of synaptic vesicle clusters in the presynaptic part of synapses in AgNP-exposed rats. In turn, the observation of the blurred structure of the postsynaptic part of synapse was found to correlate well with decreased levels of PSD-95 protein, a postsynaptic marker abundant in neuronal postsynaptic densities (PSD). This may indicate disrupted functioning of the excitatory synapses, since PSD-95 is involved in regulation of long-term synaptic plasticity (LTP) associated with NMDA glutamate receptor signaling (Keith and Ei-Husseini, 2008), as well as in positive regulation of excitatory neurotransmission. It can be assumed that alterations in the integral synaptic proteins may be responsible for disturbances in the secretory mechanism of neurotransmitters and further in neurotransmission. Morphological and biochemical indicators of synaptic destruction under conditions of AgNPs toxicity indicate possible neuronal dysfunction. In particular, synaptic degeneration in the hippocampal region of brain may result in abnormalities in glutaminergic neurotransmission, which are predictive of impairment of cognitive processes. These assumptions are supported to date by studies showing an AgNP-induced effect on the concentration of 5-hydroxytryptamine and dopamine (Hadrup et al., 2012) and impairment of cognitive and motor functions in exposed rats (Sharma and Sharma, 2007).

Another noteworthy observation that was made after prolonged oral exposure to 10 nm AgNPs is that they affect myelin sheaths. The compact multilamellar structure of myelin was found to be disturbed and accompanied by a significant (30–40%) decrease in the level of several myelin-specific proteins such as CNP, MAG, and MOG, which play a role in the stabilization of myelin structure (Dąbrowska-Bouta et al., 2015). Marked myelin degradation was also observed based on myelin basic protein (MBP) immunostaining of brains of rats exposed to different sizes of AgNPs (Sharma et al., 2013). The disintegration of myelin membranes may lead to the loss of myelinated fibers after AgNPs exposure (Sharma et al., 2009a).

Another major issue identified by Sharma et al. (2013) is that neurotoxic effects are inversely related to the size of AgNPs, as well as the age of exposed animals. The smaller AgNPs were found to produce cellular injuries (BBB, myelin, and neurons) of greater intensity in young and old age groups relative to a middle-aged group. This important observation indicates that age is a significant factor influencing AgNPs neurotoxicity. This factor should be seriously considered with regard to the safety of appliances that include AgNPs and future AgNP-based medical approaches.

6. PHYSICOCHEMICAL PARAMETERS INFLUENCING THE TOXICITY OF SILVER NANOPARTICLES

6.1 Size

Nanoformulations have a high proportion of surface area per unit mass. This influences the properties of nanoparticles. AgNPs are highly effective at penetrating cellular membranes and provide specific impacts on biological systems. The reactivity of nanoparticles, including AgNPs, is inversely proportional to the particle size (Christian et al., 2008). It is also evident that higher reactivity causes greater toxicity. Existing experimental data indicate a high correlation between the size and toxicity of AgNPs. In vitro studies comparing cytotoxicity and genotoxicity of 20, 80, and 113 nm AgNPs toward L929 fibroblasts have confirmed this relationship (Park et al., 2011). Evaluation of cytotoxic effects of AgNPs of different sizes (5, 20, and 50 nm) on various types of human cells has revealed that 5-nm AgNPs induce severe changes in cell morphology, viability, cellular membrane integrity, and oxidative stress (Liu et al., 2010). Furthermore, cytotoxic effects were found to be induced by10-nm AgNPs, but not by 40- or 75-nm citrate-coated AgNPs in human lung cells (Gliga et al., 2014). Higher cytotoxic potential of 20-nm AgNPs relative to 40-nm AgNPs has been proved in primary neuronal cultures (Haase et al., 2012).

The size-dependent toxicity was also confirmed in an animal model of exposure. In general, smaller AgNPs (20–40 nm) induce more severe damage of brain cells relative to larger AgNPs (50–60 and 130–150 nm). Greater numbers of NOS-positive neurons and a greater extent of myelin disintegration have been identified in the cortex of animals treated with smaller nanoparticles (Sharma et al., 2013). However, there are also conflicting data that do not confirm the relationship between size and toxicity (Espinosa-Cristobal et al., 2013) as well as data that fail to support the relationship between AgNPs formulations and proposed mechanisms of toxicity (for a review see Hadrup and Lam, 2014).

6.2 Shape and Coating

Bacteriotoxic activity of AgNPs has been found to depend on their shapes and concentrations. Greater effectiveness has been demonstrated for higher concentrations (above 75 μg/mL) and for triangular shapes (Pal et al., 2007), whereas nanoplates were found to exhibit higher activity than nanorods (Sadeghi et al., 2012).

Another significant parameter that may influence the characteristics of AgNPs is coating of their surfaces with different compounds. During production of AgNPs, they are coated and stabilized with compounds such as carbon, polysaccharides, PVP, or citrate to prevent agglomeration (Ahamed et al., 2008; Topuz et al., 2014). It should be noted that the type of coating agent may influence properties of nanoparticles, their behavior in physiological fluids (Braydich-Stolle et al., 2014), and their toxicity. Carbon-coated AgNPs have been shown to be less cytotoxic toward macrophages than uncoated AgNPs (Nishanth et al., 2011), whereas citrate-coated AgNPs are less cytotoxic than PVP-coated AgNPs of similar size (Kim et al., 2013). In turn, uncoated AgNPs influence the viability of mouse embryonic stem cells and mouse embryonic fibroblasts to a lesser extent than polysaccharide-coated AgNPs (Ahamed et al., 2008). Modulation of the features of nanoparticles by shaping their surfaces is under investigation and considered to be important in context of using them as drug delivery nanocarriers or diagnostic tools.

6.3 Release of Ions

Another important problem related to the toxicity of metal nanoparticles in general, including AgNPs in particular, is the release of ions from the surface of the particles. It is still discussed if the toxicity of AgNPs is mediated by a combination of mechanisms including the individual mechanisms of silver ions and of specific nanoformulation or by only one of the two mechanisms (for a review see Hadrup and Lam, 2014). Our own in vitro data indicate that NMDAR-mediated excitotoxicity is an AgNP-specific toxic effect (Ziemińska et al., 2014). Similarly, some of the observed morphological changes in synapses from exposed rat brain were found to be characteristic of the presence of nanoparticulate silver (Skalska et al., 2015). Most likely, the observed toxicity is due to the bimodal effect of the two forms of silver (Powers et al., 2011; Skalska et al., 2015; Skalska and Strużyńska, 2015).

There is substantial evidence for an AgNP-mediated "Trojan Horse effect" inside the cell (Park et al., 2010; Singh and Ramarao, 2012), where silver ions are released after surface oxidation of nanoparticles within the acidic environment of lysosomes (Setyawati et al., 2014). It has been claimed that this phenomenon depends on the size of the nanoparticles. Smaller AgNPs (<10 nm) release ions more efficiently, because of the larger surface area per unit mass (Pratsinis et al., 2013). In turn, appropriate capping of the surface area may reduce the amount of ions released as in the case of citrate-stabilized AgNPs, which are more "densely packed" than AgNPs dispersed in water (Liu et al., 2010).

7. SUMMARY

The potential benefits of AgNPs in medical applications evoke great interest and excitement. The ability to freely cross cell membranes, including the membranes of the BBB, in the context of biocidal properties of AgNPs is of particular concern in neurosciences. The majority of studies have provided interesting and promising results for both therapeutic and diagnostic applications of AgNPs in the field of treatment of neurological disorders, including cancer. However, there are increasing numbers of studies concerning adverse effects on cerebral tissues. Based on the existing findings, it should be stressed that AgNPs are not completely safe and exposure may induce neurotoxic effects. However, our present knowledge on neurotoxicity of AgNPs is derived mainly from in vitro experiments and therefore future investigations should focus on animal studies. More data should be obtained to answer questions about dose–response effects in brain. Furthermore, the more subtle effects of low-level exposure should be investigated. Applications of AgNPs as tools in targeted therapy will allow the use of smaller amounts of AgNPs, which can act at specific locations. We have to consider the tendency of AgNPs to accumulate in brain. Moreover, long-term health consequences are also worth investigating, as it can be expected that growing production and extensive usage of engineered AgNPs will increase the risk of exposure of both workers and users. In summary, although the benefits of AgNPs will motivate us to continue with promising studies on new nanotechnological solutions, it remains important to characterize the safety profile of AgNPs in various applications.

ACKNOWLEDGMENTS

This work was supported partially by statutable grant for Mossakowski Medical Research Centre, Polish Academy of Sciences and by grant nr NN 401619938 from the Polish Ministry of Science and Higher Education.

The author thanks Dr. Małgorzata Frontczak-Baniewicz for her valuable contribution to the TEM studies.

REFERENCES

Ahamed, M., Karns, M., Goodson, M., Rowe, J., Hussain, S.M., Schlager, J.J., Hong, Y., 2008. DNA damage response to different surface chemistry of silver nanoparticles in mammalian cells. Toxicol. Appl. Pharmacol. 233, 404–410.

Assaf, S.Y., Chung, S.H., 1984. Release of endogenous Zn^{2+} from brain tissue during activity. Nature 308, 734–736.

Baldi, C., Minoia, C., Di Nucci, A., Capodaglio, E., Manzo, L., 1988. Effects of silver in isolated rat hepatocytes. Toxicol. Lett. 41, 261–268.

Baram-Pinto, D., Shukla, S., Perkas, N., Gedanken, A., Sarid, R., 2009. Inhibition of herpes simplex virus type 1 infection by silver nanoparticles capped with mercaptoethane sulfonate. Bioconjug. Chem. 20, 1497–1502.

Bhaskar, S., Tian, F., Stoeger, T., Kreyling, W., de la Fuente, J.M., Grazú, V., Borm, P., Estrada, G., Ntziachristos, V., Razansky, D., 2010. Multifunctional nanocarriers for diagnostics, drug delivery and targeted treatment across blood–brain barrier: perspectives on tracking and neuroimaging. Part. Fibre Toxicol. 7, 3.

Barillo, D.J., Marx, D.E., 2014. Silver in medicine: a brief history BC 335 to present. Burns 40s, s3–s8.

Bonfoco, E., Krainc, D., Ankarcrona, M., Nicotera, P., Lepton, S.A., 1995. Apoptosis and necrosis: two distinct events induced, respectively, by mild and intense insults with N-methyl-D-aspartate or nitric oxide/superoxide in cortical cell cultures. Proc. Natl. Acad. Sci. USA 92, 7162–7166.

Braydich-Stolle, L.K., Breitner, E.K., Comfort, K.K., Schlager, J.J., Hussain, S.M., 2014. Dynamic characteristics of silver nanoparticles in physiological fluids: toxicological implications. Langmuir 30, 15309–15316.

Chaloupka, K., Malam, Y., Scifalian, A.M., 2010. Nanosilver as a new generation of nanoproduct in biomedical applications. Trends Biotechnol. 28, 580–588.

Chen, Y., Liu, L., 2012. Modern methods for delivery of drugs across the blood–brain barrier. Adv. Drug Deliv. Rev. 64, 640–665.

Chen, X., Schluesener, H.J., 2008. Nanosilver: a nanoproduct in medical application. Toxicol. Lett. 176, 1–12.

Christian, P., Von der Kammer, F., Baalousha, M., Hofmann, T.H., 2008. Nanoparticles: structure, properties preparation and behaviour in environmental media. Ecotox 17, 326–343.

Dąbrowska-Bouta, B., Sulkowski, G., Zięba, M., Orzelska, J., Strużyńska, L., 2015. The oxidative stress in myelin of the rat brain exposed to silver nanoparticles. Folia Neuropathol. 53.

De Camilli, P., Greengard, P., 1986. Synapsin I: a synaptic vesicle-associated neuronal phosphoprotein. Biochem. Pharmacol. 35, 4349–4357.

De Matteis, V., Malvindi, M.A., Galeone, A., Brunetti, V., De Luca, E., Kote, S., Kshirsagar, P., Sabella, S., Bardi, G., Pompa, P.P., 2015. Negligible particle-specific toxicity mechanism of silver nanoparticles: the role of Ag+ ion release in the cytosol. Nanomedicine 11, 731–739.

Dong, S., Duan, Y., Hu, Y., Zhao, Z., 2012. Advances in the pathogenesis of Alzheimer's disease: a re-evaluation of amyloid cascade hypothesis. Transl. Neurodegener. 1, 18.

Dos Santos, C.A., Seckler, M.M., Ingle, A.P., Gupta, I., Galdiero, S., Galdiero, M., Gade, A., Rai, M., 2014. Silver nanoparticles: therapeutical uses, toxicity, and safety issues. J. Pharm. Sci. 103, 1931–1944.

Dreaden, E.C., El-Sayed, M.A., 2012. Detecting and destroying cancer cells in more than one way with noble metals and different confinement properties on the nanoscale. Acc. Chem. Res. 45, 1854–1865.

Ebabe, E.R., Gaillet, S., Vide, J., Romain, C., Lauret, C., Rugani, N., Cristol, J.P., Rouanet, J.M., 2013. Dietary exposure to silver nanoparticles in Sprague-Dawley rats: effects on oxidative stress and inflammation. Food Chem. Toxicol. 60, 297–301.

Espinosa-Cristobal, L.F., Martinez-Castañon, G.A., Loyola-Rodriguez, J.P., Patiño-Marin, N., Reyes-Macías, J.F., Vargas-Morales, J.M., Facundo, R., 2013. Toxicity, distribution, and accumulation of silver nanoparticles in Wistar rats. J. Nanopart. Res. 15, 1702.

Fewtrell, L., 2014. Silver: water disinfection and toxicity. Centre Res. Environ. Health 1–53.

Frederickson, C.J., 1989. Neurobiology of zinc and zinc-containing neurons. Int. Rev. Neurobiol. 31, 145–238.

Gaillet, S., Rouanet, J.M., 2015. Silver nanoparticles: their potential toxic effects after oral exposure and underlying mechanisms–a review. Food Chem. Toxicol. 77, 58–63.

Gajbhiye, M.B., Kesharwani, J.G., Ingke, A.P., Gade, A.K., Rai, M.K., 2009. Fungus mediated synthesis of silver nanoparticles and its activity against pathogenic fungi in combination of fluconazole. Nanomed. NBM 5, 282–286.

Ge, L., Li, Q., Wang, M., Ouyang, J., Li, X., Xing, M.M., 2014. Nanosilver particles in medical applications: synthesis, performance, and toxicity. Int. J. Nanomed. 9, 2399–2407.

Genter, M.B., Newman, N.C., Shertzer, H.G., Ali, S.F., Bolon, B., 2012. Distribution and systemic effects of intranasally administered 25 nm silver nanoparticles in adult mice. Toxicol. Pathol. 40, 1004–1013.

Gliga, A.R., Skoglund, S., Wallinder, I.O., Fadeel, B., Karlsson, H.L., 2014. Size-dependent cytotoxicity of silver nanoparticles in human lung cells: the role of cellular uptake, agglomeration and Ag release. Part. Fibre Toxicol. 11, 11.

Gottschalk, F., Scholz, R.W., Nowack, B., 2010. Probabilistic material flow modeling for assessing the environmental exposure to compounds: methodology and an application to engineered nano-TiO_2 particles. Environ. Model. Software 25, 320–332.

Greengard, P., Valtorta, F., Czernik, A.J., Benfenati, F., 1993. Synaptic vesicle phosphoproteins and regulation of synaptic function. Science 259, 780–785.

Grunkemeier, G.L., Jin, R., Starr, A., 2006. Prosthetic heart valves: Objective Performance Criteria versus randomized clinical trial. Ann. Thorac. Surg. 82, 776–780.

Guo, D., Zhu, L., Huang, Z., Zhou, H., Ge, Y., Ma, W., Wu, J., Zhang, X., Zhou, X., Zhang, Y., Zhao, Y., Gu, N., 2013. Anti-leukemia activity of PVP-coated silver nanoparticles via generation of reactive oxygen species and release of silver ions. Biomaterials 34, 7884–7894.

Gurunathan, S., Lee, K.J., Kalishwaralal, K., Sheikpranbabu, S., Vaidyanathan, R., Eom, S.H., 2009. Antiangiogenic properties of silver nanoparticles. Biomaterials 30, 6341–6350.

Haase, A., Rott, S., Mantion, A., Graf, P., Plendl, J., Thunemann, A.F., Meier, W.P., Taubert, A., Luch, A., Reiser, G., 2012. Effects of silver nanoparticles on primary mixed neural cell cultures: uptake, oxidative stress and acute calcium responses. Toxicol. Sci. 126, 457–468.

Hadrup, N., Lam, H.R., 2014. Oral toxicity of silver ions, silver nanoparticles and colloidal silver – a review. Regul. Toxicol. Pharmacol. 68, 1–7.

Hadrup, N., Loeschner, K., Mortensen, A., Sharma, A.K., Qvortrup, K., Larsen, E.H., Lam, H.R., 2012. The similar neurotoxic effects of nanoparticulate and ionic silver in vivo and in vitro. Neurotoxicology 33, 416–423.

Hawkins, B.T., Davis, T.P., 2005. The blood–brain barrier/neurovascular unit in health and disease. Pharmacol. Rev. 57, 173–185.

Hoet, P.H., Brüske-Hohlfeld, I., Salata, O.V., 2004. Nanoparticles – known and unknown health risks. J. Nanobiotechnol. 2, 12.

Huang, C.L., Hsiao, I.L., Lin, H.C., Wang, C.F., Huang, Y.J., Chuang, C.Y., 2015. Silver nanoparticles affect on gene expression of inflammatory and neurodegenerative responses in mouse brain neural cells. Environ. Res. 136, 253–263.

Huang, Y., Li, X., Liao, Z., Zhang, G., Liu, Q., Tang, J., Peng, Y., Liu, X., Luo, Q., 2007. A randomized comparative trial between Acticoat and SD-Ag in the treatment of residual burn wounds, including safety analysis. Burns 33, 161–166.

Izumi, Y., Tokuda, K., Zorumski, C.F., 2008. Long-term potentiation inhibition by low-level N-methyl-D-aspartate receptor activation involves calcineurin, nitric oxide, and p38 mitogen-activated protein kinase. Hippocampus 18, 258–265.

Kalishwaralal, K., Banumathi, E., Ram Kumar Pandian, S., Deepak, V., Muniyandi, J., Eom, S.H., Gurunathan, S., 2009. Silver nanoparticles inhibit VEGF induced cell proliferation and migration in bovine retinal endothelial cells. Colloids Surf. B Biointerfaces 73, 51–57.

Kim, S., Ryu, D.-Y., 2013. Silver nanoparticle-induced oxidative stress, genotoxicity and apoptosis in cultured cells and animal tissues. J. Appl. Toxicol. 33, 78–89.

Kim, Y.S., Song, M.Y., Park, J.D., Song, K.S., Ryu, H.R., Chung, Y.H., Chang, H.K., Lee, J.H., Oh, K.H., Kelman, B.J., Hwang, I.K., Yu, I.J., 2010. Subchronic oral toxicity of silver nanoparticles. Part. Fibre Toxicol. 7, 20.

Kim, K.T., Truong, L., Wehmas, L., Tanguay, R.L., 2013. Silver nanoparticle toxicity in the embryonic zebrafish is governed by particle dispersion and ionic environment. Nanotechnology 24, 115101.

Kelekar, A., 2005. Autophagy. Ann. NY Acad. Sci. 1066, 259–271.

Keith, D., El-Husseini, A., 2008. Excitation control: balancing PSD-95 function at the synapse. Front. Mol. Neurosci. 1, 4.

Lansdown, A.B., 2007. Critical observations on the neurotoxicity of silver. Crit. Rev. Toxicol. 37, 237–250.

Lara, H.H., Ayala-Nuñez, N.V., Ixtepan-Turrent, L., Rodriguez-Padilla, C., 2010. Mode of antiviral action of silver nanoparticles against HIV-1. J. Nanobiotechnol. 8, 1.

Lee, J.H., Kim, Y.S., Song, K.S., Ryu, H.R., Sung, J.H., Park, J.D., Park, H.M., Song, N.W., Shin, B.S., Marshak, D., Ahn, K., Lee, J.E., Yu, J., 2013. Biopersistence of silver nanoparticles in tissues from Sprague-Dawley rats. Part. Fibre Toxicol. 10, 36.

Leite, P.E., Pereira, M.R., Granjeiro, J.M., 2015. Hazard effects of nanoparticles in central nervous system: searching for biocompatible nanomaterials for drug delivery. Toxicol. In Vitro 29, 1653–1660.

Li, L., Chin, L.S., Shupliakov, O., Brodin, L., Sihra, T.S., Hvalby, O., Jensen, V., Zheng, D., McNamara, J.O., Greengard, P., Andersen, P., 1995. Impairment of synaptic vesicle clustering and of synaptic transmission, and increased seizure propensity, in synapsin I-deficient mice. Proc. Natl. Acad. Sci. USA 92, 9235–9239.

Lin, J., Chen, R., Feng, S., Pan, J., Li, Y., Chen, G., Cheng, M., Huang, Z., Yu, Y., Zeng, H., 2011. A novel blood plasma analysis technique combining membrane electrophoresis with silver nanoparticle-based SERS spectroscopy for potential applications in noninvasive cancer detection. Nanomedicine 7, 655–663.

Liu, Y., Guan, W., Ren, G., Yang, Z., 2012a. The possible mechanism of silver nanoparticle impact on hippocampal synaptic plasticity and spatial cognition in rats. Toxicol. Lett. 209, 227–231.

Liu, W., Wu, Y., Wang, C., Li, H.C., Wang, T., Liao, C.Y., Cui, L., Zhou, Q.F., Yan, B., Jiang, G.B., 2010. Impact of silver nanoparticles on human cells: effect of particle size. Nanotoxicology 4, 319–330.

Liu, J., Zhao, Y., Guo, Q., Wang, Z., Wang, H., Yang, Y., Huang, Y., 2012b. TAT-modified nanosilver for combating multidrug-resistant cancer. Biomaterials 33, 6155–6161.

Loeschner, K., Hadrup, N., Qvortru, K., Larsen, A., Gao, X., Vogel, U., Mortensen, A., Rye Lam, H., Larsen, E.H., 2011. Distribution of silver in rats following 28 days of repeated oral exposure to silver nanoparticles or silver acetate. Part. Fibre Toxicol. 8, 18.

Lu, L., Sun, R.W., Chen, R., Hui, C.K., Ho, C.M., Luk, J.M., Lau, G.K., Che, C.M., 2008. Silver nanoparticles inhibit hepatitis B virus replication. Antivir. Ther. 13, 253–262.

Luther, E.M., Schmidt, M.M., Diendorf, J., Epple, M., Dringen, R., 2012. Upregulation of metallothioneins after exposure of cultured primary astrocytes to silver nanoparticles. Neurochem. Res. 37, 1639–1648.

Marambio-Jones, C., Hoek, E.M.V., 2010. A review of the antibacterial effects of silver nanomaterials and potential implications for human health and the environment. J. Nanopart. Res. 12, 1531–1551.

Marshall, J.P., Schneider, R.P., 1997. Systemic argyria secondary to topical silver nitrate. Arch 1077–1079.

Matsumura, Y., Yoshikata, K., Kunisaki, S., Tsuchido, T., 2003. Mode of bactericidal action of silver zeolite and its comparison with that of silver nitrate. Appl. Environ. Microbiol. 69, 4278–4281.

Meyer, D., Bonhoeffer, T., Scheuss, V., 2014. Balance and stability of synaptic structures during synaptic plasticity. Neuron 82, 430–443.

Mody, W., Siwale, R., Singh, A., Mody, H.R., 2010. Introduction to metallic nanoparticles. J. Pharm. Bioallied Sci. 2, 282–289.

Morley, K.S., Webb, P.B., Tokareva, N.V., Krasnov, A.P., Popov, V.K., Zhang, J., Roberts, C.J., Howdle, S.M., 2007. Synthesis and characterisation of advanced UHMWPE/silver nanocomposites for biomedical applications. Eur. Polym. J. 43, 307–314.

Nishanth, R.P., Jyotsna, R.G., Schlager, J.J., Hussain, S.M., Reddanna, P., 2011. Inflammatory responses of RAW 264.7 macrophages upon exposure to nanoparticles: role of ROS-NFkappaB signaling pathway. Nanotoxicology 5, 502–516.

Orrenius, S., 2007. Reactive oxygen species in mitochondria-mediated cell death. Drug Metab. Rev. 39, 443–455.

Pal, S., Tak, Y.K., Song, J.M., 2007. Does the antibacterial activity of silver nanoparticles depend on the shape of the nanoparticle? A study of the Gram-negative bacterium *Escherichia coli*. Appl. Environ. Microbiol. 73, 1712–1720.

Park, E.J., Yi, J., Kim, Y., Choi, K., Park, K., 2010. Silver nanoparticles induce cytotoxicity by a Trojan-horse type mechanism. Toxicol. In Vitro 24, 872–878.

Park, M.V., Neigh, A.M., Vermeulen, J.P., De La Fonteyne, L.J., Verharen, H.W., Briede, J.J., Van Loveren, H., De Jong, W.H., 2011. The effect of particle size on the cytotoxicity, inflammation, developmental toxicity and genotoxicity of silver nanoparticles. Biomaterials 32, 9810–9817.

Parveen, S., Misra, R., Sahoo, S.K., 2012. Nanoparticles: a boon to drug delivery, therapeutics, diagnostics and imaging. Nanomedicine 8, 147–166.

Powers, C.M., Badireddy, A.R., Ryde, I.T., Seidler, F.J., Slotkin, T.A., 2011. Silver nanoparticles compromise neurodevelopment in PC12 cells: critical contributions of silver ion, particle size, coating, and composition. Environ. Health Perspect. 119, 37–44.

Pratsinis, A., Hervella, P., Leroux, J.C., Pratsinis, S.E., Sotiriou, G.A., 2013. Toxicity of silver nanoparticles in macrophages. Small 9, 2576–2584.

Rahman, M.F., Wang, J., Patterson, T.A., Saini, U.T., Robinson, B.L., Newport, G.D., Murdock, R.C., Schlager, J.J., Hussain, S.M., Ali, S.F., 2009. Expression of genes related to oxidative stress in the mouse brain after exposure to silver-25 nanoparticles. Toxicol. Lett. 187, 15–21.

Rungby, J., Danscher, G., 1983. Localization of exogenous silver in brain and spinal cord of silver exposed rats. Acta Neuropathol. 60, 92–98.

Rejeski, D., 2011. Project on Emerging Nanotechnologies. Woodrow Wilson International Center for Scholars, Washington, DC.

Sadeghi, B., Garmaroudi, F.S., Hashemi, M., Nezhad, H.R., Nasrollahi, A., Ardalan, S., Ardalan, S., 2012. Comparison of the anti-bacterial activity on the nanosilver shapes: nanoparticles, nanorods and nanoplates. Adv. Powder Technol. 23, 22–26.

Saggu, S.K., Chotaliya, H.P., Blumbergs, P.C., Casson, R.J., 2010. Wallerian-like axonal degeneration in the optic nerve after excitotoxic retinal insult: an ultrastructural study. BMC Neurosci. 11, 97.

Saunders, N.R., Ek, C.J., Habgood, M.D., Dziegielewska, K.M., 2008. Barriers in the brain: a renaissance? Trends Neurosci. 31, 279–286.

Setyawati, M.I., Yuan, X., Xie, J., Leong, D.T., 2014. The influence of lysosomal stability of silver nanomaterials on their toxicity to human cells. Biomaterials 35, 6707–6715.

Sharma, H.S., Ali, S.F., Hussain, S.M., Schlager, J.J., Sharma, A., 2009a. Influence of engineered nanoparticles from metals on the blood–brain barrier permeability, cerebral blood flow, brain edema and neurotoxicity. An experimental study in the rat and mice using biochemical and morphological approaches. J. Nanosci. Nanotechnol. 9, 5055–5072.

Sharma, H.S., Ali, S.F., Tian, Z.R., Hussain, S.M., Schlager, J.J., Sjöquist, P.O., Sharma, A., Muresanu, D.F., 2009b. Chronic treatment with nanoparticles exacerbate hyperthermia induced blood–brain barrier breakdown, cognitive dysfunction and brain pathology in the rat. Neuroprotective effects of nanowired-antioxidant compound H-290/51. J. Nanosci. Nanotechnol. 9, 5073–5090.

Sharma, H.S., Hussain, S., Schlager, J., Ali, S.F., Sharma, A., 2010. Influence of nanoparticles on blood–brain barrier permeability and brain edema formation in rats. Acta Neurochir. Suppl. 106, 359–364.

Sharma, A., Muresanu, D.F., Lafuente, J.V., Patnaik, R., Tian, Z.R., Buzoianu, A.D., Sharma, H.S., 2015. Sleep deprivation-induced blood–brain barrier breakdown and brain dysfunction are exacerbated by size-related exposure to Ag and Cu nanoparticles. Neuroprotective effects of a 5-HT3 receptor antagonist Ondansetron. Mol. Neurobiol. 52, 867–881.

Sharma, A., Muresanu, D.F., Patnaik, R., Sharma, H.S., 2013. Size- and age-dependent neurotoxicity of engineered metal nanoparticles in rats. Mol. Neurobiol. 48, 386–396.

Sharma, H.S., Sharma, A., 2007. Nanoparticles aggravate heat stress induced cognitive deficits, blood–brain barrier disruption, edema formation and brain pathology. Prog. Brain Res. 162, 245–273.

Shin, S.H., Ye, M.K., 2012. The effect of nano-silver on allergic rhinitis model in mice. Clin. Exp. Otorhinolaryngol. 5, 222–227.

Schuettauf, F., Rejdak, R., Walski, M., Frontczak-Baniewicz, M., Voelker, M., Blatsios, G., Shinoda, K., Zagorski, Z., Zrenner, E., Grieb, P., 2004. Retinal neurodegeneration in the DBA/2J mouse – a model for ocular hypertension. Acta Neuropathol. 107, 352–358.

Sibbald, R.G., Contreras-Ruiz, J., Coutts, P., Fierheller, M., Rothman, A., Woo, K., 2007. Bacteriology, inflammation, and healing: a study of nanocrystalline silver dressings in chronic venous leg ulcers. Adv. Skin Wound Care 20, 549–558.

Singh, R.P., Ramarao, P., 2012. Cellular uptake, intracellular trafficking and cytotoxicity of silver nanoparticles. Toxicol. Lett. 213, 249–259.

Skalska, J., Frontczak-Baniewicz, M., Lenkiewicz, A., Struzynska, L., 2014. Induction of oxidative stress in rat brain after repeated oral exposure of silver nanoparticles. Folia Neuropathol. 52, 358.

Skalska, J., Frontczak-Baniewicz, M., Struzynska, L., 2015. Synaptic degeneration in rat brain after prolonged oral exposure to silver nanoparticles. Neurotoxicology 46, 145–154.

Skalska, J., Struzynska, L., 2015. Toxic effects of silver nanoparticles in mammals – does the risk of neurotoxicity exist? Folia Neuropathol. 53, 1–20.

Sung, J.H., Ji, J.H., Yoon, J.U., Kim, D.S., Song, M.Y., Jeong, J., Han, B.S., Han, J.H., Chung, Y.H., Kim, J., Kim, T.S., Chang, H.K., Lee, E.J., Lee, J.H., Yu, I.J., 2008. Lung function changes in Sprague-Dawley rats after prolonged inhalation exposure to silver nanoparticles. Inhal. Toxicol. 20, 567–574.

Sung, J.H., Ji, J.H., Park, J.D., Yoon, J.U., Kim, D.S., Jeon, K.S., Song, M.Y., Jeong, J., Han, B.S., Han, J.H., Chung, Y.H., Chang, H.K., Lee, J.H., Cho, M.H., Kelman, B.J., Yu, I.J., 2009. Subchronic inhalation toxicity of silver nanoparticles. Toxicol. Sci. 108, 452–461.

Tang, J., Xiong, L., Wang, S., Wang, J., Liu, L., Li, J., Wan, Z., Xi, T., 2008. Influence of silver nanoparticles on neurons and blood–brain barrier via subcutaneous injection in rats. Appl. Surf. Sci. 255, 502–504.

Tang, J., Xiong, L., Wang, S., Wang, J., Liu, L., Li, J., Yuan, F., Xi, T., 2009. Distribution, translocation and accumulation of silver nanoparticles in rats. J. Nanosci. Nanotechnol. 9, 4924–4932.

Tang, J., Xiong, L., Zhou, G., Wang, S., Wang, J., Liu, L., Li, J., Yuan, F., Lu, S., Wan, Z., Chou, L., Xi, T., 2010. Silver nanoparticles crossing through and distribution in the blood–brain barrier in vitro. J. Nanosci. Nanotechnol. 10, 6313–6317.

Topuz, E., Sigg, L., Talinli, I., 2014. A systematic evaluation of agglomeration of Ag and TiO_2 nanoparticles under freshwater relevant conditions. Environ. Pollut. 193, 37–44.

Trickler, W.J., Lantz, S.M., Murdock, R.C., Schrand, A.M., Robinson, B.L., Newport, G.D., Schlager, J.J., Oldenburg, S.J., Paule, M.G., Slikker Jr., W., Hussain, S.M., Ali, S.F., 2010. Silver nanoparticle induced blood–brain barrier inflammation and increased permeability in primary rat brain microvessel endothelial cells. Toxicol. Sci. 118, 160–170.

Wang, R., Chen, C., Yang, W., Shi, S., Wang, C., Chen, J., 2013. Enhancement effect of cytotoxicity response of silver nanoparticles combined with thermotherapy on C6 rat glioma cells. J. Nanosci. Nanotechnol. 13, 3851–3854.

Xu, L., Dan, M., Shao, A., Cheng, X., Zhang, C., Yokel, R.A., Takemura, T., Hanagata, N., Niwa, M., Watanabe, D., 2015. Silver nanoparticles induce tight junction disruption and astrocyte neurotoxicity in a rat blood–brain barrier primary triple coculture model. Int. J. Nanomed. 10, 6105–6119.

Xu, R., Ma, J., Sun, X., Chen, Z., Jiang, X., Guo, Z., Huang, L., Li, Y., Wang, M., Wang, C., Liu, J., Fan, X., Gu, J., Chen, X., Zhang, Y., Gu, N., 2009. Ag nanoparticles sensitize IR-induced killing of cancer cells. Cell Res. 19, 1031–1034.

Xu, F., Piett, C., Farkas, S., Qazzaz, M., Syed, N.I., 2013. Silver nanoparticles (AgNPs) cause degeneration of cytoskeleton and disrupt synaptic machinery of cultured cortical neurons. Mol. Brain 6, 29.

Yamagata, Y., 2003. New aspects of neurotransmitter release and exocytosis: dynamic and differential regulation of synapsin I phosphorylation by acute neuronal excitation in vivo. J. Pharmacol. Sci. 93, 22–29.

Yang, Z., Liu, Z.W., Allaker, R.P., Reip, P., Oxford, J., Ahmad, Z., Ren, G., 2010. A review of nanoparticle functionality and toxicity on the central nervous system. J. R. Soc. Interface 7 (Suppl. 4), S411–S422.

Yin, N., Liu, Q., Liu, J., He, B., Cui, L., Li, Z., Yun, Z., Qu, G., Liu, S., Zhou, Q., Jiang, G., 2013. Silver nanoparticle exposure attenuates the viability of rat cerebellum granule cells through apoptosis coupled to oxidative stress. Small 9, 1831–1841.

Yin, N., Yao, X., Zhou, Q., Faiola, F., Jiang, G., 2015. Vitamin E attenuates silver nanoparticle-induced effects on body weight and neurotoxicity in rats. Biochem. Biophys. Res. Commun. 458, 405–410.

van der Zande, M., Vandebriel, R.J., Van Doren, E., Kramer, E., Herrera Rivera, Z., Serrano-Rojero, C.S., Gremmer, E.R., Mast, J., Peters, R.J., Hollman, P.C., Hendriksen, P.J., Marvin, H.J., Peijnenburg, A.A., Bouwmeester, H., 2012. Distribution, elimination, and toxicity of silver nanoparticles and silver ions in rats after 28-day oral exposure. ACS Nano 6, 7427–7442.

Zhivotovsky, B., Orrenius, S., 2011. Calcium and cell death mechanisms: a perspective from the cell death community. Cell Calcium 50, 211–221.

Ziemińska, E., Stafiej, A., Strużyńska, L., 2014. The role of the glutamatergic NMDA receptor in nanosilver-evoked neurotoxicity in primary cultures of cerebellar granule cells. Toxicology 315, 38–48.

Ziemińska, E., Strużyńska, L., 2016. Zinc modulates nanosilver-induced toxicity in primary neuronal cultures. Neurotox. Res. 29, 325–343.

Chapter 8

The Applications, Neurotoxicity, and Related Mechanism of Gold Nanoparticles

S.J.S. Flora

Defence Research and Developmental Establishment, Gwalior, India

Chapter Outline

Neurotoxicity of Nanomaterials and Nanomedicine. http://dx.doi.org/10.1016/B978-0-12-804598-5.00008-8

1. INTRODUCTION

Nanoparticles are the class of engineered structures with diameter less than 100nm and can be produced via several different chemical and physical processes depending on the type of property required for its application (Gwinn and Vallyathan, 2006). They are 1/100–1/1000 of human cell. The exclusive characteristics of these particles such as high surface to mass ratio for catalytic promotions of reactions and their ability to absorb and carry other compounds such as drugs and proteins along with surfaces that might be chemically more reactive compared with their fine analogues (Borm and Kreyling, 2004) make them suitable for various applications in the fields of biomedicine, optics, and electronics.

Gold in its bulk form has been considered an inert, nontoxic, biocompatible metal used for various therapeutic purposes. But when the size of the typical objects decreases into nanoscopic dimensions, gold behaves very differently than in bulk and it is related with many important concerns in the risk assessment for humans. Also, there is rapid emergence of gold nanoparticle (AuNP) technology, which holds great promise for future applications because of their large volume specific surface areas and high diverse surface activities compared with bulk gold (Yah, 2013).

Gold is a class of noble metal that has been used for various medical purposes since 2500 BC (Antonovych, 1981; Lewis and Walz, 1982). In Indian culture also it is being used in the form of "Swarna Bhasma" for the management of various health needs. With this idea, AuNPs are designed in several forms such as gold nanospheres, nanorods, nanobelts, nanocages, nanoprisms, and nanostars to be used for various purposes (Yang and Cui, 2008), which can be synthesized by applying different strategies (Grzelczak et al., 2008; Murphy et al., 2005).

In this chapter, we have focus on a wide range of applications of AuNPs and also toxicological aspects of these nanoparticles with special focus on neurotoxicity. These aspects can be applied as measures to improve their biomedical applications and risk assessment to humans.

2. SYNTHESIS

There have been several methods for the synthesis of AuNPs, which can be broadly classified into physical or chemical methods. Examples of physical methods include γ-irradiation method, photochemical synthesis, microwave irradiation, and seed approach, whereas chemical methods are supercritical technology or biological agents (Rezende et al., 2010; Ganesh Kumar et al., 2012).

3. ADVANTAGES

The widespread applications of AuNPs for various purposes such as biomedical imaging, drug and biomolecule delivery, and cancer therapy are due to the various characteristics such as shape and size, high surface area, high biocompatibility, and noncytotoxicity (Khan et al., 2014; Pal et al., 2013).

4. PHARMACOKINETICS

Gold is a well-known and highly valued metal that has been used since histories. Owing to its availability and wide use for various purposes such as cosmetics, automobiles, beverages, etc. (Sung et al., 2011), in either bulk or nanoforms, it is important to know the pharmacokinetics of AuNPs. Absorption, distribution, metabolism, and elimination of gold are discussed and described in the following paragraphs.

4.1 Absorption

Humans can be exposed to AuNPs via ingestion, inhalation, or dermal contact (Oberdörster et al., 2005b; Walser et al., 2012; Pernodet et al., 2006). Gastrointestinal absorption may be the major source of AuNP entry into the body (Bergin and Witzmann, 2013). Inhalation has been one of the major routes for the absorption of AuNP wherein it has been capable of transporting the AuNP to the olfactory nerves leading to their entry into the brain (Balasubramanian et al., 2013). The use of AuNP in cosmetics has also led to the dermal exposure of AuNP, which has been confirmed by various studies (Freese et al., 2012, 2013; Johnston et al., 2010).

4.2 Distribution

Various studies have shown that the nanoparticles were able to distribute to various organs of the body such as blood, liver, kidney, and brain. The major factors responsible for the distribution of these nanoparticles are size, charge, and surface chemistry. The smaller nanoparticles are more prone to cross the blood–brain barrier (BBB) as compared with the larger ones (Table 8.1).

4.3 Metabolism

The metabolism is yet not clear for these nanoparticles but few studies have revealed that they may be taken up by the body and get deposited into various tissues (Loeschner et al., 2011; van der Zande et al., 2012; Aaseth et al., 1981).

4.4 Elimination

Studies conducted by Schleh et al. (2012) revealed that urine could be the major route of elimination for these nanoparticles (Fig. 8.1).

5. APPLICATIONS

Owing to a wide range of unique properties and advantages, these particles are used for various purposes (Fig. 8.2) as described in the following sections.

5.1 Electronics

It has been widely used as conductors, which can be used to connect the resistors, conductors of an electronic chip.

TABLE 8.1 Available Literature Suggesting Distribution of Gold Nanoparticles in Brain

#	Nature and Size of Nanoparticle	Surface Coating	Model	Dose, Route, and Duration	References
1	CG-4, 10, 28, 58 nm	–	Mice	200 mg/mL in drinking water for 7 days	Hillyer and Albrecht (2001)
2	^{198}Au-15–20 nm	Gum arabic	Mice	10–20 μCi, intravenous (iv) for 1, 4, 24 h	Katti et al. (2006)
3	Au-dendrimer (nm) PAMAM, 5, 11, 22	PAMAM	Mice	0.8–1.6 mg/g, iv, 5 min, 1 h, 1, 4, 7 days	Balogh et al. (2007)
4	GNS SiO_2/Au 110/10 nm	PEG	Mice	10 mg/g iv for 4 h, 1, 4, 7, 14, 21, 28 days	James et al. (2007)
5	CG-10, 50, 100, 250 nm	–	Rats	0.3–0.5 mg/g iv 24 h	De Jong et al. (2008)
6	GNS Au/Au_2S 20–50 nm	Mercaptoundecanoic acid + cisplatin	Mice	10 mg/g iv for 1 and 7 days	Huang et al. (2008)
7	GNR-65, 11 nm	PEG-SH Phosphatidylcholine	Mice	1.3–6 mg/g iv for 6 min	Niidome et al. (2008)
8	Gold clusters-1.4, CG-18 nm	Na salt of monosulfonated triphenylphosphine	Rats	0.01–0.1 mg/g iv for 24 h	Semmler-Behnke et al. (2008)
9	CG-15, 50, 100, 200 nm	Na alginate	Mice	1000 mg/g iv for 24 h	Sonavane et al. (2008)
10	CG-13 nm	PEG-SH	Mice	0.85, 4.26 mg/g iv for 5, 30 min, 4, 24 h, 7 days	Cho et al. (2009)

11	CG-15, 50, GNS SiO_2/Au 120/20 nm	PEG-SH	Rats	1.3 mg/g for 24 h	Terentyuk et al. (2009)
12	CG-20 nm	–	Rats	0.01–0.015 mg/g, iv 1, 7 days, 1 and 2 months	Balasubra-manian et al. (2010)
13	GNR-56 × 13 nm	CTAB, BSA	Rats	0.560 mg/g for 0.5, 1–24 h, 3–28 days	Wang et al. (2010)
14	AuNP	Polyelectrolyte	Mice	AuNP uptake in cortex, hippocampus, thalamus, hypothalamus, spinal cord	Sousa et al. (2010)
15	AuNP-7, 20 nm	–	Rats	Inhalation exposure, 1×10^6 particles/cm^3 for 15 days	Balasubra-manian et al. (2013)
16	AuNP-6, 35 nm	–	Rats	100 μg/kg oral for 28 or 14 days	Buzulukov et al. (2014)
17	AuNP-15	PEG COOH NH_2	Mice	1 mg/kg iv, Au was measured in organs at 30 min, 4, 24 h, 7, 14 days, 1, 3, and 6 months	Lee et al. (2016)

AuNP, gold nanoparticle; *BSA*, bovine serum albumin; *CG*, colloidal gold; *CTAB*, cetyltrimethyl ammonium bromide; *GNR*, gold nanorod; *GNS*, gold nanosphere; *PAMAM*, polyamidoamine; *PEG*, polyethylene glycol.

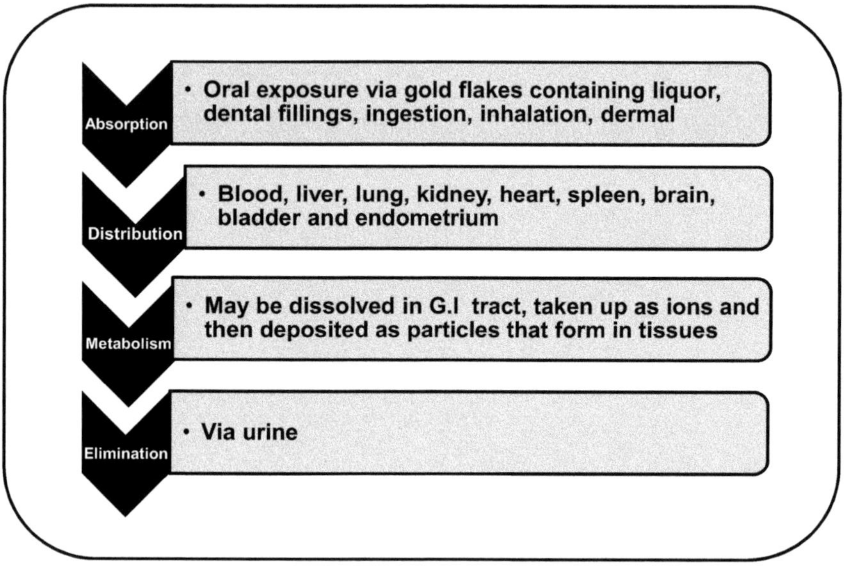

FIGURE 8.1 Pharmacokinetics of gold nanoparticle. *GI*, gastrointestinal.

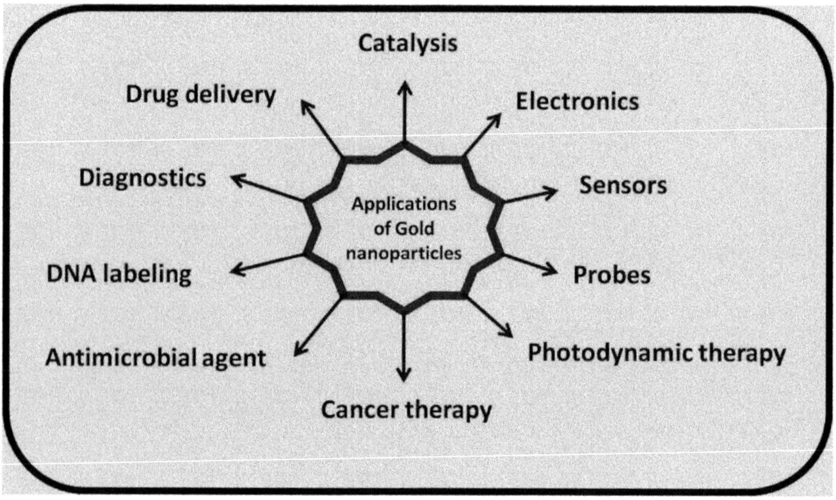

FIGURE 8.2 Applications of gold nanoparticle.

5.2 For Labeling

AuNPs have the ability to strongly absorb and scatter visible light particularly at the surface plasmon resonance frequency. Owing to its high absorption cross-section, it can be used as an attractive contrast agent. This absorption of light causes heating of the particle that further leads to the heating of the environment (Csáki et al., 2002).

5.3 Sensors

Various signals can be used to build sensors using AuNPs based on the principles of electrochemical response, photoluminescence, UV–Vis absorption, and Raman scattering.

5.4 As Delivery Vehicle

Owing to their ability to easily conjugate with ligands (Brown et al., 2010), they can efficiently be used for the delivery of molecules that are adsorbed on the surface of the AuNP into the desired cell of interest either via gene guns or by particle ingestion.

5.5 Probes

AuNPs possess high density and atomic weight, which enable them to be used in transmission electron microscopy. They also have a high ability of scattering X-ray, which provides contrast in X-ray imaging (Hainfeld et al., 2006). They have the ability to scatter light and produce large numbers of colors, which make them useful as probes in dark field microscopy.

5.6 As Heat Source

Gold particles which are composed of free electrons get excited on absorbing light and this excitation occurs at the plasmon resonance frequency wherein the excited electrons relax and the thermal energy is transferred to the lattice, which produces heat from the AuNPs dissipating into the surrounding environment (Bendix et al., 2010) leading to cell death due to rise in temperature.

5.7 Catalysis

Catalysts are the class of compounds that increases the rate of a reaction by lowering the activation energy and thereby increasing the rate of a chemical reaction. The AuNP surface can be used for selective oxidation.

6. MECHANISM OF CELLULAR UPTAKE

Different studies have been conducted by various groups to postulate the uptake of these nanoparticles into the cell, which can be classified into various forms as discussed in the following sections (Fig. 8.3).

6.1 Phagocytosis

Phagocytosis takes place by a class of dendritic cells such as macrophages and monocytes for up to particle size of 500 nm by the formation of phagosomes that

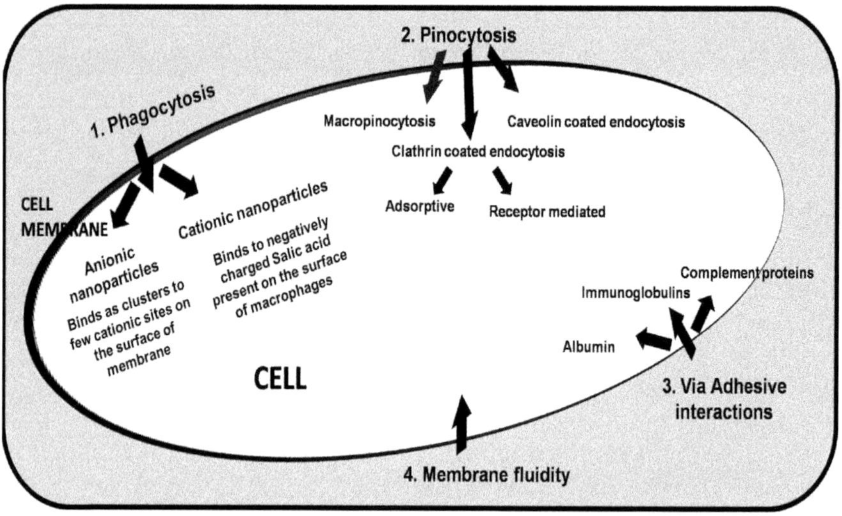

FIGURE 8.3 Mechanism of cellular uptake of gold nanoparticle.

result in a series of fusion events with lysosomes and formation of phagolysosomes (Chu and Liu, 2000; Kettiger et al., 2013).

6.2 Pinocytosis

It is a mode of internalization wherein smaller particles are invigilated inside the cell. It can be further subdivided into the following types.

6.2.1 Macropinocytosis

It is one of the forms of endocytosis that allows the nonselective uptake of solute molecules or foreign particles. Studies have revealed that it is an important mechanism for the uptake of positively charged nanoparticles (Zhao et al., 2011).

6.2.2 Clathrin-Mediated Endocytosis

It mediates the internalization of 120–150 nm sized nanoparticles within the clathrin-coated vesicles, which enters the endosomal/lysosomal trafficking route and causes the endocytosis of the nanoparticle (Kettiger et al., 2013). Chithrani and Chan (2007) found that this process mediates AuNP uptake within the cells with half-lives of ~2 h. It can also be of two types shown in Fig. 8.3.

6.3 Caveolae-Dependent Endocytosis

Caveolae are flask-shaped structures rich in protein and lipids and are invaginations of 50–100 nm in the plasma membrane. These vesicles detach from the membrane to bring their content into caveosomes, the compartments formed by them (Panariti et al., 2012).

6.4 Adhesive Interactions

Nanoparticles have the ability to interact with various serum proteins and immunoglobulins, which can be adsorbed on the surface of nanoparticles and allow them to undergo receptor-mediated endocytosis (Kettiger et al., 2013).

The cellular uptake of AuNPs depends on the shape and size of the nanoparticles. Also, few studies on the mechanism of AuNP uptake have been reported (Chithrani et al., 2006; Chithrani and Chan, 2007).

7. GENERAL MECHANISM OF TOXICITY

Although gold is known to be inert and a less toxic metal, several studies have been conducted that have revealed the major mechanisms leading to toxicity caused by the growing use of AuNPs (Gerber et al., 2013). The conflicting data in the literature regarding AuNP bioactivity on the basis of differences in laboratory research protocols make it difficult to evaluate and generalize important aspects and do not allow one to reach a definitive conclusion and shared opinion about AuNP cytotoxicity (Fratoddi et al., 2015a). Sabella et al. (2011) reviewed the toxicity of AuNPs in in vitro and in vivo models. They concluded that, beyond the wide variability of the experimental conditions like physical, chemical, and other characteristics, AuNPs are significantly toxic both in vitro and in vivo, whereas appropriate coating may partially prevent their harmful effects. Some of the major mechanisms by which they cause toxicity are listed in the following sections (Fig. 8.4).

7.1 Oxidative Stress

Oxidative stress is a mechanism that is described as the excessive generation of reactive oxygen species (ROS), which causes an imbalance between the prooxidant and antioxidant levels in the body (Flora, 2011; Agrawal et al., 2015). Study conducted by Li et al. (2010) stated that AuNPs, when exposed to MRC-5 human lung fibroblasts cells in vitro, showed high levels of lipid peroxidation indicative of oxidative stress in lung fibroblast cells. A similar study conducted by Mateo et al. (2014) also stated that AuNPs when exposed to human leukemia (HL-60) and hepatoma (HepG2) cell lines were responsible in depleting the glutathione (GSH) levels thereby disturbing the antioxidant status of the body leading to oxidative stress.

7.2 Disruption of Lipid Bilayer

Plasma membrane is the outermost covering of the cell, which is composed of a phospholipids bilayer, highly selective in nature thereby allowing only selective substances to enter inside the cell. Therefore it is very important to maintain the

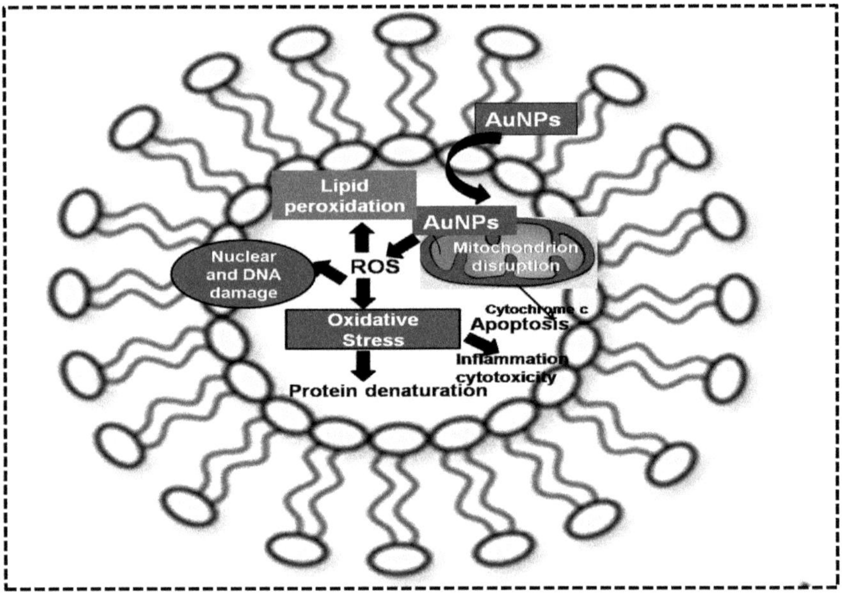

FIGURE 8.4 General mechanism of toxicity due to gold nanoparticle (AuNP). *ROS*, reactive oxygen species.

integrity of this membrane to protect the cell. AuNPs disrupt this integrity of the cellular membrane leading to toxic manifestations in the cell. This was confirmed by studies conducted by Leroueil et al. (2008), Camesano and D'Angelo (2011), and Camesano et al. (2014), which stated that AuNPs were able to form holes and lead to membrane thinning, causing membrane disruption. It was also found that this disintegrity of the membrane was caused by smaller-sized nanoparticles (2–10 nm), whereas the larger-sized nanoparticles up to 20 nm were unable to cause so. This membrane disruption was also dependent on the charge of the nanoparticles. The small positively charged AuNPs were much successful in penetrating inside the negatively charged membrane causing holes in the membrane (Lin et al., 2011).

7.3 Necrosis/Apoptosis

Necrosis and apoptosis are considered as the forms of cell death where necrosis is a common phenomenon in which the cellular membrane disrupts, which causes the depletion of ATP leading to metabolic collapse thereby causing cell swelling followed by rupture marked by inflammation. On the contrary, apoptosis leads to membrane blabbing, shrinkage of cell, and nuclear fragmentation and condensation followed by the formation of apoptotic bodies, which are engulfed by the white blood cells. Studies have revealed the reason of cell death due to AuNP. Imperatore et al. (2015) used both in vivo and in vitro models to study the effects of AuNPs. In the C57BL/6 mice model, these particles showed typical hallmarks of apoptosis in the brain region. Similarly in an in vitro study too, in SH-SY5Y they found the onset of neuronal death by apoptosis. Few studies conducted by Schaeublin et al. (2011) and Pan et al. (2007) have revealed that the size and charge of AuNPs have been the major determining factors for the fate of cell death pathway to be chosen.

7.4 Mitochondrial Dysfunction

Oxidative stress is a potent mechanism behind mitochondrial damage. The small size of these AuNPs enables them to cross the inner mitochondrial membrane. Therefore these AuNPs have known to cause mitochondrial damage too. The intrinsic mitochondrial apoptotic pathway plays a major role in nanoparticle-induced oxidative stress. High levels of ROS generated due to AuNPs could lead to a small proportion of electrons that escapes the mitochondrial chain thereby interacting with O_2, which either forms H_2O_2 or OH (Manke et al., 2013). Administration of 8-nm AuNP in HL7702 cell lines showed prominent depletion in cytosolic and mitochondrial GSH with marked increase in the levels of H_2O_2 leading to apoptosis (Gao et al., 2011).

7.5 DNA Damage

AuNPs cause DNA damage leading to genotoxicity majorly because of two reasons (Rivera et al., 2010). Nanoparticles with the size of about 1.4 nm correspond to the size of the major DNA grooves enabling them to interact with the DNA. Also, due to the highly electronegative nature of AuNPs they have the ability to attract DNA grooves due to their negative environment. Several studies are associated with AuNP-induced DNA damage, such as Li et al. (2008) on 20 nm and Auffan et al. (2009) on 8 nm. Studies by Cardoso et al. (2014) also confirmed the generation of caspase-3, interferon-c, and 8-hydroxydeoxyguanosine, leading to inflammation and DNA damage.

7.6 Endocrine Disruption

Any molecule that disrupts the endocrine gland of the body and leads to hormonal or reproductive defects is termed as endocrine disruptor. The biodistribution of AuNPs to testis has been reported by few investigators (De Jong et al., 2008; Cho et al., 2009; Balasubramanian et al., 2010). Stelzer and Hutz (2009) confirmed that AuNPs increased the accumulation of estrogens in ovarian granulose cells. A similar increase in the levels of testosterone was also seen in mice exposed to AuNPs (Li et al., 2013).

8. FACTORS AFFECTING TOXICITY

Several factors determine the toxicity of nanoparticles (summarized in Fig. 8.5).

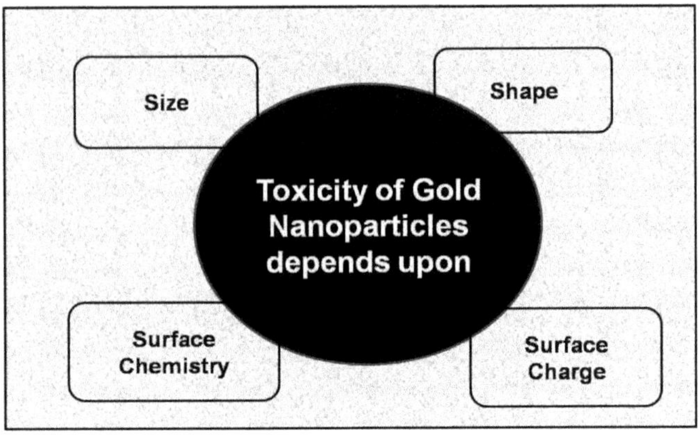

FIGURE 8.5 Factors affecting the toxicity of gold nanoparticle.

8.1 Size

Various studies have been conducted to determine the effects of size of nanoparticles on toxicity. De Jong and Borm (2008) reported that the smaller AuNPs caused more toxicity than the larger particles when injected intravenously into mice. A similar study conducted by Trickler et al. (2011) suggested that only the smallest (3 nm) nanoparticles were able to induce signs of cellular toxicity in primary rat brain microvessel endothelial cells (rBMECs) compared with other large nanoparticles. Size has also been a major factor in crossing the BBB leading to possible neurotoxicity (Sonavane et al., 2008). Only the 15- and 50-nm AuNPs were able to cross the BBB because of their small size compared with the large 100- and 200-nm AuNPs.

8.2 Shape

Shape has been another attribute to determine the degree of toxicity. Schaeublin et al. (2012) showed that nanorods have a high degree of toxicity in human keratinocytes, HaCaT cell line compared with gold nanospheres on the basis of cell viability, ROS generation, and altered gene expression. Similarly, Wang et al. (2008) reported that cetyltrimethyl ammonium bromide (CTAB)-coated gold nanorods (AuNRs) showed a higher degree of toxicity than spherical AuNPs. Hutter et al. (2010) on the other hand suggested that gold spheres coated by CTAB were more toxic to N9 glial cells than polyethylene glycol (PEG)-coated spheres and CTAB- or PEG-coated nanorods or CTAB- and PEG-coated nanourchins. They also found that only PEG-coated spheres and nanorods were responsible in increasing the levels of N9 granulocyte macrophage colony-stimulating factor not the PEG/CTAB-coated nanourchins. On the contrary only these nanourchins were responsible in increasing interleukin 1α levels.

8.3 Surface Charge

Charge also influences the toxicity of nanoparticles. Goodman et al. (2004) studied both positively and negatively charged AuNPs for their toxicity on Cos-1 cells, red blood cells, and *Escherichia coli* bacteria and reported that positively charged AuNPs were more toxic than negatively charged AuNPs of the same size. Another study by Schaeublin et al. (2012) showed that charged nanoparticles of the same size showed more DNA damage in HaCaT cells than the neutral ones.

8.4 Surface Chemistry

Treuel et al. (2010) studied the influence of surface chemistry and reported an interaction between functionalized AuNPs as well as polystyrene nanoparticles with bovine serum albumin (BSA). The study confirmed that citrate-coated AuNPs exhibited much stronger interaction with BSA than the polymeric-coated nanoparticles. Nanoparticles when added to the protein solution destroyed the helical structure due to their surface adsorption. Alkilany et al. (2009) also studied a similar phenomenon using AuNPs with different surface agents such as CTAB, polyacrylic acid (PAA), and poly (allylamine) hydrochloride (PAH). It was seen that the PAA overcoating (PAA-CTAB-AuNRs) eliminated the observed cytotoxicity compared with CTAB-AuNRs, as the polymer masked the exposure of the highly toxic free CTAB molecule to the HT-29 cells. Hutter et al. (2010) showed that both CTAB- and PEG-coated gold nanourchins showed a similar and much greater content of gold in N9 microglia compared with CTAB/PEG-coated gold spheres or rods. Fratoddi et al. (2015b) suggested that differently functionalized AuNPs behave similarly in Hela cells and that the different surface coating of the different nanoparticles considered defines the range of particle concentration where toxic effects begin. The authors suggested that this kind of analysis could furnish some albeit preliminary suggestions toward an appropriate method to study AuNP toxicity.

9. NEUROTOXICITY

The brain is an important component of the central nervous system, which is composed of several structures to protect it from any kind of external insult. AuNP-induced neurotoxicity is one of the least focused and explored topic. Very few studies have been conducted on the toxicity associated with AuNPs (Oberdörster et al., 2009). The nanoparticles translocation into the central nervous system due to the external insult to the BBB leads to the activation of cerebral epithelial cells thereby leading to the generation of oxidative stress (Oberdörster et al., 2009; Leitea et al., 2015).

9.1 Neuronal Uptake

Neuronal uptake of AuNPs can take place via two routes.

9.1.1 Neuronal Uptake via Olfactory Nerves

Neuronal uptake of the nanoparticles may be via several sensory nerve endings present in the nasal and the trachea-bronchial regions (Oberdörster et al., 2005b). The silver-coated AuNPs (50 nm) instilled intranasally in monkeys migrated to the olfactory nerves and bulb with an axonal transport velocity of 2.5 mm/h (Oberdörster et al., 2005a) and were found to be located in the mitochondria leading to toxicity. Activity of nanogold in the CSF could be attributed to the perineural transport of these particles, which can also be described as a very rapid translocation route from the nose to the brain (Illum, 2000). Another study by Segal (2000) also states that the nanogold particles after axonal transport to the olfactory bulb cross into the CSF and get distributed into the various regions of the brain. Studies have also revealed that only a very small fraction of AuNPs deposited by intranasal instillation onto the olfactory mucosa was translocated to the olfactory bulb in rats, which depended on the size of the nanoparticles where larger 20-nm gold particle was less able to translocate when compared with the smaller particle, that is, 2 and 5 nm (Rinderknecht et al., 2007).

9.1.2 Neuronal Uptake via Blood–Brain Barrier

The BBB is a barrier made up of endothelial cells also known as brain microvascular endothelial cells (BMECs), which is maintained by tight junctions present between the BMECs to allow the selective transport of molecules across it (Zlokovic, 2008). Some studies (Khlebtsov and Dykman, 2011) that determined the passage of AuNPs across the BBB have been summarized in Table 8.1. One study by Lasagna-Reeves et al. (2010) reported that 12.5-nm AuNP on intraperitoneal administration at the dose of 40, 200, and 400 μg/kg/day for 8 days also was present in the neural tissue as it was successful in crossing the BBB.

9.2 General Neurotoxicity

Very few studies are available in the literature regarding AuNP-induced neurotoxicity. The available literature reveals that AuNPs are capable of neuronal uptake both via olfactory nerves and by crossing the BBB. Isolated studies are also available on the passage of AuNPs into the brain, which have already been described earlier. Thus there is a need to study in detail the toxicity associated with these nanoparticles on CNS and neurobehavioral variables (Table 8.2).

In an in vitro conducted by Trickler et al. (2011) on primary rBMECs, AgNPs at a concentration of 0.8–50 μg/mL and size 3, 5, 7, 10, 30, and 60 nm were administered for 24 h, but there was no morphological change in the culture. The smallest nanoparticle of 3 nm was found to be slightly cytotoxic, whereas no signs of cytotoxicity were observed for other larger nanoparticles. Another in vitro study conducted by Jung et al. (2014) stated that administration of 5-nm nanoparticles at the concentration of 1.1×10^{11} NPs/mL and 40-nm nanoparticles at the concentration of 1.44×10^{8} NPs/mL in brain slice

TABLE 8.2 Neurotoxicity of AuNP

#	Model	Administration Dose/ Route	Duration	Study Highlight	References
1	Microglial cells in vitro	10^9 NP/mL	6 or 24 h	Alterations in microglial activation. Increased TLR-2 expression, IL-1a, and GM-CSF levels	Hutter et al. (2010)
2	Male C57/BL6 mice	40, 200, and 400 μg/kg/day 12.5 nm AuNP, i.p.	Daily, 8 days	AuNP crossed BBB and accumulated in the neural tissues	Lasagna-Reeves et al. (2010)
3	BALB/c mice	0.5–14.6 mg/kg 17 and 37 nm AuNP, i.p.	Single	AuNP damaged cognition, size-dependent, and entered hippocampus	Chen et al. (2010)
4	rBMEC	0.8–50 μg/mL (3, 5, 7, 10, 30, and 60 nm)	24 h	Smallest NP tested (3 nm) induced mild signs of cellular toxicity	Trickler et al. (2011)
5	Male Wistar rats	20 μg/kg body weight of 20-nm AuNP, i.p.	3 days	Impaired brain GPx, Generated 8-OHdG, caspase-3, and Hsp70, IFN-γ, DNA damage	Siddiqi et al. (2012)
6	Wistar rat	10 and 30 nm of 70 mg/L of NP	Single or 28 days i.p.	DNA damage in cerebral cortex. Chronic exposure caused more damage	Cardoso et al. (2014)
7	Brain slices of mouse	1.1×10^{11} NPs/mL (5 nm) and 1.44×10^8 NPs/mL (40 nm)	10 min	Increased cellular excitability, decreased threshold, increased action potentials, aggravated seizure activity	Jung et al. (2014)
8	pBMECs	15 μg/mL of AuNP-3, 5 nm	–	Proinflammatory response influenced integrity of the BBB	Trickler et al. (2014)

Continued

TABLE 8.2 Neurotoxicity of AuNP—cont'd

#	Model	Administration Dose/ Route	Duration	Study Highlight	References
9	Female Sprague–Dawley	100 nm AuNP and 5 nm AuNP unilateral injection in right cerebral cortex	Single injection	5 nm AuNP more nestin expression than 100 nm AuNP	Lee et al. (2016)
10	(i) In vivo, C57BL/6 mice; (ii) in vitro, SH-SY5Y cell line	0.1 mg/mL (AuNP)-mmi NP	1, 2, 4 h	(i) Detected in caudate putamen area mainly in the striatal neurons suggesting apoptosis (ii) Led to the onset of neuronal death by apoptosis	Imperatore et al. (2015)
11	Albino rat	400 μg/kg/day AuNP, orally	8 weeks once daily	Led to its deposition in rat brain with various degrees of astrogliosis	El-Drieny et al. (2015)

8-OHdG, 8-hydroxydeoxyguanosine; *AuNP*, gold nanoparticle; *BBB*, blood–brain barrier; *GM-CSF*, granulocyte-macrophage colony-stimulating factor; *GPx*, glutathione peroxidase; *Hsp*, heat shock protein; *i.p.*, intraperitoneal; *IFN*, interferon; *IL*, interleukin; *NP*, nanoparticle; *pBMEC*, porcine brain microvessel endothelial cell; *rBMEC*, rat brain microvessel endothelial cells; *TLR*, Toll-like receptors.

of mouse for 10 min led to an increased excitability levels in hippocampus and seizure activity induced by low Mg^{2+}. The intracellular AuNPs altered the intrinsic properties of neurons toward increasing their excitability and had deleterious effect such as seizure under pathological conditions. Few in vivo studies have also been conducted to see the effects of these nanoparticles on brain tissues. Cardoso et al. (2014) studied the effects of 70 mg/L of 10- and 30-nm AuNPs, which were injected intraperitoneally for both single acute and chronic exposure for 28 days in Wistar rats. It was found that both the nanoparticles were potent in inducing DNA damage in the cerebral cortex and chronic exposure was more potent in causing the damage than acute exposure of these nanoparticles. Another study describing the neurotoxicity was conducted by El-Drieny et al. (2015), which stated that when the albino rats were administered 400 μg/kg/day of AuNPs by gastric tube for 8 weeks once daily, it led to the deposition of the particles in the brain with various degrees of astrogliosis proving that these particles were potent enough to generate neurotoxicity. Lee et al. (2015) studied the effect of a single injection of 100-nm AuNP and the 5-nm AuNP unilateral injection into the right cerebral cortex in female Sprague–Dawley rats. The authors reported that a 5-nm AuNP showed greater nestin expression than 100-nm AuNP at short-term implantation into the cerebral cortex of the brain. Study conducted by Imperatore et al. (2015) in an in vivo model also revealed that AuNPs when injected in the caudal vein of C57BL/6 mice were found to cross the BBB and were detected in the caudate putamen area of mice mainly in the striatal neurons, which showed typical hallmarks of apoptosis. The in vitro study conducted by them on SH-SY5Y human neuroblastoma cells also revealed that the administration of the lowest dose of 0.1 mg/mL (AuNP)-mmi nanoparticles also led to the onset of neuronal death by apoptosis (Fig. 8.6).

9.3 Neurological Pathology

The ability of AuNPs to cross the BBB enables them to cause several neurological pathological symptoms as described in the following sections.

9.3.1 Astrogliosis

This condition is often referred to as "reactive astrocytosis" wherein there is an abnormal increase in the number of astrocytes due to the destruction of nearby neurons. Astrogliosis may cause morphological and molecular changes in the expression of astrocytes leading to scar formation and inhibition of axon regeneration (Fig. 8.7). El-Drieny et al. (2015) studied the histological effects of long-term exposure to AuNPs on the brain of adult male albino rats exposed to 400 μg/kg/day AuNPs by gastric tube once daily for 8 weeks. Immunohistochemical studies revealed the widespread distribution of the AuNPs in the neurons of cerebral cortex and hippocampus along with the choroid plexus showing hyalinization of the wall of some blood vessels and disruption of capillaries.

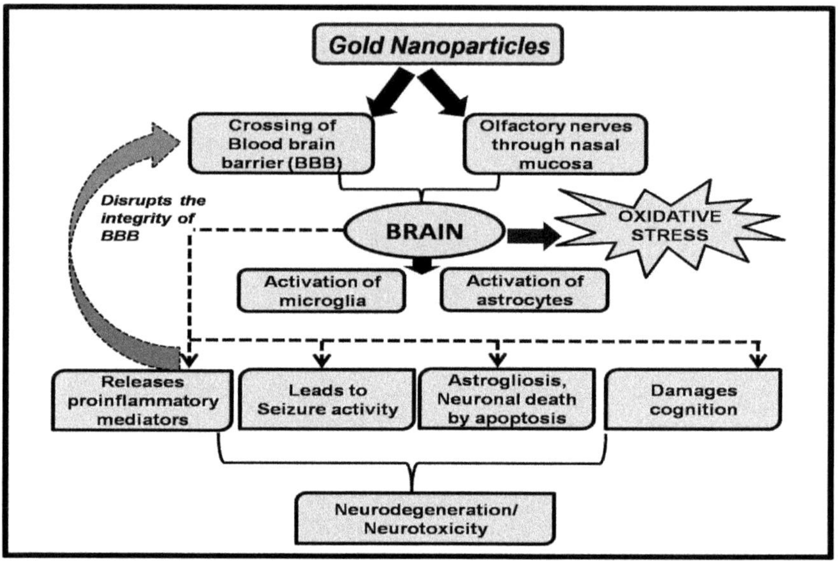

FIGURE 8.6 Schematic diagram showing neurotoxicity due to gold nanoparticle.

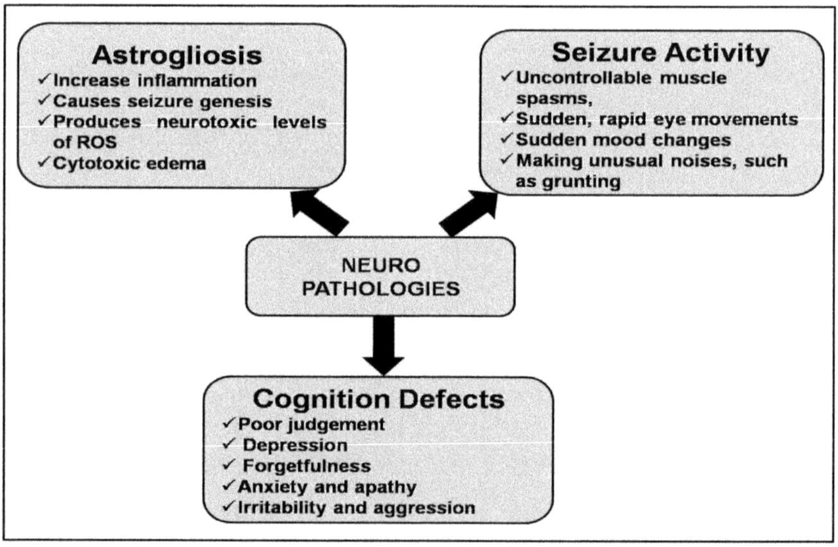

FIGURE 8.7 Various symptoms of neurological pathologies due to gold nanoparticle. *ROS*, reactive oxygen species.

Astrogliosis was marked by astrocytic proliferation and increased size of their cell body with increase in the number and length of their processes. The study concluded that repeated exposure of AuNPs to adult male albino rats leads to deposition of AuNPs in the brain and histological alterations and various degrees of astrogliosis.

9.3.2 Generation of Seizure Activity

Seizure may be defined as any change in the brain's electrical activity. Violent shaking and loss of control are two widely recognized symptoms of seizure activity. Epilepsy is the major cause of the same. Several extracranial and intracranial causes may be responsible for seizure. Several symptoms associated with it are listed in Fig. 8.7. Jung et al. (2014) studied the effect of AuNPs on a single neuron under physiological and pathological conditions in vitro. In this study AuNPs were intracellularly applied to hippocampal CA1 neurons from the mouse brain. Intracellular application of AuNPs increased both the number of action potentials (APs) and input resistance, whereas the threshold and duration of APs and the after hyperpolarization were decreased by the intracellular AuNPs. This study demonstrated that intracellular AuNPs alter the intrinsic properties of neurons toward increasing their excitability, and may have deleterious effects on neurons under pathological conditions, such as seizure.

9.3.3 Cognition Defect

Knowledge, attention, memory, and judgment are all the set of mental abilities and processes that forms the collective basis of cognition. Cognition can also be assumed to be a mode of judging a person's ability and psychology. Studies of AuNP intake in animal models have shown AuNPs to impart cognition defects. Chen et al. (2010) performed a study in BALB/c mice to explore the size-dependent impairment of cognition caused by the injection of AuNPs. Different sizes of AuNPs, 17 and 37 nm in diameter, were injected intraperitoneally into mice at doses ranging from 0.5 to 14.6 mg/kg. Results obtained from this study showed that the ability of AuNPs to damage cognition in mice was size dependent and associated with the ability of the particles to invade the hippocampus.

10. CONCLUSION

Data on effects of nanoparticles in the brain tissue and cells both in vitro and in vivo are scarce and two main features were observed, namely, nanoparticle-mediated oxidative stress and inflammatory response. This chapter provides a detailed insight about the usage of AuNPs and the toxicity associated with them. There are very sporadic reports available for the neurotoxic effects of AuNP and thus warrants comprehensive investigation. Also, there is dearth of information on AuNP health effects and no regulatory safety guidelines related to AuNP toxicity. Knowledge about size-, charge-, and surface chemistry–dependent in vivo toxicity is also needed to predict the hazard potential of AuNPs for humans. However, for better understanding of AuNP toxicity and its health effects, there must be recommendations of methods for testing of reproductive toxicity, genotoxicity, or carcinogenicity effects that have been made available by the Organization for Economic Cooperation and Development. Also, more in vitro and in vivo AuNP studies are highly recommended to augment the limited current health effects data on humans. Furthermore, in-depth research should be done on the size, shape, and surface activities of AuNPs because they exhibit different properties both in vitro and in vivo. This chapter focuses on providing some evidence for the ability of the AuNP to cross the BBB leading to various toxic manifestations including neurotoxicity. These effects generally depend on the size, shape, and surface properties of gold nanostructures. These evidences will be of paramount importance for establishing regulatory safety guidelines in the relatively new field of gold nanotechnology. This chapter provides an updated account of the data available for the AuNP-induced neurotoxic manifestations.

LIST OF ABBREVIATIONS

8-OHdG 8-Hydroxydeoxyguanosine
ATP Adenosine triphosphate
AuNRs Gold nanorods
BBB Blood–brain barrier
BSA Bovine serum albumin
CG Colloidal gold
CSF Cerebrospinal fluid
CTAB Cetyltrimethyl ammonium bromide
GNR Gold nanorods
GNS Gold nanospheres
PEG Polyethylene glycol
rBMEC Primary rat brain microvessel endothelial cells

REFERENCES

Aaseth, J., Olsen, A., Halse, J., Hovig, T., 1981. Argyria-tissue deposition of silver as selenide. Scand. J. Clin. Lab. Invest. 41, 247–251.

Agrawal, S., Bhatnagar, P., Flora, S.J., 2015. Changes in tissue oxidative stress, brain biogenic amines and acetylcholinesterase following co-exposure to lead, arsenic and mercury in rats. Food Chem. Toxicol. 86, 208–216.

Alkilany, M., Nagaria, P.K., Hexel, C.R., Shaw, T.J., Murphy, C.J., Wyatt, M.D., 2009. Cellular uptake and cytotoxicity of gold nanorods: molecular origin of cytotoxicity and surface effects. Small 6, 701–708.

Antonovych, T.T., 1981. Gold nephropathy. Ann. Clin. Lab. Sci. 11 (5), 386–391.

Auffan, M., Rose, J., Orsiere, T., De Meo, M., Thill, A., Zeyons, O., Proux, O., Masion, A., Chaurand, P., Spalla, O., 2009. CeO_2 nanoparticles induce DNA damage towards human dermal fibroblasts in vitro. Nanotoxicology 3, 161–171.

Balasubramanian, S.K., Jittiwat, J., Manikandan, J., Ong, C.N., Yu, L.E., Ong, W.Y., 2010. Biodistribution of gold nanoparticles and gene expression changes in the liver and spleen after intravenous administration in rats. Biomaterials 31, 2034–2042.

Balasubramanian, S.K., Poh, K.W., Ong, C.N., Kreyling, W.G., Ong, W.Y., Yu, L.E., 2013. The effect of primary particle size on biodistribution of inhaled gold nano-agglomerates. Biomaterials 34, 5439–5452.

Balogh, L., Nigavekar, S.S., Nair, B.M., Lesniak, W., Zhang, C., Sung, L.Y., Kariapper, M.S., El-Jawahri, A., Llanes, M., Bolton, B., Mamou, F., Tan, W., Hutson, A., Minc, L., Khan, M.K., 2007. Significant effect of size on the in vivo biodistribution of gold composite nanodevices in mouse tumor models. Nanomedicine 3, 281–296.

Bendix, P.M., Reihani, S.N., Oddershede, L.B., 2010. Direct measurements of heating by electromagnetically trapped gold nanoparticles on supported lipid bilayers. ACS Nano 4, 2256–2262.

Bergin, I.L., Witzmann, F.A., 2013. Nanoparticle toxicity by the gastrointestinal route: evidence and knowledge gaps. Int. J. Biomed. Nanosci. Nanotechnol. 3.

Borm, P.J., Kreyling, W., 2004. Toxicological hazards of inhaled nanoparticles–potential implications for drug delivery. J. Nanosci. Nanotechnol. 4, 521–531.

Brown, S.D., Nativo, P., Smith, J.A., Stirling, D., Edwards, P.R., Venugopal, B., Flint, D.J., Plumb, J.A., Graham, D., Wheate, N.J., 2010. Gold nanoparticles for the improved anticancer drug delivery of the active component of oxaliplatin. J. Am. Chem. Soc. 132, 4678–4684.

Buzulukov, Y.P., Arianova, E.A., Demin, V.F., Safenkova, I.V., Gmoshinski, I.V., Tutelyan, V.A., 2014. Bioaccumulation of silver and gold nanoparticles in organs and tissues of rats by neutron activation analysis. Izv. Akad. Nauk Ser. Biol. 3, 286–295.

Camesano, T.A., D'Angelo, S.M., 2011. Interactions of model cell membranes with nanoparticles. In: 85th ACS Colloid and Surface Science Symposium, Montreal, QC, Canada.

Camesano, T.A., Kamaloo, E., Bailey, C., Waterman, K., 2014. Abstracts of Papers. In: 247th ACS National Meeting & Exposition, Dallas, TX, United States.

Cardoso, E., Rezin, G.T., Zanoni, E.T., de Souza Notoya, F., Leffa, D.D., Damiani, A.P., Daumann, F., Rodriguez, J.C., Benavides, R., da Silva, L., Andrade, V.M., da Silva Paula, M.M., 2014. Acute and chronic administration of gold nanoparticles cause DNA damage in the cerebral cortex of adult rats. Mutat. Res. 766/767, 25–30.

Chen, Y.S., Hung, Y.C., Lin, L.W., Liau, I., Hong, M.Y., Huang, G.S., 2010. Size-dependent impairment of cognition in mice caused by the injection of gold nanoparticle. Nanotechnology 21, 485102.

Chithrani, B.D., Chan, W.C.W., 2007. Elucidating the mechanism of cellular uptake and removal of protein-coated gold nanoparticles of different sizes and shapes. Nano Lett. 7, 1542–1550.

Chithrani, B.D., Ghazani, A.A., Chan, W.C.W., 2006. Determining the size and shape dependence of gold nanoparticle uptake by mammalian cells. Nano Lett. 6, 662–668.

Cho, W.S., Cho, M., Jeong, J., Choi, M., Cho, H.Y., Han, B.S., Kim, S.H., Kim, H.O., Lim, Y.T., Chung, B.H., Jeong, J., 2009. Acute toxicity and pharmacokinetics of 13 nm-sized PEG-coated gold nanoparticles. Toxicol. Appl. Pharmacol. 236, 16–24.

Chu, B., Liu, T., 2000. Characterization of nanoparticles by scattering techniques. J. Nanopart. Res. 2, 29–41.

Csáki, A., Möller, R., Fritzsche, W., 2002. Gold nanoparticles as novel label for DNA diagnostics. Expert Rev. Mol. Diagn. 2, 187–193.

De Jong, W.H., Borm, P., 2008. Drug delivery and nanoparticles: applications and hazards. Int. J. Nanomed. 3, 133–149.

De Jong, W.H., Hagens, W.I., Krystek, P., Burger, M.C., Sips, A.J., Geertsma, R.E., 2008. Particle size-dependent organ distribution of gold nanoparticles after intravenous administration. Biomaterials 29, 1912–1919.

El-Drieny, E.A.E.A., Sarhan, N.I., Bayomy, N.A., Elsherbeni, S.A.E., Momtaz, R., Mohamed, H.E.D., 2015. Histological and immunohistochemical study of the effect of gold nanoparticles on the brain of adult male albino rat. J. Microsc. Ultrastruct. http://dx.doi.org/10.1016/j.jmau.2015.05.001.

Flora, S.J.S., 2011. Arsenic-induced oxidative stress and its reversibility. Free Radic. Biol. Med. 51, 257–281.

Fratoddi, I., Venditti, I., Cametti, C., Russo, M.V., 2015a. How toxic are gold nanoparticle? The state of the art. Nano Res. 8(6), 1–38.

Fratoddi, I., Venditti, I., Camettib, C., Russo, M.V., 2015b. The puzzle of toxicity of gold nanoparticles. The case-study of HeLa cells. Toxicol. Res. 4, 796–800.

Freese, C., Gibson, M.I., Klock, H.A., Unger, R.E., Kirkpatrick, C.J., 2012. Size- and coating-dependent uptake of polymer-coated gold nanoparticles in primary human dermal microvascular endothelial cells. Biomacromolecules 13, 1533–1543.

Freese, C., Unger, R.E., Deller, R.C., Gibson, M.I., Brochhausen, C., Klok, H.A., Kirkpatrick, C.J., 2013. Uptake of poly(2-hydroxypropylmethacrylamide)-coated gold nanoparticles in microvascular endothelial cells and transport across the blood–brain barrier. Biomater. Sci. 1, 824–833.

Ganesh Kumar, M., Sastry, T.P., Sathish Kumar, M., Dinesh, M.G., Kannappan, S., Suguna, L., 2012. Sun light mediated synthesis of gold nanoparticles as carrier for 6-mercaptopurine: preparation, characterization and toxicity studies in zebra fish embryo model. Mater. Res. Bull. 47, 2113–2119.

Gao, W., Xu, K., Ji, L., Tang, B., 2011. Effect of gold nanoparticles on glutathione depletion-induced hydrogen peroxide generation and apoptosis in HL7702 cells. Toxicol. Lett. 205, 86–95.

Gerber, A., Bundschuh, M., Klingelhofer, D., Groneberg, D.A., 2013. Gold nanoparticles: recent aspects for human toxicology. J. Occup. Med. Toxicol. 8, 32.

Goodman, C.M., McCusker, C.D., Yilmaz, T., Rotello, V.M., 2004. Toxicity of gold nanoparticles functionalized with cationic and anionic side chains. Bioconjug. Chem. 15, 897–900.

Grzelczak, M., Perez-Juste, J., Mulvaney, P., Liz-Marzan, L.M., 2008. Shape control in gold nanoparticle synthesis. Chem. Soc. Rev. 37, 1783–1791.

Gwinn, M.R., Vallyathan, V., 2006. Nanoparticles: health effects—pros and cons. Environ. Health Perspect. 114, 1818–1825.

Hainfeld, J.F., Slatkin, D.N., Focella, T.M., Smilowitz, H.M., 2006. Gold nanoparticles: a new X-ray contrast agent. Br. J. Radiol. 79, 248–253.

Hillyer, J.F., Albrecht, R.M., 2001. Gastrointestinal persorption and tissue distribution of differently sized colloidal gold nanoparticles. J. Pharm. Sci. 90, 1927–1936.

Huang, X.L., Zhang, B., Ren, L., Ye, S.F., Sun, L.P., Zhang, Q.Q., Tan, M.C., Chow, G.M., 2008. In vivo toxic studies and biodistribution of near infrared sensitive Au-Au(2)S nanoparticles as potential drug delivery carriers. J. Mat. Sci. Mater. Med. 19, 2581–2588.

Hutter, E., Boridy, S., Labrecque, S., Lalancette-Hebert, M., Kriz, J., Winnik, F.M., Maysinger, D., 2010. Microglial response to gold nanoparticles. ACS Nano 4, 2595–2606.

Illum, L., 2000. Transport of drugs from the nasal cavity to the central nervous system. Eur. J. Pharm. Sci. 11, 1–18.

Imperatore, R., Carotenuto, G., di Grazia, M.A., Ferrandino, I., Palomba, L., Mariotti, R., Vitale, E., de Nicola, S., Longo, A., Cristino, L., 2015. Imidazole-stabilized gold nanoparticles induce neuronal apoptosis: an in vitro and in vivo study. J. Biomed. Mater. Res. 103, 1436–1446.

James, W.D., Hirsch, L.R., West, J.L., Neal, P.D.O., Payne, J.D., 2007. Application of INAA to the build-up and clearance of gold nanoshells in clinical studies in mice. J. Radioanal. Nucl. Chem. 271, 455–459.

Johnston, H.J., Hutchison, G., Christensen, F.M., Peters, S.S., Hankin, V., 2010. A review of the in vivo and in vitro toxicity of silver and gold particulates: particle attributes and biological mechanisms responsible for the observed toxicity. Crit. Rev. Toxicol. 40, 328–346.

Jung, S., Bang, M., Kim, B.S., Lee, S., Kotov, N.A., Kim, B., Jeon, D., 2014. Intracellular gold nanoparticles increase neuronal excitability and aggravate seizure activity in the mouse brain. PLoS One 9, e91360.

Katti, K.V., Raghuraman, K., Rahing, V., Cutler, C., Pandrapragada, R., Katti, K.K., Kattumuri, V., David Robertson, J., Stan, J.C., Jurisson, S., Boote, E., 2006. Hybrid nanoparticles in molecular imaging and therapy. In: Czech. J. Phys – Proc. 15th Radiochemical Conference, 56 (Suppl. D).

Kettiger, H., Schipanski, A., Wick, P., Huwyler, J., 2013. Engineered nanomaterial uptake and tissue distribution: from cell to organism. Int. J. Nanomed. 8, 3255–3269.

Khan, A.K., Rashid, R., Murtaza, G., Zahra, A., 2014. Gold nanoparticles: synthesis and applications in drug delivery. Trop. J. Pharm. Res. 13, 1169–1177.

Khlebtsov, N., Dykman, L., 2011. Biodistribution and toxicity of engineered gold nanoparticles: a review of in vitro and in vivo studies. Chem. Soc. Rev. 40, 1647–1671.

Lasagna-Reeves, C., Gonzalez-Romero, D., Barria, M.A., Olmedo, I., Clos, A., Sadagopa Ramanujam, V.M., Urayama, A., Vergara, L., Kogan, M.J., Soto, C., 2010. Bioaccumulation and toxicity of gold nanoparticles after repeated administration in mice. Biochem. Biophys. Res. Commun. 393, 649–655.

Lee, J.K., Kim, T.S., Bae, J.Y., Jung, A.Y., Lee, S.M., Seok, J.H., Roh, H.S., Song, C.W., Choi, M.J., Jeong, J., Chung, B.H., Lee, Y.G., Jeong, J., Cho, W.S., 2015. Organ-specific distribution of gold nanoparticles by their surface functionalization. J. Appl. Toxicol. 35, 573–580.

Lee, U., Yoo, C.J., Kim, Y.J., Yoo, Y.M., 2016. Cytotoxicity of gold nanoparticles in human neural precursor cells and rat cerebral cortex. J. Biosci. Bioeng. 121, 341–344.

Leitea, P.E.C., Pereira, M.R., Granjeiroa, J.M., 2015. Hazard effects of nanoparticles in central nervous system: searching for biocompatible nanomaterials for drug delivery. Toxicol. In Vitro 29, 1653–1660.

Leroueil, P.R., Berry, S.A., Duthie, K., Han, G., Rotello, V.M., McNerny Daniel, Q., Baker, J.R., Orr, B.G., Banaszak Holl, M.M., 2008. Wide varieties of cationic nanoparticles induce defects in supported lipid bilayers. Nano Lett. 8, 420–424.

Lewis, A.J., Walz, D.T., 1982. Immunopharmacology of gold. Med. Chem. 19, 1–58.

Li, J.J., Zou, L., Hartano, D., Ong, C.N., Bay, B.H., Lanry Yung, L.Y., 2008. Gold nanoparticles induce oxidative damage in lung fibroblasts in vitro. Adv. Mater. 20, 138–142.

Li, J.J., Hartono, D., Ong, C.N., Bay, B.H., Yung, L.Y., 2010. Autophagy and oxidative stress associated with gold nanoparticles. Biomaterials 31, 5996–6003.

Li, W.Q., Wang, F., Liu, Z.M., Wang, Y.C., Wang, J., Sun, F., 2013. Gold nanoparticles elevate plasma testosterone levels in male mice without affecting fertility. Small 9, 1708–1714.

Lin, J.Q., Zheng, Y.G., Zhang, H.W., Chen, Z.A., 2011. Simulation study on nanoscale holes generated by gold nanoparticles on negative lipid bilayers. Langmuir 27, 8323–8332.

Loeschner, K., Hadrup, N., Qvortrup, K., 2011. Distribution of silver in rats following 28 days of repeated oral exposure to silver nanoparticles or silver acetate. Part. Fibre Toxicol. 8, 18.

Manke, A., Wang, L., Rojanasakul, Y., 2013. Mechanisms of nanoparticle-induced oxidative stress and toxicity. BioMed Res. Int. 942916.

Mateo, D., Morales, P., Ávalos, A., Haza, A.I., 2014. Oxidative stress contributes to gold nanoparticle-induced cytotoxicity in human tumor cells. Toxicol. Mech. Methods 24, 161–172.

Murphy, C.J., Sau, T.K., Gole, A.M., Orendorff, C.J., Gao, J., Gou, L., Hunyadi, S.E., Li, T.J., 2005. Anisotropic metal nanoparticles: synthesis, assembly, and optical applications. Phys. Chem. B 109, 13857–13870.

Niidome, T., Akiyama, Y., Shimoda, K., Kawano, T., Mori, T., Katayama, Y., Niidome, Y., 2008. In vivo monitoring of intravenously injected gold nanorods using near-infrared light. Small 4, 1001–1007.

Oberdörster, G., Maynard, A., Donaldson, K., Castranova, V., Fitzpatrick, J., Ausman, K., Carter, J., Karn, B., Kreyling, W., Lai, D., Olin, S., Monteiro-Riviere, N., Warheit, D., Yang, H., 2005a. Principles for characterizing the potential human health effects from exposure to nanomaterials: elements of a screening strategy. Part. Fibre Toxicol. 6 (2), 8.

Oberdörster, G., Oberdorster, E., Oberdorster, J., 2005b. Nanotoxicology: an emerging discipline evolving from studies of ultrafine particles. Environ. Health Perspect. 113, 823–839.

Oberdörster, G., Elder, A., Rinderknecht, A., 2009. Nanoparticles and the brain: cause for concern? J. Nanosci. Nanotech. 9, 4996–5007.

Pal, R., Panigrahi, S., Bhattacharyya, D., Chakraborti, A.S., 2013. Characterization of citrate capped gold nanoparticle quercetin complex: experimental and quantum chemical approach. J. Mol. Struct. 1046, 153–163.

Pan, Y., Neuss, S., Leifert, A., Fischler, M., Wen, F., Simon, U., Schmid, G., Brandau, W., Jahnen-Dechent, W., 2007. Size-dependent cytotoxicity of gold nanoparticles. Small 3, 1941–1949.

Panariti, A., Miserocchi, G., Rivolta, I., 2012. The effect of nanoparticle uptake on cellular behavior: disrupting or enabling functions? Nanotechnol. Sci. Appl. 5, 87–100.

Pernodet, N., Fang, X., Sun, Y., Bakhtina, A., Ramakrishnan, A., Sokolov, J., Ulman, A., Rafailovich, M., 2006. Adverse effects of citrate/gold nanoparticles on human dermal fibroblasts. Small 2, 766–773.

Rezende, T.S., Andrade, G.R.S., Barreto, L.S., Costa Jr., N.B., Gimenez, I.F., Almeida, L.E., 2010. Facile preparation of catalytically active gold nanoparticles on a thiolated chitosan. Mater. Lett. 64, 882–884.

Rinderknecht, A., Elder, A., Prud'homme, R., Gindy, M., Harkema, J., Oberdörster, G., 2007. Surface functionalization affects the role of nanoparticle disposition. Am. J. Respir. Crit. Care Med. 175, A246.

Rivera, G.P., Huhn, D., del Mercato, L.L., Sasse, D., Parak, W.J., 2010. Nanopharmacy: inorganic nanoscale devices as vectors and active compounds. Pharmacol. Res. 62, 115–125.

Sabella, S., Galeone, A., Vecchio, G., Cingolani, R., Pompa, P.P., 2011. AuNPs are toxic in vitro and in vivo: a review. J. Nanosci. Lett. 1, 145–165.

Schaeublin, N.M., Braydich-Stolle, L.K., Schrand, A.M., Miller, J.M., Hutchison, J., Schlager, J.J., Hussain, S.M., 2011. Surface charge of gold nanoparticles mediates mechanism of toxicity. Nanoscale 3, 410–420.

Schaeublin, N.M., Braydich-Stolle, L.K., Maurer, E.I., Park, K., MacCuspie, R.I., Nabiul-Afrooz, A.R.M., Vaia, R.A., Saleh, N.B., Hussain, S.M., 2012. Does shape matter? Bioeffects of gold nanomaterials in a human skin cell model. Langmuir 28, 3248–3258.

Schleh, C., Semmler-Behnke, M., Lipka, J., et al., 2012. Size and surface charge of gold nanoparticles determine absorption across intestinal barriers and accumulation in secondary target organs after oral administration. Nanotoxicology 6, 36–46.

Segal, M.B., 2000. The choroid plexuses and the barriers between the blood and the cerebrospinal fluid. Cell. Mol. Neurobiol. 20, 183–196.

Semmler-Behnke, M., Kreyling, W.G., Lipka, J., Fertsch, S., Wenk, A., Takenaka, S., Schmid, G., Brandau, W., 2008. Biodistribution of 1.4- and 18-nm gold particles in rats. Small 4, 2108–2111.

Siddiqi, N.J., Abdelhalim, M.A., El-Ansary, A.K., Alhomida, A.S., Ong, W.Y., 2012. Identification of potential biomarkers of gold nanoparticle toxicity in rat brains. J. Neuroinflamm. 12, 123.

Sonavane, G., Tomoda, K., Makino, K., 2008. Biodistribution of colloidal gold nanoparticles after intravenous administration: effect of particle size. Colloids Surf. B Biointerfaces 66, 274–280.

Sousa, F., Mandal, S., Garrovo, C., Astolfo, A., Bonifacio, A., Latawiec, D., Menk, R.H., Arfelli, F., Huewel, S., Legname, G., Galla, H.J., Krol, S., 2010. Functionalized gold nanoparticles: a detailed in vivo multimodal microscopic brain distribution study. Nanoscale 2, 2826–2834.

Stelzer, R., Hutz, R.J., 2009. Gold nanoparticles enter rat ovarian granulosa cells and subcellular organelles, and alter in-vitro estrogen accumulation. J. Reprod. Dev. 55, 685–690.

Sung, J.H., Ji, J.H., Park, J.D., Song, M.Y., Song, K.S., Ryu, H.R., Yoon, J.U., Jeon, K.S., Jeong, J., Han, B.S., Chung, Y.H., Chang, H.K., Lee, J.H., Kim, D.W., Kelman, B.J., Yu, I.J., 2011. Subchronic inhalation toxicity of gold nanoparticles. Part. Fibre Toxicol. 14 (8), 16.

Terentyuk, G.S., Maslyakova, G.N., Suleymanova, L.V., Khlebtsov, B.N., Kogan, B.Y., Akchurin, G.G., Shantrocha, A.V., Maksimova, I.L., Khlebtsov, N.G., Tuchin, V.V., 2009. Circulation and distribution of gold nanoparticles and induced alterations of tissue morphology at intravenous particle delivery. J. Biophotonics 2, 292–302.

Treuel, L., Malissek, M., Gebauer, J.S., Zellner, R., 2010. The influence of surface composition of nanoparticles on their interactions with serum albumin. ChemPhysChem 11, 3093–3099.

Trickler, W.J., Lantz, S.M., Murdock, R.C., Schrand, A.M., Robinson, B.L., Newport, G.D., Schlager, J.J., Oldenburg, S.J., Paule, M.G., Slikker Jr., W., Hussain, S.M., Ali, S.F., 2011. Brain microvessel endothelial cells responses to gold nanoparticles: in vitro pro-inflammatory mediators and permeability. Nanotoxicology 5, 479–492.

Trickler, W.J., Lantz-McPeak, S.M., Robinson, B.L., Paule, M.G., Slikker Jr., W., Biris, A.S., Schlager, J.J., Hussain, S.M., Kanungo, J., Gonzalez, C., Ali, S.F., 2014. Porcine brain microvessel endothelial cells show pro-inflammatory response to the size and composition of metallic nanoparticles. Drug Metab. Rev. 46, 224–231.

Walser, T., Hellweg, S., Juraske, R., Luechinger, N.A., Wang, J., Fierz, M., 2012. Exposure to engineered nanoparticles: model and measurements for accident situations in laboratories. Sci. Total Environ. 420, 119–126.

Wang, S.G., Lu, W.T., Tovmachenko, O., Rai, U.S., Yu, H.T., Ray, P.C., 2008. Challenge in understanding size and shape dependent toxicity of gold nanomaterials in human skin keratinocytes. Chem. Phys. Lett. 463, 145–149.

Wang, L., Li, Y.F., Zhou, L., Liu, Y., Meng, L., Zhang, K., Wu, X., Zhang, L., Li, B., Chen, C., 2010. Characterization of gold nanorods in vivo by integrated analytical techniques: their uptake, retention, and chemical forms. Anal. Bioanal. Chem. 396, 1105–1114.

Yah, C.S., 2013. The toxicity of Gold Nanoparticles in relation to their physiochemical properties. Biomed. Res. 24, 400–413.

Yang, D.P., Cui, D.X., 2008. Advances and prospects of gold nanorods. Chem. Asian J. 3, 2010–2022.

van der Zande, M., Vandebriel, R.J., Van, D.E., 2012. Distribution, elimination, and toxicity of silver nanoparticles and silver ions in rats after 28-day oral exposure. ACS Nano 6, 7427–7442.

Zhao, F., Zhao, Y., Liu, Y., Chang, X., Chen, C., Zhao, Y., 2011. Cellular uptake, intracellular trafficking, and cytotoxicity of nanomaterials. Small 7, 1322–1337.

Zlokovic, B.V., 2008. The blood–brain barrier in health and chronic neurodegenerative disorders. Neuron 57, 178–201.

Chapter 9

The Applications, Neurotoxicity, and Related Mechanisms of Manganese-Containing Nanoparticles

K. Ikeda, T. Tachibana, Y. Manome
The Jikei University School of Medicine, Tokyo, Japan

Chapter Outline

Neurotoxicity of Nanomaterials and Nanomedicine. http://dx.doi.org/10.1016/B978-0-12-804598-5.00009-X

1. CHEMICAL PROPERTIES AND APPLICATIONS OF MANGANESE

1.1 Chemical Properties of Manganese

Manganese, with the chemical symbol Mn and atomic number of 25, is a hard and brittle metal belonging to group 7 and d-block in the periodic table (Atkins et al., 2014; Barceloux, 1999). Manganese compounds have both oxidizing and antioxidizing properties with the following oxidation states and chemical characteristics:

1. Manganese exhibits oxidation states from −3 to +7, the most common being +2, +3, +4, +6, and +7.
2. The manganese compounds with the +2 oxidation state (Mn (II)) are manganese chloride ($MnCl_2$), manganese carbonate ($MnCO_3$), and manganese sulfate ($MnSO_4$).
3. The manganese compounds with an oxidation state of +3 are manganese acetate ($Mn(OAc)_3$) and manganese (III) oxide (Mn_2O_3).
4. Manganese dioxide (MnO_2), which catalyzes H_2O_2 into $H_2O + O_2$, is a well-known manganese compound with a +4 oxidation state.
5. Permanganate, with an oxidative state of +7 (Mn (VII) compounds), is purple in color and used to make glasses violet (Patrick, 1998). The commonly known permanganate compounds are potassium permanganate ($KMnO_4$), sodium permanganate ($NaMnO_4$), and barium permanganate ($BaMn_2O_8$). Another manganese compound with the +7 oxidation state is manganese oxide (Mn_2O_7).

The oxidation state of manganese contributes to manganese toxicity. For example, manganese (II) compounds can be both nontoxic and toxic, whereas manganese (VII) compounds are highly oxidizing and poisonous (Atkins et al., 2014; Barceloux, 1999; Carmona et al., 2010, 2014).

1.2 Applications of Manganese in Industrial, Research, and Medical Purposes

Manganese is widely used in daily life, especially in combination with iron, eg, to construct steel and aluminum alloys and as a component of magnetic substances. Manganese per se is rather brittle, but stronger and more corrosion resistant as an iron or aluminum alloy (Atkins et al., 2014). Owing to these features, manganese is used to produce alloys with steel with varying manganese contents and aluminum drink cans, among other applications. Moreover, manganese is often used as manganese dioxide (MnO_2) and permanganate (MnO_4^-). MnO_2 exerts catalytic activity on H_2O_2 and is used in cathodes of zinc-carbon, alkaline, and lithium batteries (Dell, 2000; Liu et al., 2013; Rutz, 2009). Especially, MnO_2 nanoparticle for lithium batteries was synthesized in different forms and considered to have high theoretical capacity, environmental benignity, low cost, and special properties (Liu et al., 2013). Manganese was used as a coloring agent for ceramics and glasses (Atkins et al., 2014; Patrick, 1998). Manganese is also added to nonlead gasoline as an organic compound, methylcyclopentadienyl manganese tricarbonyl (MMT), to improve antiknocking performance. Other usage of manganese nanoparticles is a nanosponge Mn_2O_3 adsorbent for preconcentration of palladium (II) and rhodium (III) in sea water, waste water, rock, street sediment, and catalytic converter samples (Yavuz et al., 2014). In addition, manganese ferrite nanoparticles ($MnFe_2O_4$) strengthen the adsorption of sulfamethoxazole from water and higher regeneration efficiency by H_2O_2 (Wan et al., 2014).

Manganese is involved in various physiological functions in humans and animals. The metal acts as a cofactor for numerous enzymes in living organisms, such as the oxygen-evolving complex in photosynthesis (Najafpour et al., 2014), manganese catalase, which catalyzes hydrogen peroxide, manganese superoxide dismutase (Mn-SOD), which converts O_2^- into H_2O_2 and decreases oxidative stress, arginase generating ornithine and urea from arginine, and manganese ribonucleotide reductase, which converts RNA to DNA (Bresciani et al., 2015; Huang et al., 2014; Law et al., 1998). In addition, manganese contributes to certain metabolic and psychological states. One of the major toxic mechanisms in biological systems is oxidative stress, characterized by an imbalance in reactive oxygen species (ROS) production and antioxidant defense. Oxidative stress leads to cellular and tissue damage; reaction with DNA, lipids, and proteins; and various pathological states, such as senescence, colon cancer, cardiovascular disease, and Alzheimer's disease (Bresciani et al., 2015; Hybertson et al., 2011; Tao et al., 2014). Oxygen is converted to superoxide anion (O_2^-), also known as "free radicals" or ROS, in mitochondria during energy production or ionizing radiation and in activated leukocytes during inflammation (Hosseinimehr, 2015; Mokgobu et al., 2012). Mn-SOD enzymes catalyze superoxide, which is mainly produced in mitochondria and triggers oxidative stress in animal cells, thus constituting an essential defense against superoxide-induced oxidative stress (Indo et al., 2015). Manganese is one of the main trace elements with a known protective effect against radiation-induced DNA damage (Hosseinimehr, 2015).

Dietary manganese deficiency appears to affect manganese homeostasis in the brain, followed by alteration of neural activity (Takeda, 2003). Experimentally, an enhanced manganese diet was shown to protect neural cells against apoptosis in a scrapie-infected Tga20 mouse model (Bolea et al., 2010). Manganese is required for the synthetic pathway of cholesterol and bone metabolism (Food and Nutrition Board, 2001), the biosynthesis of mucopolysaccharides in bone matrix formation, and as a cofactor for several enzymes in bone tissue (Palacios, 2006). Long-term manganese and copper deficiencies may result in decreased osteogenesis and osteoclast activity (Strause et al., 1987), and decreased plasma manganese concentration is possibly related to osteoporosis (Food and Nutrition Board, 2001; Palacios, 2006). Reports showed that manganese supplementation improves the calcium content in adult antlers (Cappelli et al., 2015), supporting the theory that manganese may be required to maintain normal bone metabolism. However, Whelan et al. (2006) suggested that evidence on specific effects of manganese on osteoporosis is incomplete, because the results were obtained from trials on manganese supplementation in combination with other minerals. Moreover, manganese supplementation improved both hypocholesterolemia in humans and the atherogenic index in Wistar rats fed a high-salt diet (Food and Nutrition Board, 2001; Muhammad et al., 2012).

In addition to the above-mentioned disorders, manganese supplementation is reported to improve salt-induced hypertension, insulin resistance, and type 2 diabetes mellitus via enhancing insulin secretion (Lee et al., 2013; Muhammad et al., 2012). Psychologically, intake of low-dose manganese (1 mg/day) increased mood and pain symptoms during the premenstrual phase, implying a symptomatological role in menstrual distress (Bendich, 2000; Penland and Johnson, 1993).

Manganese nanoparticles are also employed as a magnetic resonance imaging (MRI) probe (Briley-Saebo et al., 2014; Rutz, 2009; Yang et al., 2012; Zhen and Xie, 2012). One study advocated the usage of manganese-containing nanoparticles as an intracellular image enhancer for MRI of human atherosclerotic lesions (Briley-Saebo et al., 2014). In addition, manganese nanoparticle compound with gold (gold-manganese oxide nanoparticle) is suggested as a unique enhanced positive contrast agent for in vitro or in vivo applications (Li et al., 2014). Another potential medical usage of manganese nanoparticle is to apply for the magnetic fluid hyperthermia against cancer because it is reported that nanosized manganese zinc ferrite, which is used for the magnetic fluid hyperthermia, may have unique beneficial effects, such as universality of applicable tumor cells, high specific absorption rate, thermal bystander effects, small size of targeting effects, hyperthermia temperature uniformity, automatic temperature control, and constant temperature effect (Lin et al., 2014). Furthermore, another report showed that generation 4 PEGylated poly(amidoamine) dendrimer with gold or manganese nanoparticle protected delivered RNA from RNase degradation and improved anticancer RNA delivery in experimental cancer treatment (Reyes-Reveles et al., 2013). These reports indicate that manganese nanoparticle may be a useful agent in future anticancer therapies.

In another usage of manganese, potassium permanganate is applied as a fixative for light and electron microscopy (Luft, 1956). In agriculture, manganese is used as an organic compound, the fungicidal pesticide Maneb (International Union of Pure and Applied Chemistry name: manganese, [[2-[(dithiocarboxy)amino]ethyl]carbamodithioato(2-)-kS, kS']; chemical name: manganese ethylenebis[dithiocarbamate]). Maneb is often used in combination with *N,N′*-dimethyl-4,4′-bipyridinium dichloride (Paraquat) to generate an animal model of Parkinson's disease, owing to its blood–brain barrier (BBB) permeability and ability to induce apoptosis through Bak activation (Cicchetti et al., 2009). Another unusual application of manganese is minting coins. Because of the decrease in key metals for coin minting, such as nickel (used for stainless steel) and copper (used for ammunition casings), the US Mint changed the composition of the Jefferson nickel in 1942 to copper, silver, and manganese during the period of World War II (Kuwahara et al., 2001).

2. ENVIRONMENTAL AND OCCUPATIONAL EXPOSURE TO MANGANESE

2.1 Overview

As discussed earlier, manganese has numerous wide-ranging applications, for instance, as a component of steel, aluminum alloys, and an additive of gasoline. Manganese exposure and ensuing toxicities occur in a variety of environmental settings, such as welding plants, manganese mines, expressways exposed to MMT exhaust fumes, nutritional sources, contaminated foods, infant formulae, water, soil, and air with natural or man-made contaminants (Bouchard et al., 2011; Elsner and Spangler, 2005; Erikson et al., 2008; O'Neal and Zheng, 2015; Zatta et al., 2003). Environmental exposure to airborne manganese may be attributable to emission of combusted exhaust fumes of MMT additive-containing gasoline (Lynam et al., 1999), working in manganese mines (Gonzalez-Cuyar et al., 2014), and fumes of metal arc welding in manganese alloy production plants and the construction industry (Meeker et al., 2007; Pesch et al., 2012; Taube, 2013).

The US Occupational Safety and Health Administration (OSHA, https://www.osha.gov/) and National Institute of Occupational Safety and Health (http://www.cdc.gov/niosh/) have established guidelines for manganese exposure. The permissible exposure limit (PEL) for manganese and its compounds is a ceiling level of 5 mg/m^3 as fumes, and PEL-time-weighted average (TWA) of 1 mg/m^3 and PEL-short-term exposure limit (STEL) of 3 mg/m^3 as dust. The recommended exposure limit (REL)-TWA for manganese compounds and fumes is 1 mg/m^3 and REL-STEL is 3 mg/m^3. The level of manganese immediately dangerous to life and health has been established as 500 mg/m^3 by the OSHA (Barceloux, 1999).

2.2 Environmental Exposure in Industry Concerning Manganese

In welding plants, the diameter of median airborne particles by arc welding is between 0.2 and 0.6 μm, with less than 10% or 30% of the fume mass being larger than 1 μm (Taube, 2013). A case of airborne manganese exposure was reported in a gun shooting range (Lach et al., 2015). Airborne manganese particles, which can be deposited in the alveolar range, may be generated in such conditions and become toxic at diameters 40 nm in size. On emissions from such sources, airborne manganese particles may exert toxicity. Elevated levels of manganese have been detected in the blood and axillary hair of manganese-exposed workers (Bader et al., 1999). Previous studies reported neurotoxic effects of manganese fumes in shipyard and welding workers (Halatek et al., 2005; He and Niu, 2004). Consistent with these findings, Criswell et al. (2012) provided evidence of potential neurotoxicity of manganese in arc welding workers in shipyards using MRI and showed elevated indices associated with diffusion-weighted abnormalities in the pallidum and anterior putamen, suggestive of neurotoxicity in these regions.

Mean microglial density was increased in the globus pallidus external and internal segments of manganese mine workers, compared with those of non-manganese mine workers. The number of years spent working in the manganese mines was significantly correlated with microglial density in the globus pallidus internal segment. In addition, the ratio of astrocytes to microglia in each brain region (caudate, putamen, the globus pallidus external and internal segments) was lower in manganese than non-manganese mine workers, indicating a trend toward lower astrocyte and neuron densities in the striatum and inflammatory microglial response in the globus pallidus of manganese-exposed mine workers (Gonzalez-Cuyar et al., 2014).

Photocopiers in commercial photocopy centers present another means of occupational exposure to nanoparticles containing manganese (Martin et al., 2015). Photocopiers may emit high levels of nanoparticles in real workplace settings. Toner particles carry airborne ultrafine particulate matter with diameters less than 0.1 μm ($PM_{0.1}$). Elemental analysis of copy toner particles revealed iron as the most abundant metal element, along with the presence of other elements, such as titanium, sulfur, manganese, zinc, copper, nickel, niobium, and cerium. The ratio of manganese included in toner particles ranges from 0.01% to 0.12%.

2.3 Environmental Exposure Caused by MMT Exhaust

Usage of MMT, an additive of nonlead gasoline as an octane enhancer in automobiles, leads to increased airborne manganese exposure in cities and areas along expressways (Loranger et al., 1996; Lynam et al., 1999). The majority of manganese emitted from tailpipes of cars using MMT-containing gasoline is associated with particles smaller than 2.5 μm in size ($PM_{2.5}$). However, exposure of the general population to manganese, eg, in Toronto, is well within the safe limits determined by the US Environmental Protection Agency and other standard setting organizations worldwide (Lynam et al., 1999).

3. MANGANESE-ASSOCIATED NEURODISORDERS AND PATHOPHYSIOLOGY OF MANGANESE NEUROTOXICITY

3.1 Manganese and Neurological Disorders

Manganese has been identified as the causative agent of neuronal disorders, including manganism (or manganese-induced parkinsonism), Huntington's disease, amyotrophic lateral sclerosis (ALS), Alzheimer's disease, Parkinson's disease (Bowman et al., 2011; Chin-Chan et al., 2015; Farina et al., 2013; Guilarte, 2011; Kwakye et al., 2015; Normandin and Hazell, 2002; O'Neal and Zheng, 2015; Olanow, 2004; Takeda, 2003; Zatta et al., 2003), and psychiatric syndrome (compulsive behavior, emotional lability, hallucinations) (Barceloux, 1999). In addition, overexposure of manganese can lead to adverse cognitive, intellectual, motor, and behavioral effects during childhood (Bouchard et al., 2011; Claus Henn et al., 2014; Grandjean and Landrigan, 2014; Zoni and Lucchini, 2013). Potassium permanganate, which includes manganese at an oxidation state of +7, is known to exert systemic and fatal toxic effects (Kwakye et al., 2015).

3.2 Mechanisms of Neurological Disorders by Manganese

Manganese neurotoxicity mechanisms involve stress to the endoplasmic reticulum (Xu et al., 2013), accumulation of manganese in Golgi apparatus (Carmona et al., 2010, 2014), disruption of the glutamine/glutamate-γ-aminobutyric acid (GABA) cycle neurotransmitter systems in human and nonhuman primates (Burton and Guilarte, 2009; Sidoryk-Wegrzynowicz and Aschner, 2013a,b), which influence GABA release in the striatum (Takeda et al., 2003), and alterations in glutamate transporter expression in astrocytes (Karki et al., 2015). Astrocytes, which contribute to neural function and survival by production of neurophilic factors, express glutamate transporters, such as excitatory amino acid transporter (EAAT) 1 and 2 (also known as glutamate aspartate transporter [GLAST] and glutamate transporter 1 [GLT-1] in rodents, respectively), which take up excess glutamate from the extracellular space. Knockout of GLAST and GLT-1 in rats, corresponding to EAAT 1 and 2 in humans, resulted in decreased glutamate transport and a marked increase in glutamate concentrations in dialysates of extracellular fluid of brain, along with manifestation of motor syndrome and seizure, implying progressive motor impairment (Rothstein et al., 1996). Activity of EAAT 2, in particular, accounts for more than 90% extracellular glutamate clearance, and its reduction is associated with numerous neurodisorders, including ALS, Alzheimer's disease, Parkinson's disease, schizophrenia, and epilepsy. Manganese has been shown to suppress EAAT 1 and 2 expression in astrocytes in an epigenetic manner via upregulation of the transcription factor, YY1, subsequently leading to reduced glutamine uptake from the extracellular fluid (Karki et al., 2015).

Manganese concentrations in brain were elevated in experimental rats receiving end-to-side portacaval anastomosis and exposed to manganese phosphate via inhalation as well as patients with hepatic cirrhosis who died of hepatic coma (Rose et al., 1999; Salehi et al., 2001), indicating that manganese exerts neurotoxicity in chronic liver disease with portacaval anastomosis. Butterworth (2014) reviewed the association of manganese with hepatic encephalopathy in patients with alcoholic liver cirrhosis. Their study revealed an increase in plasma concentrations of manganese. A highly consistent MRI finding in cirrhotic patients was bilateral symmetric hyperintensities in globus pallidus on T1-weighted imaging, which may be the result of pallidal manganese deposition. This hyperintensity in globus pallidus on T1-weighted imaging is associated with the tissue content of manganese in globus pallidus samples obtained at autopsy from patients with liver cirrhosis, Alagille syndrome, an autosomal dominant disorder characterized by cholestasis, intrahepatic bile duct paucity, and end-stage liver disease, compared with age-matched controls. In these patients, excretion of manganese, which normally occurs through the hepatobiliary route, is markedly impaired. Similar pallidal signal hyperintensities in brain MRI have been described in patients receiving total parental nutrition, by which a large amount of manganese has been supplied. Consequently, manganese that accumulates in the brain contributes to basal ganglia dysfunction and is a strong contender for the basis of associated extrapyramidal abnormalities (Butterworth, 2014; Guilarte, 2011; Hazell, 2002). Mutations in one of the manganese exporters, SLC30A10 (solute carrier family 30, member 10), which transports excess manganese from the cytoplasm to extracellular space, result in manganese-induced toxicity. Mutant SLC30A10 was previously identified in patients with hepatic cirrhosis, dystonia, polycythemia, and hypermanganesemia, concomitant with high brain manganese levels. Although none of the patients containing mutant SLC30A10 genes were exposed to excessive manganese, they manifested symptoms consistent with Parkinson's disease. In addition, overexpression of wild-type SLC30A10 resulted in protection against manganese-induced lethality and dopaminergic neurodegeneration as well as locomotion deficits in *Caenorhabditis elegans* (Chen et al., 2014; Kwakye et al., 2015), indicating that genetic impairment constitutes one of the mechanisms of manganese-induced neurotoxicity.

The majority of manganese in brain is bound to glutamine synthetase predominantly expressed in astrocytes that catalyzes the conversion of glutamate to glutamine and is a key enzyme in the metabolism of the major excitatory neurotransmitter, glutamate. In addition, manganese may exist in synaptic vesicles in glutamatergic neurons (Kwakye et al., 2015; Takeda, 2003). Reduced glutamine synthetase activity possibly enhances glutamate trafficking and glutamatergic signaling, leading to excitotoxicity (eg, excessive glutamate at the synapse). Manganese deposition in the basal ganglia and manganese intoxication results in alteration of glutamatergic neurotransmission, hyperactivity of corticostriatal neurons and subsequent enhancement of the frequency and amplitude of spontaneous excitatory postsynaptic potential in the striatum, elevated synaptic glutamate levels, and reduced glycine interactions with the *N*-methyl-D-aspartate subtype of glutamate receptors in mouse brain (Kwakye et al., 2015).

One of the potential mechanisms underlying neurotoxicity to dopaminergic cells is oxidative stress. Earlier in vitro studies revealed that exposure of manganese chloride triggers increased ROS accumulation in rat neonatal astrocytes. Oxidative stress induces DNA damage as well as lipid and protein oxidation and is an important feature of neurodegeneration, suggesting that endogenous antioxidant protective pathways are inadequate or overwhelmed (Giordano et al., 2014), leading to increased oxidative DNA damage in the brains of affected patients (Coppedè and Migliore, 2015).

Manganese-induced neurotoxicity may also involve forkhead winged-helix transcription factor O (FoxO) levels and its translocation, an important regulator of the cellular stress response that promotes antioxidant defense (Klotz et al., 2015), into nuclei of astrocytes (Exil et al., 2014). FoxO is critical for cell survival and involved in direct transcriptional regulation of the antioxidant Mn-SOD, leading to reduction of oxidative stress (Kops et al., 2002). Oxidative stress triggers FoxO activation, which promotes the transcription of known antioxidant proteins, selenoprotein P and Mn-SOD, via inhibition of wingless protein signaling, consistent with colocalization of selenoprotein P with neurofibrillary tangles and β-amyloid plaques (eg, Alzheimer's lesions) and increased Mn-SOD expression in neurons under oxidative stress conditions. However, inhibition of wingless signaling may result in activation of glycogen synthase kinase 3 (GSK-3). Increased GSK-3 activity is associated with neuronal dystrophy and generation of neurofibrillary tangles, and may trigger increased τ protein phosphorylation, β-amyloid protein synthesis, and neuronal apoptosis (Bellinger et al., 2008; Manolopoulos et al., 2010). Anantharam et al. (2002) reported that MMT exposure induces a rapid increase in ROS generation within 5–15 min, followed by release of mitochondrial cytochrome C into the cytoplasm and subsequent activation of the cysteine proteases caspase-9 (two- to threefold) and caspase-3 (15- to 25-fold), but not caspase-8, in a time- and dose-dependent manner. Both peroxidase and catalase activities in substantia nigra were decreased in autopsy cases of Parkinson's disease (Ambani et al., 1975), indicating that ROS activity in the brain is enhanced in these cases. Studies have shown that *N*-acetylcysteine amide exerts protective effects against manganese-induced neurotoxicity by inhibiting lipid peroxidation, scavenging ROS, and preserving intracellular glutathione, the most abundant and crucial intracellular thiol antioxidant in the central nervous system, and mitochondrial membrane potential (Maddirala et al., 2015). These results verify the involvement of ROS in manganese-induced neurotoxicity. The groups of Latronico et al. (2013) and Exil et al. (2014) reported a potentially important role of manganese (II) in neurotoxicity as an oxidant in the central nervous system, despite its known antioxidant property. Briefly, exposure of rat astrocyte cells to manganese chloride induced expression of matrix metalloproteinase-9 in a dose-dependent manner, resulting in a significant increase in ROS and decrease in SOD activity (Latronico et al., 2013) or increased oxidative stress with no decrease in SOD-2 protein and glutathione peroxidase levels (Exil et al., 2014), signifying increased oxidative stress in the central nervous system on overexposure of astrocytes to manganese.

Clinically, one of the putative causes of manganese-induced parkinsonism is failure of conversion of levodopa to dopamine in dopaminergic neurons. Degeneration of dopaminergic neurons is one of the well-known causes of parkinsonism and Parkinson's disease. Interestingly, patients with extrapyramidal deficits caused by chronic manganese intoxication failed to respond to levodopa, despite the dramatic response of patients with idiopathic Parkinson's disease (Lu et al., 1994). In addition, Tong et al. (2014) reported significant correlation of manganese with Mini-Mental State Examination and Clinical Dementia Rating Scale scores, and showed increased plasma β-amyloid peptide levels with elevated manganese, indicating that high plasma concentrations of manganese are an essential pathogenic factor in the progression of Alzheimer's disease. In addition, although the plasma concentrations of manganese did not differ significantly between patients with ALS and controls (Garzillo et al., 2014), Roos et al. (2013) reported increased concentrations of manganese and other metals in blood–cerebrospinal fluid (CSF) of patients with ALS. Another study on an animal model of ALS, TDP-43 mice carrying the disease-causing A315T mutation ($TDP\text{-}43^{A315T}$), also showed increased levels of zinc, copper, and manganese in the spinal cord, but not brain, supporting the theory that the manganese level in CSF potentially contributes to pathogenesis of ALS.

4. NEUROTOXICITY MECHANISMS OF MANGANESE NANOPARTICLES

4.1 Manganese Nanoparticles

Nanotechnology has led to the development of nanosized particles or nanoparticles (1–1000 nm in size) that have been successfully applied in medical contexts (De Jong and Borm, 2008; Feng et al., 2015b). Nanoparticles are used in industrial products, including medical as well as dental materials, such as light polymerization composite resins and bonding systems, coating materials for dental implants, bioceramics, endodontic sealers, and mouthwashes (Feng et al., 2015a). The wide usage of nanoparticles in daily life can lead to various problems, because engineered nanoparticles have unusual physicochemical properties due to their small size (surface area and size distribution), chemical composition (purity, crystallinity, electronic properties), surface structure (surface reactivity, surface groups, inorganic or organic coating), solubility, shape, and aggregation (Nel et al., 2006). Metal oxide nanoparticles, such as titanium dioxide, silicon dioxide, zinc oxide, and copper oxide, and silver, manganese, copper, and iron nanoparticles often induce neurotoxicity or neuroinflammation (Migliore et al., 2015; Wang et al., 2009).

A typical image of manganese nanoparticles (MnO_2) synthesized according to a previous study (Luo, 2007) is shown in Fig. 9.1. Briefly, potassium permanganate (Wako Pure Chemical Industry Ltd., Tokyo, Japan) solution was prepared in ultrapure water and poly(allylamine hydrochloride) (PAH, Mw 450,000, Sigma–Aldrich Co., St. Louis, MO, USA) solution at a concentration of 20 mmol/L. Then the PAH solution was further diluted by 30 times to adjust the total amount of PAH as the previous report (Luo, 2007); 20 mmol/L potassium permanganate and diluted PAH solution were mixed with a 1:4 volume

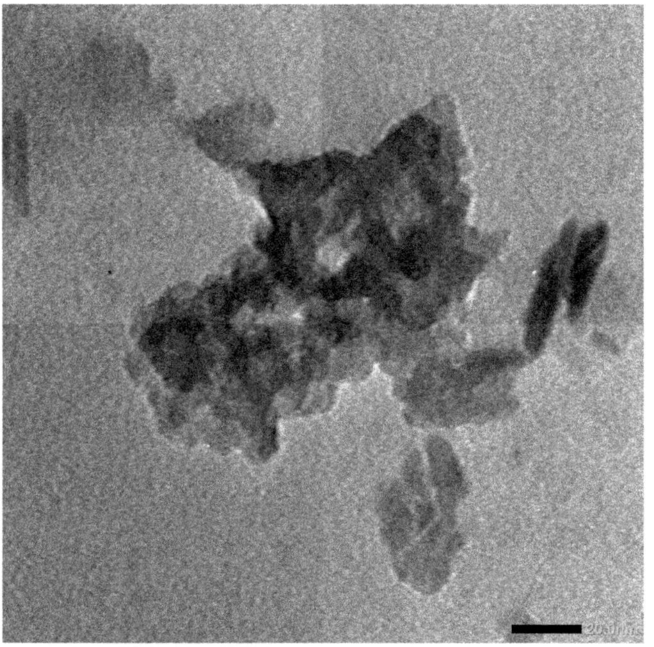

FIGURE 9.1 Typical image of MnO_2 nanoparticles under a transmission electron microscope synthesized according to a previous report. The size bar indicates 20 nm.

ratio. Samples were incubated at room temperature for 5 min and further diluted 10 times with ultrapure water. A drop of sample colloid was placed on the carbon-coated copper grid and air dried at room temperature. The prepared sample was examined under a transmission electron microscope (JEM-1400Plus, JEOL Ltd., Tokyo, Japan) operated at 100 kV, and nanoparticle images acquired.

4.2 Transport of Manganese Into Central Nervous System

The BBB, a specialized structure composed of endothelial cells, astrocyte end feet, pericytes, and tight junctions between the cerebral endothelial cells, forms a diffusion barrier (referred to as zonulae occludentes) that selectively excludes most blood-borne substances from entering the brain, impedes influx of the majority of compounds from blood to brain, and plays a pivotal role in homeostasis of the central nervous system, fibrinolysis, coagulation, vasotonus regulation, and blood cell activation and migration during physiological and pathological processes. Substance transport of BBB is different from that of other organs in that metabolic barriers, such as deficiency in pinocytic vesicles causing low pinocytosis of specific compounds, and a series of intra- and extracellular enzymes limit transport through the BBB. Nanoparticles are transferred across by their carriers, which are called nanocarriers, without any damage to the BBB. Various factors, such as the type of polymer or surfactant used, particle size, and the drug molecule, also affect the transport via the BBB (Jain, 2012). Metal ions are conveyed into brain through BBB by the transporters, divalent metal transporter 1, divalent cation transporter 1, and natural resistance-associated macrophage protein 2, which are specific for divalent ions, but not other oxidation states. In addition, for some metal ions, especially iron (Fe^{2+} and Fe^{3+}), transportation through the BBB may be mediated by transferrin receptor–mediated endocytosis, which is also believed to play a role in the transport of other trivalent metal ions through the BBB, such as aluminum and manganese. However, it seems unlikely that divalent metal transporter 1 is significantly involved in manganese transport into brain (Yokel, 2006). Efflux systems, such as p-glycoprotein of endothelial cells, play an important role in the elimination of harmful molecules (Ballabh et al., 2004; Feng et al., 2015a; Zheng et al., 2003). For most solutes and macromolecules, permeability across the BBB depends on lipophilicity and size. Small hydrophilic materials, such as nutrients or metal ions, and macromolecules required for proper brain function cross the BBB through specific carrier mechanisms or facilitated diffusion (Zheng et al., 2003). In addition to the BBB, substances entering the brain are restricted by the blood–CSF barrier consisting of choroid plexus epithelial cells, which possess tight junctions similar to BBB. However, tightness of the choroid plexus in blood–CSF barrier is not as complete as that for cerebral capillary endothelial cells, leading to passage of very hydrophilic drugs and toxicants to the brain via the choroid plexus (Feng et al., 2015a; Zheng et al., 2003). Despite stringent protective mechanisms, essential hydrophilic metal ions, such as calcium, copper, magnesium, manganese, iron, zinc, cobalt, and molybdenum, are imported via the BBB or blood–CSF barrier for maintaining homeostasis of the central nervous system. Therefore a series of active or receptor-mediated transport systems function to control the transport of each metal ion into the brain and maintain their optimal concentrations (Albrecht et al., 2007). Nanoparticles of such metals, in particular, manganese, also translocate into brain tissues directly through the BBB via the olfactory route or from disrupted neural cell membranes, and often cause inflammatory responses in vivo and oxidative stress (Feng et al., 2015b; Hanley et al., 2009; Karmakar et al., 2014; Mofid et al., 1997; Win-Shwe and Fujimaki, 2011).

In a study by Máté et al. (2016), inhalation of manganese nanoparticles resulted in marked accumulation of the metal in blood, brain, and lung, with significant correlation between the time spent with immobility during the open field test and brain manganese levels, and abnormal electrocorticogram in rats. Previously, Takács et al. (2012) investigated manganese accumulation in male Wistar rats wearing chronically implanted cortical electrodes and exposed to manganese-containing nanoparticles (MnO_2, particle size: 10–65 nm, average size: 30.9 ± 9.9 nm in diameter) via the airway. Instillation of manganese in rat trachea resulted in an elevation of manganese concentrations in blood and significant accumulation in brain tissue, indicating direct transfer through the BBB. In the central nervous system, divalent and trivalent manganese are supplied to the brain via both the BBB and blood–CSF barrier through transferrin-receptor mediated transporter system or other mechanisms. Manganese may be absorbed as a divalent metal in the gut and oxidized by ceruloplasmin in the plasma. Absorbed manganese is stored in the liver and supplied to the brain via two putative pathways through the BBB. One pathway is via receptor-mediated endocytosis of transferrin-bound manganese. Manganese translocated in brain is further taken up via transferrin receptor–mediated endocytosis or a manganese-specific transporter (Takeda, 2003). A previous study showed that manganese nanoparticles are transferred from olfactory epithelium to brain through the olfactory bulb in addition to the transport pathway via the BBB (Moshkin et al., 2014; Takeda, 2003), indicating that exposure to airborne manganese nanoparticles may induce neurotoxicity. However, the transmission route of manganese nanoparticles to brain via the olfactory bulb is controversial (Oszlánczi et al., 2010).

4.3 Manganese Nanoparticles and Neurotoxicity and Perspectives of Manganese Nanoparticle

Chronic exposure to manganese compounds, such as manganese phosphate from MMT, used in nonlead gasoline and sources of airborne manganese (Lynam et al., 1999), leads to accumulation of manganese in the olfactory bulb, central nervous system, and other peripheral tissues, especially lung (Normandin et al., 2002). Moreover, chronic exposure to manganese sulfate produced by combustion of MMT is reported to induce a dose-dependent decrease in spatial learning, memory ability, and spontaneous activity; overt morphological changes in hippocampi; cellular degeneration and death; and significant increase in Hsp70 mRNA levels of hippocampal cells in rats (Su et al., 2015). Overexposure to manganese tetraoxide (Mn_3O_4) resulted in significant accumulation in pigeon brain, but with no changes in various biochemical parameters (Sierra et al., 1998).

Owing to their widespread usage, nanomaterials have potentially hazardous effects on the central nervous system (Feng et al., 2015b). One possible mechanism of manganese nanoparticle–mediated cytotoxicity was reported in the RAW264.7 murine macrophage cell line using charged manganese ferrite nanoparticles (Yang et al., 2012). Briefly, cationic manganese ferrite nanoparticles promote contact with the cellular membrane, leading to increased cellular uptake, suggesting that charges may be involved in cytotoxicity. Manganese nanoparticles (40 nm) may also be toxic to dopamine-producing cells, in view of the finding that tyrosine hydroxylase and park2 mRNA (encoding Parkin protein) are suppressed and Sna mRNA is upregulated. Because tyrosine hydroxylase is a key enzyme in dopamine production in PC12 rat pheochromocytoma cells, suppression of its expression results in depletion of dopamine (Hussain et al., 2006; Wang et al., 2009). Parkin protein is an E3 ubiquitin ligase, a component of the ubiquitin-proteasome system that identifies and targets misfolded proteins, such as a-synuclein and Gpr37, to the proteasome for degradation and protects cells from unfolded protein stress (Imai et al., 2000; Wang et al., 2009), possibly leading to loss of dopaminergic neural cells. In addition, manganese chloride, manganese sulfate, and MMT, with oxidation states of +2, +2, and +1, respectively, but not Mn_2O_3, exerted similar dose-dependent toxicities on accumulation in the Golgi apparatus of PC12 rat pheochromocytoma cells. Maneb, with a manganese oxidation state of +2, exerts high toxicity in PC12 rat pheochromocytoma cells (Carmona et al., 2010, 2014). These toxic manganese compounds are water soluble. Accordingly, Carmona et al. (2014) speculated that the toxic actions of manganese are attributed to its solubility, and consequent accumulation of manganese in Golgi apparatus may be the result of detoxification via exocytosis. Furthermore, another study revealed that manganese nanoparticle was taken up into the rat dopaminergic cell, N27 cells, and uptake of manganese nanoparticles resulted in upregulation of transferrin, a transporter protein of manganese; induction of oxidative stress and autophagy of dopaminergic neuronal cells; dopaminergic cell death; decreased number of dopaminergic neurons; and reduction of neurite cell length (Afeseh Ngwa et al., 2011). The data collectively indicate that the pathophysiological cause of manganese-induced Parkinson's disease may be partially attributable to alterations in expression of genes associated with the dopaminergic system and protection against unfolded protein stress in dopaminergic cells.

As mentioned earlier, manganese nanoparticles have both beneficial and adverse properties, such as neurotoxicity, which may cause the neurodegenerative disorders. But some studies showed the possibility of more beneficial use of manganese nanoparticles, eg, usage for anticancer treatment (Lin et al., 2014). Therefore investigation on the mechanism of toxicity of manganese nanoparticles and the development of prevention or therapeutic methods against its toxic effects is required.

5. SUMMARY AND CONCLUSIONS

Manganese, with the chemical symbol Mn and atomic number 25, is a metal element belonging to d-block and group 7 in the periodic table. Manganese ore is mainly mined as pyrolusite ($Mn(IV)O_2$) in South Africa. Manganese in organic or inorganic compounds has various oxidation states, from −3 to +7, the most common being +2, +3, +4, +6, and +7, and is used in several industrial products in daily life, including medical and dental applications. Manganese nanoparticles, including hybrid nanoparticles, such as manganese-zinc ferrite and gold-manganese nanoparticles, are used as a part of devices, ie, cathodes of lithium batteries, pollutant absorption material, MRI probes, and anticancer agent.

Manganese compound emissions are harmful to individuals who work in places contaminated with manganese fumes and dust, and those who live along expressways exposed to MMT exhaust fumes. Environmental or occupational overexposure to manganese (including nanoparticles) from automobiles combusting MMT, an organic manganese compound, fumes and dust from welding plants, the construction industry, manganese mines, and commercial photocopy centers may lead to health problems.

Manganese compounds act as both oxidants and antioxidants in vivo, and therefore sometimes exert toxicity to organs, especially components of the central nervous system. Neurotoxicity of manganese may be attributed to increased oxidative stress, direct cell accumulation, direct cytotoxic effects on neural cells (especially dopaminergic cells), interference with enzymatic activity of neural cells, such as glutamate synthetase, and disruption of glutamate transportation to astrocytes. Manganese nanoparticles may be translocated through the BBB and blood–CSF barrier or directly transferred to brain via the olfactory bulb, and trigger neurotoxicity leading to neural degeneration, parkinsonism, Alzheimer's disease, ALS, and other disorders.

Owing to the beneficial properties and neurotoxicity of manganese compounds, further investigation on adverse effects of manganese nanosized compounds for usage in daily and medical purpose is required.

ACKNOWLEDGMENTS

The authors thank Yuki Takemura for excellent technical assistance.

REFERENCES

Afeseh Ngwa, H., Kanthasamy, A., Gu, Y., Fang, N., Anantharam, V., Kanthasamy, A.G., 2011. Manganese nanoparticle activates mitochondrial dependent apoptotic signaling and autophagy in dopaminergic neuronal cells. Toxicol. Appl. Pharmacol. 256, 227–240.

Albrecht, J., Sonnewald, U., Waagepetersen, H.S., Schousboe, A., 2007. Glutamine in the central nervous system: function and dysfunction. Front. Biosci. 12, 332–343.

Ambani, L.M., Van Woert, M.H., Murphy, S., 1975. Brain peroxidase and catalase in Parkinson disease. Arch. Neurol. 32, 114–118.

Anantharam, V., Kitazawa, M., Wagner, J., Kaul, S., Kanthasamy, A.G., 2002. Caspase-3-dependent proteolytic cleavage of protein kinase Cdelta is essential for oxidative stress-mediated dopaminergic cell death after exposure to methylcyclopentadienyl manganese tricarbonyl. J. Neurosci. 22, 1738–1751.

Atkins, P.W., Overton, T.L., Rourke, J.P., Weller, M.T., Armstrong, F.A., 2014. Inorganic Chemistry, sixth ed. Oxford University Press, Oxford, UK.

Bader, M., Dietz, M.C., Ihrig, A., Triebig, G., 1999. Biomonitoring of manganese in blood, urine and axillary hair following low-dose exposure during the manufacture of dry cell batteries. Int. Arch. Occup. Environ. Health 72, 521–527.

Ballabh, P., Braun, A., Nedergaard, M., 2004. The blood–brain barrier: an overview: structure, regulation, and clinical implications. Neurobiol. Dis. 16, 1–13.

Barceloux, D.G., 1999. Manganese. J. Toxicol. Clin. Toxicol. 37, 293–307.

Bellinger, F.P., He, Q.P., Bellinger, M.T., Lin, Y., Raman, A.V., White, L.R., Berry, M.J., 2008. Association of selenoprotein P with Alzheimer's pathology in human cortex. J. Alzheimers Dis. 15, 465–472.

Bendich, A., 2000. The potential for dietary supplements to reduce premenstrual syndrome (PMS) symptoms. J. Am. Coll. Nutr. 19, 3–12.

Bolea, R., Hortells, P., Martín-Burriel, I., Vargas, A., Ryffel, B., Monzón, M., Badiola, J.J., 2010. Consequences of dietary manganese and copper imbalance on neuronal apoptosis in a murine model of scrapie. Neuropathol. Appl. Neurobiol. 36, 300–311.

Bouchard, M.F., Sauvé, S., Barbeau, B., Legrand, M., Brodeur, M.È., Bouffard, T., Limoges, E., Bellinger, D.C., Mergler, D., 2011. Intellectual impairment in school-age children exposed to manganese from drinking water. Environ. Health Perspect. 119, 138–143.

Bowman, A.B., Kwakye, G.F., Herrero Hernández, E., Aschner, M., 2011. Role of manganese in neurodegenerative diseases. J. Trace Elem. Med. Biol. 25, 191–203.

Bresciani, G., da Cruz, I.B., González-Gallego, J., 2015. Manganese superoxide dismutase and oxidative stress modulation. Adv. Clin. Chem. 68, 87–130.

Briley-Saebo, K., Yeang, C., Witztum, J.L., Tsimikas, S., 2014. Imaging of oxidation-specific epitopes with targeted nanoparticles to detect high-risk atherosclerotic lesions: progress and future directions. J. Cardiovasc. Transl. Res. 7, 719–736.

Burton, N.C., Guilarte, T.R., 2009. Manganese neurotoxicity: lessons learned from longitudinal studies in nonhuman primates. Environ. Health Perspect. 117, 325–332.

Butterworth, R.F., 2014. Hepatic encephalopathy in alcoholic cirrhosis. Handb. Clin. Neurol. 125, 589–602.

Cappelli, J., Garcia, A., Ceacero, F., Gomez, S., Luna, S., Gallego, L., Gambin, P., Landete-Castillejos, T., 2015. Manganese supplementation in deer under balanced diet increases impact energy and contents in minerals of antler bone tissue. PLoS One 10, e0132738.

Carmona, A., Devès, G., Roudeau, S., Cloetens, P., Bohic, S., Ortega, R., 2010. Manganese accumulates within golgi apparatus in dopaminergic cells as revealed by synchrotron X-ray fluorescence nanoimaging. ACS Chem. Neurosci. 1, 194–203.

Carmona, A., Roudeau, S., Perrin, L., Veronesi, G., Ortega, R., 2014. Environmental manganese compounds accumulate as Mn(II) within the Golgi apparatus of dopamine cells: relationship between speciation, subcellular distribution, and cytotoxicity. Metallomics 6, 822–832.

Chen, P., Parmalee, N., Aschner, M., 2014. Genetic factors and manganese-induced neurotoxicity. Front. Genet. 5, 265.

Chin-Chan, M., Navarro-Yepes, J., Quintanilla-Vega, B., 2015. Environmental pollutants as risk factors for neurodegenerative disorders: Alzheimer and Parkinson diseases. Front. Cell. Neurosci. 9, 124.

Cicchetti, F., Drouin-Ouellet, J., Gross, R.E., 2009. Environmental toxins and Parkinson's disease: what have we learned from pesticide-induced animal models? Trends Pharmacol. Sci. 30, 475–483.

Claus Henn, B., Coull, B.A., Wright, R.O., 2014. Chemical mixtures and children's health. Curr. Opin. Pediatr. 26, 223–229.

Coppedè, F., Migliore, L., 2015. DNA damage in neurodegenerative diseases. Mutat. Res. 776, 84–97.

Criswell, S.R., Perlmutter, J.S., Huang, J.L., Golchin, N., Flores, H.P., Hobson, A., Aschner, M., Erikson, K.M., Checkoway, H., Racette, B.A., 2012. Basal ganglia intensity indices and diffusion weighted imaging in manganese-exposed welders. Occup. Environ. Med. 69, 437–443.

De Jong, W.H., Borm, P.J., 2008. Drug delivery and nanoparticles: applications and hazards. Int. J. Nanomed. 3, 133–149.

Dell, R.M., 2000. Batteries. Solid State Ionics 134, 139–158.

Elsner, R.J., Spangler, J.G., 2005. Neurotoxicity of inhaled manganese: public health danger in the shower? Med. Hypotheses 65, 607–616.

Erikson, K.M., Dorman, D.C., Lash, L.H., Aschner, M., 2008. Duration of airborne-manganese exposure in rhesus monkeys is associated with brain regional changes in biomarkers of neurotoxicity. Neurotoxicology 29, 377–385.

Exil, V., Ping, L., Yu, Y., Chakraborty, S., Caito, S.W., Wells, K.S., Karki, P., Lee, E., Aschner, M., 2014. Activation of MAPK and FoxO by manganese (Mn) in rat neonatal primary astrocyte cultures. PLoS One 9, e94753.

Farina, M., Avila, D.S., da Rocha, J.B., Aschner, M., 2013. Metals, oxidative stress and neurodegeneration: a focus on iron, manganese and mercury. Neurochem. Int. 62, 575–594.

Feng, X., Chen, A., Zhang, Y., Wang, J., Shao, L., Wei, L., 2015a. Application of dental nanomaterials: potential toxicity to the central nervous system. Int. J. Nanomed. 10, 3547–3565.

Feng, X., Chen, A., Zhang, Y., Wang, J., Shao, L., Wei, L., 2015b. Central nervous system toxicity of metallic nanoparticles. Int. J. Nanomed. 10, 4321–4340.

Food and Nutrition Board IoM: Manganese, 2001. Dietary Reference Intakes for Vitamin A, Vitamin K, Boron, Chromium, Copper, Iodine, Iron, Manganese, Molybdenum, Nickel, Silicon, Vanadium, and Zinc. National Academies Press, Washington, DC, pp. 394–419.

Garzillo, E.M., Lamberti, M., Genovese, G., Pedata, P., Feola, D., Sannolo, N., Daniele, L., Trojsi, F., Monsurro, M.R., Miraglia, N., 2014. Blood lead, manganese, and aluminum levels in a regional Italian cohort of ALS patients: does aluminum have an influence? J. Occup. Environ. Med. 56, 1062–1066.

Giordano, S., Darley-Usmar, V., Zhang, J., 2014. Autophagy as an essential cellular antioxidant pathway in neurodegenerative disease. Redox Biol. 2, 82–90.

Gonzalez-Cuyar, L.F., Nelson, G., Criswell, S.R., Ho, P., Lonzanida, J.A., Checkoway, H., Seixas, N., Gelman, B.B., Evanoff, B.A., Murray, J., Zhang, J., Racette, B.A., 2014. Quantitative neuropathology associated with chronic manganese exposure in South African mine workers. Neurotoxicology 45, 260–266.

Grandjean, P., Landrigan, P.J., 2014. Neurobehavioural effects of developmental toxicity. Lancet Neurol. 13, 330–338.

Guilarte, T.R., 2011. Manganese and Parkinson's disease: a critical review and new findings. Cien Saude Colet 16, 4549–4566.

Halatek, T., Sinczuk-Walczak, H., Szymczak, M., Rydzynski, K., 2005. Neurological and respiratory symptoms in shipyard welders exposed to manganese. Int. J. Occup. Med. Environ. Health 18, 265–274.

Hanley, C., Thurber, A., Hanna, C., Punnoose, A., Zhang, J., Wingett, D.G., 2009. The influences of cell type and ZnO nanoparticle size on immune cell cytotoxicity and cytokine induction. Nanoscale Res. Lett. 4, 1409–1420.

Hazell, A.S., 2002. Astrocytes and manganese neurotoxicity. Neurochem. Int. 41, 271–277.

He, S.C., Niu, Q., 2004. Subclinical neurophysiological effects of manganese in welding workers. Int. J. Immunopathol. Pharmacol. 17, 11–16.

Hosseinimehr, S.J., 2015. The protective effects of trace elements against side effects induced by ionizing radiation. Radiat. Oncol. J. 33, 66–74.

Huang, M., Parker, M.J., Stubbe, J., 2014. Choosing the right metal: case studies of class I ribonucleotide reductases. J. Biol. Chem. 289, 28104–28111.

Hussain, S.M., Javorina, A.K., Schrand, A.M., Duhart, H.M., Ali, S.F., Schlager, J.J., 2006. The interaction of manganese nanoparticles with PC-12 cells induces dopamine depletion. Toxicol. Sci. 92, 456–463.

Hybertson, B.M., Gao, B., Bose, S.K., McCord, J.M., 2011. Oxidative stress in health and disease: the therapeutic potential of Nrf2 activation. Mol. Aspects Med. 32, 234–246.

Imai, Y., Soda, M., Takahashi, R., 2000. Parkin suppresses unfolded protein stress-induced cell death through its E3 ubiquitin-protein ligase activity. J. Biol. Chem. 275, 35661–35664.
Indo, H.P., Yen, H.C., Nakanishi, I., Matsumoto, K., Tamura, M., Nagano, Y., Matsui, H., Gusev, O., Cornette, R., Okuda, T., Minamiyama, Y., Ichikawa, H., Suenaga, S., Oki, M., Sato, T., Ozawa, T., Clair, D.K., Majima, H.J., 2015. A mitochondrial superoxide theory for oxidative stress diseases and aging. J. Clin. Biochem. Nutr. 56, 1–7.
Jain, K.K., 2012. Nanobiotechnology-based strategies for crossing the blood–brain barrier. Nanomedicine (Lond.) 7, 1225–1233.
Karki, P., Smith, K., Johnson Jr., J., Aschner, M., Lee, E.Y., 2015. Genetic dys-regulation of astrocytic glutamate transporter EAAT2 and its implications in neurological disorders and manganese toxicity. Neurochem. Res. 40, 380–388.
Karmakar, A., Zhang, Q., Zhang, Y., 2014. Neurotoxicity of nanoscale materials. J. Food Drug Anal. 22, 147–160.
Klotz, L.O., Sánchez-Ramos, C., Prieto-Arroyo, I., Urbánek, P., Steinbrenner, H., Monsalve, M., 2015. Redox regulation of FoxO transcription factors. Redox Biol. 6, 51–72.
Kops, G.J., Dansen, T.B., Polderman, P.E., Saarloos, I., Wirtz, K.W., Coffer, P.J., Huang, T.T., Bos, J.L., Medema, R.H., Burgering, B.M., 2002. Forkhead transcription factor FOXO3a protects quiescent cells from oxidative stress. Nature 419, 316–321.
Kuwahara, R.T., Skinner, R.B.I., Skinner, R.B.J., 2001. Nickel coinage in the United States. The history of a common contact allergen. West J. Med. 175, 112–114.
Kwakye, G.F., Paoliello, M.M., Mukhopadhyay, S., Bowman, A.B., Aschner, M., 2015. Manganese-induced parkinsonism and Parkinson's disease: shared and distinguishable features. Int. J. Environ. Res. Public Health 12, 7519–7540.
Lach, K., Steer, B., Gorbunov, B., Mička, V., Muir, R.B., 2015. Evaluation of exposure to airborne heavy metals at gun shooting ranges. Ann. Occup. Hyg. 59, 307–323.
Latronico, T., Branà, M.T., Merra, E., Fasano, A., Di Bari, G., Casalino, E., Liuzzi, G.M., 2013. Impact of manganese neurotoxicity on MMP-9 production and superoxide dismutase activity in rat primary astrocytes. Effect of resveratrol and therapeutical implications for the treatment of CNS diseases. Toxicol. Sci. 135, 218–228.
Law, N.A., Caudle, T.M., Pecoraro, V.L., 1998. Manganese redox enzymes and model systems: properties, structures, and reactivity. Adv. Inorg. Chem. 46, 305–440.
Lee, S.H., Jouihan, H.A., Cooksey, R.C., Jones, D., Kim, H.J., Winge, D.R., McClain, D.A., 2013. Manganese supplementation protects against diet-induced diabetes in wild type mice by enhancing insulin secretion. Endocrinology 154, 1029–1038.
Li, L., Xu, J., Zheng, X., Ma, C., Song, X., Ge, S., Yu, J., Yan, M., 2014. Growth of gold-manganese oxide nanostructures on a 3D origami device for glucose-oxidase label based electrochemical immunosensor. Biosens. Bioelectron. 61, 76–82.
Lin, M., Huang, J., Sha, M., 2014. Recent advances in nanosized Mn-Zn ferrite magnetic fluid hyperthermia for cancer treatment. J. Nanosci. Nanotechnol. 14, 792–802.
Liu, X., Chen, C., Zhao, Y., Jia, B., 2013. A review on the synthesis of manganese oxide nanomaterials and their applications on lithium-ion batteries. J. Nanomater. 2013, 1–7.
Loranger, S., Tétrault, M., Kennedy, G., Zayed, J., 1996. Manganese and other trace elements in urban snow near an expressway. Environ. Pollut. 92, 203–211.
Lu, C.S., Huang, C.C., Chu, N.S., Calne, D.B., 1994. Levodopa failure in chronic manganism. Neurology 44, 1600–1602.
Luft, J.H., 1956. Permanganate – a new fixative for electron microscopy. J. Biophys. Biochem. Cytol. 2, 799–802.
Luo, Y., 2007. Preparation of MnO_2 nanoparticles by directly mixing potassium permanganate and polyelectrolyte aqueous solutions. Mater. Lett. 61, 1893–1895.

Lynam, D.R., Roos, J.W., Pfeifer, G.D., Fort, B.F., Pullin, T.G., 1999. Environmental effects and exposures to manganese from use of methylcyclopentadienyl manganese tricarbonyl (MMT) in gasoline. Neurotoxicology 20, 145–150.

Maddirala, Y., Tobwala, S., Ercal, N., 2015. N-acetylcysteineamide protects against manganese-induced toxicity in SHSY5Y cell line. Brain Res. 1608, 157–166.

Manolopoulos, K.N., Klotz, L.O., Korsten, P., Bornstein, S.R., Barthel, A., 2010. Linking Alzheimer's disease to insulin resistance: the FoxO response to oxidative stress. Mol. Psychiatry 15, 1046–1052.

Martin, J., Bello, D., Bunker, K., Shafer, M., Christiani, D., Woskie, S., Demokritou, P., 2015. Occupational exposure to nanoparticles at commercial photocopy centers. J. Hazard. Mater. 298, 351–360.

Máté, Z., Horváth, E., Kozma, G., Simon, T., Kónya, Z., Paulik, E., Papp, A., Szabó, A., 2016. Size-dependent toxicity differences of intratracheally instilled manganese oxide nanoparticles: conclusions of a subacute animal experiment. Biol. Trace Elem. Res. 156–166.

Meeker, J.D., Susi, P., Flynn, M.R., 2007. Manganese and welding fume exposure and control in construction. J. Occup. Environ. Hyg. 4, 943–951.

Migliore, L., Uboldi, C., Di Bucchianico, S., Coppedè, F., 2015. Nanomaterials and neurodegeneration. Environ. Mol. Mutagen 56, 149–170.

Mofid, M.M., Thomposon, R.C., Pardo, C.A., Manson, P.N., Vander Kolk, C.A., 1997. Biocompatibility of fixation materials in the brain. Plast. Reconstr. Surg. 100, 14–20.

Mokgobu, M.I., Anderson, R., Steel, H.C., Cholo, M.C., Tintinger, G.R., Theron, A.J., 2012. Manganese promotes increased formation of hydrogen peroxide by activated human macrophages and neutrophils in vitro. Inhal. Toxicol. 24, 634–644.

Moshkin, M.P., Petrovski, D.V., Akulov, A.E., Romashchenko, A.V., Gerlinskaya, L.A., Ganimedov, V.L., Muchnaya, M.I., Sadovsky, A.S., Koptyug, I.V., Savelov, A.A., Troitsky, S.Y., Moshkn, Y.M., Bukhtiyarov, V.I., Kolchanov, N.A., Sagdeev, R.Z., Fomin, V.M., 2014. Nasal aerodynamics protects brain and lung from inhaled dust in subterranean diggers, *Ellobius talpinus*. Proc. Biol. Sci. 281.

Muhammad, S.A., Bilbis, L.S., Saidu, Y., Adamu, Y., 2012. Effect of antioxidant mineral elements supplementation in the treatment of hypertension in albino rats. Oxid. Med. Cell. Longev. 2012 134723.

Najafpour, M.M., Ghobadi, M.Z., Haghighi, B., Eaton-Rye, J.J., Tomo, T., Shen, J.R., Allakhverdiev, S.I., 2014. Nano-sized manganese-calcium cluster in photosystem II. Biochemistry (Mosc) 79, 324–336.

Nel, A., Xia, T., Mädler, L., Li, N., 2006. Toxic potential of materials at the nanolevel. Science 311, 622–627.

Normandin, L., Carrier, G., Gardiner, P.F., Kennedy, G., Hazell, A.S., Mergler, D., Butterworth, R.F., Philippe, S., Zayed, J., 2002. Assessment of bioaccumulation, neuropathology, and neurobehavior following subchronic (90 days) inhalation in Sprague-Dawley rats exposed to manganese phosphate. Toxicol. Appl. Pharmacol. 183, 135–145.

Normandin, L., Hazell, A.S., 2002. Manganese neurotoxicity: an update of pathophysiologic mechanisms. Metab. Brain Dis. 17.

O'Neal, S.L., Zheng, W., 2015. Manganese toxicity upon overexposure: a decade in review. Curr. Environ. Health Rep. 2, 315–328.

Olanow, C.W., 2004. Manganese-induced parkisonism and Parkison's disease. Ann. N.Y. Acad. Sci. 1012, 209–223.

Oszlánczi, G., Vezér, T., Sárközi, L., Horváth, E., Szabó, A., Horváth, E., Kónya, Z., Papp, A., 2010. Metal deposition and functional neurotoxicity in rats after 3–6 weeks nasal exposure by two physicochemical forms of manganese. Environ. Toxicol. Pharmacol. 30, 121–126.

Palacios, C., 2006. The role of nutrients in bone health, from A to Z. Crit. Rev. Food Sci. Nutr. 46, 621–628.

Patrick, M.W., 1998. Glassmaking in renaissance Italy: the innovation of venetian cristallo. J. Miner. Metal Mater. Soc. 50, 14.

Penland, J.G., Johnson, P.E., 1993. Dietary calcium and manganese effects on menstrual cycle symptoms. Am. J. Obstet. Gynecol. 168, 1417–1423.

Pesch, B., Weiss, T., Kendzia, B., Henry, J., Lehnert, M., Lotz, A., Heinze, E., Käfferlein, H.U., Van Gelder, R., Berges, M., Hahn, J.U., Mattenklott, M., Punkenburg, E., Hartwig, A., Brüning, T., 2012. Levels and predictors of airborne and internal exposure to manganese and iron among welders. J. Expo. Sci. Environ. Epidemiol. 22, 291–298.

Reyes-Reveles, J., Sedaghat-Herati, R., Gilley, D.R., Schaeffer, A.M., Ghosh, K.C., Greene, T.D., Gann, H.E., Dowler, W.A., Kramer, S., Dean, J.M., Delong, R.K., 2013. mPEG-PAMAM-G4 nucleic acid nanocomplexes: enhanced stability, RNase protection, and activity of splice switching oligomer and poly I: C RNA. Biomacromolecules 14, 4108–4115.

Roos, P.M., Vesterberg, O., Syversen, T., Flaten, T.P., Nordberg, M., 2013. Metal concentrations in cerebrospinal fluid and blood plasma from patients with amyotrophic lateral sclerosis. Biol. Trace Elem. Res. 151, 159–170.

Rose, C., Butterworth, R.F., Zayed, J., Normandin, L., Todd, K., Michalak, A., Spahr, L., Huet, P.M., Pomier-Layrargues, G., 1999. Manganese deposition in basal ganglia structures results from both portal-systemic shunting and liver dysfunction. Gastroenterology 117, 640–644.

Rothstein, J.D., Dykes-Hoberg, M., Pardo, C.A., Bristol, L.A., Jin, L., Kuncl, R.W., Kanai, Y., Hediger, M.A., Wang, Y., Schielke, J.P., Welty, D.F., 1996. Knockout of glutamate transporters reveals a major role for astroglial transport in excitotoxicity and clearance of glutamate. Neuron 16, 675–686.

Rutz, A., 2009. Synthesis and properties of manganese oxide nanoparticles for environmental applications. In: The 2009 NNIN REU Research Accomplishments, pp. 98–99.

Salehi, F., Carrier, G., Normandin, L., Kennedy, G., Butterworth, R.F., Hazell, A., Therrien, G., Mergler, D., Philippe, S., Zayed, J., 2001. Assessment of bioaccumulation and neurotoxicity in rats with portacaval anastomosis and exposed to manganese phosphate: a pilot study. Inhal. Toxicol. 13, 1151–1163.

Sidoryk-Wegrzynowicz, M., Aschner, M., 2013a. Manganese toxicity in the central nervous system: the glutamine/glutamate-gamma-aminobutyric acid cycle. J. Intern. Med. 273, 466–477.

Sidoryk-Wegrzynowicz, M., Aschner, M., 2013b. Role of astrocytes in manganese mediated neurotoxicity. BMC Pharmacol. Toxicol. 14, 23.

Sierra, P., Chakrabarti, S., Tounkara, R., Loranger, S., Kennedy, G., Zayed, J., 1998. Bioaccumulation of manganese and its toxicity in feral pigeons (*Columba livia*) exposed to manganese oxide dust (Mn_3O_4). Environ. Res. 79, 94–101.

Strause, L., Saltman, P., Glowacki, J., 1987. The effect of deficiencies of manganese and copper on osteoinduction and on resorption of bone particles in rats. Calcif. Tissue Int. 41, 145–150.

Su, C., Chen, K., Zou, Y., Shen, Y., Xia, B., Liang, G., Lv, Y., Wang, F., Huang, D., Yang, X., 2015. Chronic exposure to manganese sulfate leads to adverse dose-dependent effects on the neurobehavioral ability of rats. Environ. Toxicol. http://dx.doi.org/10.1002/tox.22161 (Epub ahead of print).

Takács, S.Z., Szabó, A., Oszlánczi, G., Pusztai, P., Sápi, A., Kónya, Z., Papp, A., 2012. Repeated simultaneous cortical electrophysiological and behavioral recording in rats exposed to manganese-containing nanoparticles. Acta Biol. Hung. 63, 426–440.

Takeda, A., 2003. Manganese action in brain function. Brain Res. Rev. 41, 79–87.

Takeda, A., Sotogaku, N., Oku, N., 2003. Influence of manganese on the release of neurotransmitters in rat striatum. Brain Res. 965, 279–282.

Tao, R., Vassilopoulos, A., Parisiadou, L., Yan, Y., Gius, D., 2014. Regulation of MnSOD enzymatic activity by Sirt3 connects the mitochondrial acetylome signaling networks to aging and carcinogenesis. Antioxid. Redox Signal. 20, 1646–1654.

Taube, F., 2013. Manganese in occupational arc welding fumes–aspects on physiochemical properties, with focus on solubility. Ann. Occup. Hyg. 57, 6–25.

Tong, Y., Yang, H., Tian, X., Wang, H., Zhou, T., Zhang, S., Yu, J., Zhang, T., Fan, D., Guo, X., Tabira, T., Kong, F., Chen, Z., Xiao, W., Chui, D., 2014. High manganese, a risk for Alzheimer's disease: high manganese induces amyloid-beta related cognitive impairment. J. Alzheimers Dis. 42, 865–878.

Wan, J., Deng, H., Shi, J., Zhou, L., Su, T., 2014. Synthesized magnetic manganese ferrite nanoparticles on activated carbon for sulfamethoxazole removal. CLEAN – Soil, Air, Water 42, 1199–1207.

Wang, J., Rahman, M.F., Duhart, H.M., Newport, G.D., Patterson, T.A., Murdock, R.C., Hussain, S.M., Schlager, J.J., Ali, S.F., 2009. Expression changes of dopaminergic system-related genes in PC12 cells induced by manganese, silver, or copper nanoparticles. Neurotoxicology 30, 926–933.

Whelan, A.M., Jurgens, T.M., Bowles, S.K., 2006. Natural health products in the prevention and treatment of osteoporosis: systematic review of randomized controlled trials. Ann. Pharmacother. 40, 836–849.

Win-Shwe, T.T., Fujimaki, H., 2011. Nanoparticles and neurotoxicity. Int. J. Mol. Sci. 12, 6267–6280.

Xu, B., Shan, M., Wang, F., Deng, Y., Liu, W., Feng, S., Yang, T.Y., Xu, Z.F., 2013. Endoplasmic reticulum stress signaling involvement in manganese-induced nerve cell damage in organotypic brain slice cultures. Toxicol. Lett. 222, 239–246.

Yang, S.H., Heo, D., Park, J., Na, S., Suh, J.S., Haam, S., Park, S.W., Huh, Y.M., Yang, J., 2012. Role of surface charge in cytotoxicity of charged manganese ferrite nanoparticles towards macrophages. Nanotechnology 23, 505702.

Yavuz, E., Tokalıoğlu, Ş., Şahan, H., Patat, Ş., 2014. Nano sponge Mn_2O_3 as a new adsorbent for the preconcentration of Pd(II) and Rh(III) ions in sea water, wastewater, rock, street sediment and catalytic converter samples prior to FAAS determinations. Talanta 128, 31–37.

Yokel, R.A., 2006. Blood–brain barrier flux of aluminum, manganese, iron and other metals suspected to contribute to metal-induced neurodegeneration. J. Alzheimers Dis. 10, 223–253.

Zatta, P., Lucchini, R., van Rensburg, S.J., Taylor, A., 2003. The role of metals in neurodegenerative processes: aluminum, manganese, and zinc. Brain Res. Bull. 62, 15–28.

Zhen, Z., Xie, J., 2012. Development of manganese-based nanoparticles as contrast probes for magnetic resonance imaging. Theranostics 2, 45–54.

Zheng, W., Aschner, M., Ghersi-Egea, J.F., 2003. Brain barrier systems: a new frontier in metal neurotoxicological research. Toxicol. Appl. Pharmacol. 192, 1–11.

Zoni, S., Lucchini, R.G., 2013. Manganese exposure: cognitive, motor and behavioral effects on children: a review of recent findings. Curr. Opin. Pediatr. 25, 255–260.

Chapter 10

The Application, Neurotoxicity, and Related Mechanism of Silica Nanoparticles

C. Zhong, M. He, K. Lou, F. Gao
East China University of Science and Technology, Shanghai, China

Chapter Outline

1. INTRODUCTION

It is well known that silica nanoparticles (NPs) are of increasing importance due to their applications in drug delivery, bioimaging, catalysis, optical devices, cosmetics, polymeric fillers, and other biomedical fields. Thus it is necessary to characterize these nanomaterials and optimize their properties during the development process to eliminate or ameliorate identified toxic characteristics. Considering the extensive applications of silica NPs, especially in the food industry, performing various in vitro and in vivo experiments for clarifying their toxic effect and mechanism is essential to lower the risk of various potential diseases

Neurotoxicity of Nanomaterials and Nanomedicine. http://dx.doi.org/10.1016/B978-0-12-804598-5.00010-6

such as pulmonary and neurodegenerative disorders [ie, Alzheimer's disease (AD) and Huntington's diseases]. Some in vitro studies have found that silica NPs induce cytotoxicity by exploring various kinds of cells including human lung cancer cells, human liver cell line HepG2, human umbilical vein endothelial cells (HUVECs), human bronchial epithelial cell Beas-2B, human monocytic leukemia cell line THP-1, human alveolar epithelial cell line A549, and human breast cancer cell line MCF-7, as well as animal cells such as mouse embryonic stem cells, RAW 264.7cell line, and a rodent fibroblast cell line. Researchers have already found that cytotoxicity of silica NPs is closely linked to the size, dose, cell type, time, surface area, and structural discrimination. There are many potential complicated factors affecting toxicity of silica NPs that need to be studied further. The mechanisms toward toxic effects of silica NPs are mainly associated with oxidative stress, which is a common index for assessing toxicity of synthetic NPs. Several cellular and physiological reactions including apoptosis, inflammation, and DNA damage are attributed to reactive oxygen species (ROS) such as hydroxyl radical (HO•), superoxide anion (O_2^-), and hydrogen peroxide (H_2O_2) generated by oxidative stress. Oxidative stress plays an important role in increasing ROS, lipid peroxidation (LPO), and depletion of glutathione (GSH). Now that existing studies have investigated the toxic effect of silica NPs on human tissues, particularly lung, it is important to explore whether silica NPs have neurotoxicity. Researchers have found that single-walled carbon nanotubes (Wang et al., 2008), quantum dots (QDs) (Yang et al., 2007), and other nanomaterials could cross the blood–brain barrier (BBB) and thus enter the brain. Existing evidence has shown that titanium dioxide and manganese oxide NPs (Wang et al., 2008; Elder et al., 2006) are able to translocate along the olfactory nerve pathway to the brain after intranasal instillation exposure, followed by accumulation in the olfactory bulb, cortex, as well as cerebellum. Consequently, there is an urgent need to discuss whether silica NPs are capable of passing the BBB and pose a consequential influence on the brain. A series of in vitro and in vivo studies showed that silica NPs indeed exhibit toxicity on neurons and central nervous system (CNS) in some conditions by experiments on neural cells and animals, respectively. To achieve exact and scientific research results, enough careful and deliberate experiments are needed to be planned and conducted.

2. APPLICATIONS OF SILICA NANOPARTICLES

With the popularity of nanotechnology, nanomaterials have been widely applied in several fields, such as drug delivery, bioimaging, gene delivery, and engineering (Lanone and Boczkowski, 2006; Mulder et al., 2009; Sahoo and Labhasetwar, 2003; Farokhzad and Lansar, 2006; Caruthers et al., 2007; Shi et al., 2010). Due to the unique properties of porous silica NPs such as high surface area and pore volume, controllable particle size, and desirable biocompatibility, numerous studies have explored these nanomaterials (Wang et al., 2013;

Kilpeläinen et al., 2009). Table 10.1 lists some of these studies, including material, morphology, and size of NPs and their application. Especially, mesoporous silica nanoparticles (MSNs) have been widely applied due to their unique mesoporous structure that possess surface functionality, chemical stability, and biocompatibility, assuring the controlled release, and target drug delivery of a variety of drug molecules (Quan et al., 2015; Tao et al., 2014; Mishra et al., 2014; Zhang and Kong, 2015).

2.1 Drug Delivery

With the continuous development of nanomedicine in the recent years, the necessity to enhance drug delivery efficiency while reducing its toxicity has been solved by engineering nanoscaled structure and function adjustable devices (David, 2013; Re et al., 2012). Among different devices, NP technology is rapidly advancing. NPs are objects sized between 1 and 100 nm that work as a whole unit in terms of transport and properties (Youns et al., 2011). NPs are favorable for biological and medical applications possibly due to their capability of conjugating either hydrophilic or hydrophobic pharmaceuticals and are able to be administered with various routes such as oral, inhalational, and parenteral (Petkar et al., 2011). Furthermore, different targeting ligands including proteins, antibodies, and aptamers could be conjugated to the NP's surface to improve their targeting yield. The large surface-area-to-volume ratio of NPs permits multiple types of ligands to be attached to their surfaces that could dramatically augment their binding affinity (Xavier et al., 2006). Among those NP drug delivery systems, MSNs have several significant attractive properties such as ordered porous structure, large surface area, controllable particle size, as well as interior and exterior dual-functional surfaces. Different types of MSNs and their modifications are listed in Table 10.1. Studies with inorganic NPs modified with functional groups have demonstrated improvement in drug delivery into the cancer cells. Owing to the presence of specific receptors on cancer cells, surface modification of MSNs has been well studied and used. Different drugs that were loaded into the MSNs are also described in Table 10.1.

Magali et al. (Gary-Bobo et al., 2012) synthesized MSNs that covalently encapsulated fluorescein or a photosensitizer and were functionalized with galactose on the surface. The MSNs functionalized with photosensitizer and galactose were loaded with the anticancer drug camptothecin. Those MSNs combining drug delivery and photodynamic therapy were tested on three cancer cell lines and showed a dramatic enhancement of cancer cell death compared to separate treatments. Hanafi-Bojd et al. (2015) prepared, characterized, and evaluated MSNs as carriers for epirubicin hydrochloride to improve the antitumor efficacy of this drug. The results indicated that this formulation MSN-polyethylenimine (PEI)-polyethylene glycol (PEG) could be a promising drug delivery system for cancer therapy. Quan et al. (2015) developed a hepatoma targeting drug delivery system by conjugating lactose to MSNs as the

TABLE 10.1 Applications of Silica Nanoparticles

Application	Material	Chemical Moiety	Structure	Particle Size (nm)	References
Biotechnological	SiO_2-NPs	None	Nanorods	15	Show et al. (2015)
Rapid detection of *Entamoeba histolytica*	FSNP	Fluorescent	Spherical particles	60	Hemadi et al. (2015)
Biomedical	Biogenic silica nanoparticles	None	Spherical particles	10–30	Alshatwi et al. (2014)
	FSNP	None	Spherical particles	67	Nooney et al. (2015)
	MNP@SiO_2	Fe_3O_4	Spherical particles	14	Kuzminska et al. (2015)
Tissue engineering	MSNs	PLGA	Nanospheres	70	Mehrasa et al. (2015)
Labeling	MSNs	^{18}F	Spherical particles	60–90	Rojas et al. (2015)
Gene delivery	MSNs	DNA	Spherical particles	230, 100	Dengler et al. (2013) and Lin et al. (2015)
	Gadolinium oxide-doped silica NPs	DNA		40–50	Bhakta et al. (2011)
	MSNs	siRNA		47	Ngamcherdtrakul et al. (2015)
Bioimaging	MSNs	Fluorescent	Spherical particles	255	Kempen et al. (2015)
	Silica-coated gold nanorods	None	Nanorods	50	Qian et al. (2010)
	MSNs	DOX	Spherical particles	106, 150	Dai et al. (2015) and Chen et al. (2015b)

Aerosol droplet delivery	MSNs	Fluorescent	Spherical particles	70	Li et al. (2015)
Drug delivery	MSNs	DOX, CPT, PTX	Nanospheres	80–150	Ge et al. (2015), Dai et al. (2015), Zhang and Kong (2015), Chen et al. (2015a,b), Zhang et al. (2012), Mishra et al. (2014), Gary-Bobo et al. (2012), and Jia et al. (2015)
		DTX		100	Quan et al. (2015)
		CpG		75	Tao et al. (2014)

CPT, Camptothecin; *DOX*, doxorubicin; *DTX*, docetaxel; *FSNP*, fluorescent silica nanoparticle; *MNP*, magnetic nanoparticle; *MSN*, mesoporous silica nanoparticle; *NP*, nanoparticle; *PLGA*, poly(lactic-co-glycolic acid); *PTX*, paclitaxel; *siRNA*, small interfering RNA.

active targeting agent that can specifically target asialoglycoprotein receptor-positive HepG2 and SMMC7721 cells. The study showed that the internalization of functionalized MSNs into the hepatoma cells is an energy-consuming process and predominated by clathrin-mediated endocytosis. Another significant outcome of this study is that water-insoluble anticancer drug, docetaxel, was effectively loaded in the pores of lactosaminated MSN, showing significantly enhanced antitumor efficacy. A potential cytosine-phosphate-guanosine oligodeoxynucleotides (CpG ODN) delivery system has been developed by binding of CpG ODN into aminated MSNs with small particle size. The structure of amine-functionalized MSNs is well designed to acquire a high CpG ODN loading capacity due to the contribution of high surface area and mesoporous structure (Tao et al., 2014). In another study, Chinese traditional medicine "tetran-drine" and conventional antitumor drug "paclitaxel" were first codelivered by the multifunctional MSNs to overcome multidrug resistance of MCF-7/ADR cells (Jia et al., 2015), indicating that binding drugs to MSNs with small particle size would be promising for enhancing the delivery efficiency of those drugs.

The drug release of silica NPs could be well moderated by the microenvironment of the tumor it situated on, such as pH value or oxidation environment. Chen et al. (2015b) developed functionalized MSNs with desirable biocompatibility by using fluorescence-conjugated β-cyclodextrin as acid-activated gatekeepers to cap the mesopores of MSNs. This targeted MSN-based drug delivery with excellent biocompatibility has great potential for cancer diagnosis and therapy. The pH-operated MSNs exhibited intracellular pH sensitivity in acidic conditions mimicking the endosomal/lysosomal compartments, with a high cellular proliferation inhibition efficacy (Chen et al., 2015a). Administration of some anticancer drugs is limited by their dose-related side effects and nonselective delivery to healthy cells. This issue can be addressed by using stimuli-responsive targeted drug delivery systems. Selective transport of anticancer drugs to the tumor cells and their release inside the tumor microenvironment can reduce their adverse effects because of the lower required dosage (Mishra et al., 2014). Zhang and Kong (2015) showed a drug delivery system based on doxorubicin (DOX)-tethered fluorescent silica NPs. DOX was conjugated to the silica NPs through a hydra zone linkage. After uptake into the acidic environment of cancer cells, DOX was released from the surfaces of silica NPs because of the breakage of the pH-sensitive hydrazine bond. NPs were effectively delivered into the cytoplasm of HeLa cells and DOX was released in the nucleus according to the results. In another study (Palanikumar et al., 2015), a robust and simple one-pot route is developed to synthesize polymer-gatekeeper MSNs by noncovalent capping of the pores of drug-loaded nanocontainers with disulfide cross-linkable polymers. The drug molecules loaded in the nanocontainers can be released by the degradation of the polymer shell in the intracellular reducing microenvironment, which consequentially induces cell death.

There are increasing novel methods that MSNs had been applied to improve drug delivery. In one study, a highly versatile scheme that connects MSNs with an aerosol technology is designed to achieve direct nanoscale delivery to the respiratory tract (Li et al., 2015). This novel method can deposit MSNs through the entire respiratory tract, including nasal, tracheobronchial, and pulmonary regions using a water-based aerosol. The MSNs used have the potential for carrying and delivering therapeutic agents to highly specific target sites of the respiratory tract. The approach provides a critical foundation for developing therapeutic treatment protocols for a wide range of diseases where aerosol delivery to the respiratory system would be desirable. It is worthy to note that a new study takes advantage of silica NPs to achieve in vitro data for predicting NP–protein interactions in vivo and then possibly provides more efficient NPs-based drug delivery systems for disease treatment (Mirshafiee et al., 2015, 2016).

2.2 Gene Delivery

Silica NPs are also a promising candidate for the development of safe and efficient nonviral gene carriers, owing to their controlled morphologies, potential of facile surface modification, excellent biocompatibility, and in vivo biodegradability. In one particular study (Lin et al., 2015), the design of a series of novel gene carriers by employing polycation modified silica NPs with five different morphologies and the method of preparation of gadolinium oxide-doped silica NPs whose surface has been functionalized to anchor DNA onto this silica NP were investigated. Gajadhar et al. (Bhakta et al., 2011) observed that the surface-modified silica NPs such as gadolinium oxide–doped silica NPs were able to facilitate the transfer of plasmid DNA into mammalian cells in vitro. Considering the chemical inertness and biocompatibility of silica, this technology may provide a new platform for nonviral gene delivery as well as a contrasting agent in magnetic imaging. Ngamcherdtrakul et al. (2015) have developed a new generation of small interfering RNA (siRNA) in vivo delivery system comprising MSNs coated with PEI polymer to enhance the delivery, a PEG layer to protect the siRNA, and an antibody on the NP's surface for targeted delivery. Dengler et al. (2013) studied the in vitro cell stress and in vivo physiological profile of liposomes fused on MSNs, and examined their use as gene therapeutic vectors delivered to the spinal cord.

2.3 Bioimaging

Biomedical imaging could help to diagnose fatal diseases such as cancer in their early stages that could lead to more effective treatments. There are certain medical imaging techniques including optical imaging (OI), magnetic resonance imaging (MRI), positron emission computed tomography (Gambhir, 2002; Ametamey et al., 2008), computed tomography (CT), and conventional radiography applied clinically (Margolis et al., 2007). With the advent of molecular imaging, the accuracy of imaging in diagnosis and treatment dramatically increases (Pysz et al., 2010). Molecular imaging includes visualizing, characterizing, and measuring biological parameters at the molecular and cellular level in human tissues (Mankoff, 2007). Applying NP for imaging purposes could significantly help with disease diagnosis and monitoring (Sharma et al., 2006; Rosenblum et al., 2010). Numerous NPs have been applied for biological imaging such as magnetic NPs in MRI (Wang et al., 2001; Bulte and Kraitchman, 2004; Thorek et al., 2006; Lee and Hyeon, 2012), gold NPs in CT (Kim et al., 2007; Popovtzer et al., 2008; Alric et al., 2008; Peng et al., 2012), and QDs in OI (Michalet et al., 2005; Gao et al., 2004; Medintz et al., 2005; Smith et al., 2008; Jin and Hildebrandt, 2012). Silica NPs, which are used for contrast agents, have caused significant developments in molecular imaging. NPs such as QDs, gold NPs, silica NPs, and carbon nanotubes have many special physical and chemical properties, and can be used as favorable exogenous agents for optical bioimaging (Prasad and Masters, 2004). Some NPs are dispersible and stable in aqueous solutions (Murphy et al., 2005; Tan et al., 2004), and compatible to the biological environment at the same time. Some NPs can be modified and functionalized with biomolecules for different biomedical applications (Yun et al., 2007). Qian et al. (2010) introduced briefly the properties of NPs such as QDs, gold nanorods (GNRs), and organically modified silica (ORMOSIL) NPs for optical bioimaging, and they simultaneously utilized silica to encapsulate GNRs for cell imaging. ORMOSIL NPs usually have bigger pores than normal silica NPs by using tetraethoxy silane as the precursor. Furthermore, hydrocarbon groups usually have functional groups such amine in aminopropyl triethoxysilane and thiol in (3-mercaptopropyl) trimethoxysilane, making the ORMOSIL NPs easy to be functionalized for further bioconjugation. ORMOSIL NPs have some unique advantages over common silica NPs in biological applications (Jain et al., 1998; Shraboni et al., 2002). Compared with common silica, their mesoporous and bigger pores in matrix can facilitate controlled release of encapsulated biomolecules like drugs and proteins. Dai et al. (2015) explored dendrimer like magnetic nanoparticles with hierarchical pores to be used for drug delivery, bioimaging, and in vivo targeted tumor therapy, which provides an alternative drug delivery carrier for imaging targeted tumor therapy.

2.4 Cosmetics

The applications of nanotechnology has expanded to various areas, from electronics to medicine and now cosmetic industries due to the enhanced characteristics including color, transparency, and solubility of the particles at the nanoscale. The different types of nanomaterials that are employed in cosmetics include nanosomes, liposomes, fullerenes, and solid lipid NPs. Silica NPs have been widely used for cosmetics because of their nanoscaled size that enables them to pass through human skin. Therein synthetic amorphous silica NPs have been applied in toothpaste and cosmetics (Merget et al., 2002; Zhu et al., 2010; Yan et al., 2013).

3. NEUROTOXICITY OF SILICA NANOPARTICLES

Nowadays, applications of silica NPs are not just limited to engineering, food, and cosmetics, but also extended to extensive biomedical field, including targeted drug delivery, cancer therapy, OI, and gene delivery. However, there are also increased concerns about the safety issue with the use of silica NPs due to their unique physicochemical properties such as small size, large surface area, and high reactivity. Considering potential release of these NPs into the environment and accumulation in human tissues, especially in the case of occupational exposure to manufacturing premises, there is an urgent need to figure out toxicity especially neurotoxicity related to silica NPs based on well-designed experiments. Although numerous in vitro and in vivo studies have demonstrated that silica NPs exerted cytotoxic effects on cells such as human lung carcinoma cells (Gonzalez et al., 2014), human vascular endothelial cells (Yang et al., 2013), mouse peritoneal macrophages (Park and Park, 2009), microglia (Choi et al., 2009), and human skin HaLa T-cells (Gong et al., 2012; Yang et al., 2010), the neurotoxic outcomes of silica NPs remain relatively unexplored. Early studies have observed that the inhaled NPs penetrate the lungs and deposit themselves in the extrapulmonary tissues, such as heart, bone marrow, liver, kidney, and CNS (Chen et al., 2006; Kleinman et al., 2008; Nemmar et al., 2004). NPs that are deposited on the nasal olfactory area may escape from the BBB, get absorbed into the cerebrospinal fluid, enter the CNS, and cause potential damages to the CNS (Chen et al., 2006; Kleinman et al., 2008; Nemmar et al., 2004; Oberdorster et al., 2004). More seriously, NPs can cross the placenta and enter the fetus (Wick et al., 2010; Kulvietis et al., 2011; Refuerzo et al., 2011). Indeed, in vivo studies observed that the silica NPs are able to pass through the placental barrier and consequently reach the placenta, fatal brain, and fetal liver. Here, we selectively overview several studies on tissues susceptible to silica NPs, the main potential route through which silica NPs enter the CNS, as well as underlying mechanisms for neurotoxicity of silica NPs.

A selection of studies on NP toxicity on a number of biological systems would be used to illustrate these points, and also advocate the need to understand these interactions in greater detail. As for the neurotoxicity, the general interaction between physical properties and biological functionalities has been closely associated with certain biological issues, such as CNS and neurons, liver, blood, lung, skin, and spleen (Oberdorster et al., 2004).

Neurons are nerve cells that constitute the nervous tissue together with neuroglial cells. A neuron cell consists of a cell body, axon, and dendrites. Neurons receive nerve signals or action potentials, and integrate and transmit the signals to other neurons. Although the human nervous system is much more specialized and complicated than those of other animals, the structure and function of neurons is essentially the same as most animals. Hence, it is feasible to explore the neurotoxic effects of NPs on the nervous system by the aid of in vivo studies (Oberdorster et al., 2004).

The CNS, one of the most delicate microenvironments of the body, is protected by the BBB for maintaining its homeostasis. The BBB is a complex structure, which consists of endothelial cells, pericytes, astroglia, and perivascular mast cells (Wahl et al., 1988) and allows movement of certain small molecules and ions and a limited number of macromolecules from the blood to the brain to protect brain from injuries and diseases. The tightness of the BBB is attributed mainly to the vascular layer of brain capillary endothelial cells, which are interconnected side by side by tight and adherent junctions. These tight junctions perform two functions (Masserini, 2013): (1) They prevent the passage of small molecules and ions through the barrier between cells, so that their passage must occur by entering the cells by diffusion or active transport. This pathway controls the type and amount of substances that are allowed to pass. (2) They prevent the movement of integral membrane proteins between the apical and basolateral membranes of the cell, so that each of the cell membrane surfaces preserves its peculiar functions, for example, receptor-mediated endocytosis at the apical surface and exocytosis at the basolateral surface.

3.1 The Absorption, Distribution, Metabolism, and Excretion of Silica Nanoparticles

Humans are exposed to NPs mainly through dermal absorption, ingestion, and inhalation. Potential pathways of NPs-induced neurotoxicity are shown in Fig. 10.1. One well-studied pathway is the nose-to-brain transport through the olfactory epithelium. Olfactory sensory neurons (OSNs) reside in the olfactory epithelium and have direct contact with the outside environment. Thus NPs might be able to enter the brain through OSNs and their projections to the olfactory bulb and subsequently move to other regions in the brain (Chu et al., 2010; Calderón-Garcidueñas et al., 2002). Another possible nose-to-brain transport can occur through the trigeminal nerve, one largest nerve among the cranial nerves. Afferent neurons from the trigeminal nerve pass directly through the nasal mucosa and enter the brainstem at the level of the pons (Mistry et al., 2009). For example, in rats, intranasal administration of the neurotrophic factor insulin-like growth factor-1 (IGF-1) resulted in the delivery of IGF-1 to both the olfactory bulb and brainstem areas (Thorne et al., 2004). In addition, NPs can be synthetically engineered to bypass the BBB and gain direct entry into the brain depending on their sizes and engineered chemical compositions (Lockman et al., 2004). The other emerging concern about NPs is that NPs could cross the placenta barrier in pregnant mice and lead to neurotoxicity in their offspring, and more severely maternal inflammatory cytokines induced by NPs can also cross the placenta and affect fetal brain development (Jonakait, 2007; Meyer et al., 2009). Li et al. (Yang et al., 2014) have found that intravenous injection of silica (70 nm) and titanium dioxide NPs (35 nm) in pregnant BALB/c mice significantly increased fetal resorption rates. In a similar study, Yamashita et al. (2011) investigated the neurotoxicity of 70-nm silica NPs and 35-nm titanium dioxide NPs. The silica NPs and titanium dioxide NPs were detected in the placenta,

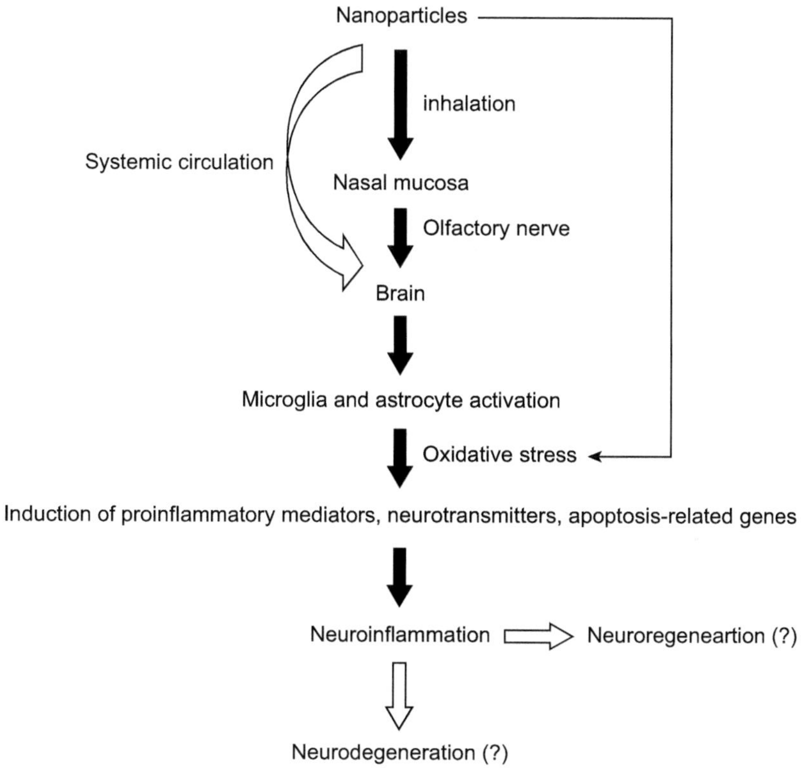

FIGURE 10.1 Potential pathways of nanoparticle-induced neurotoxicity. *Adapted from Win-Shwe, T.T., et al., 2011. Nanoparticles and Neurotoxicity. Int. J. Mol. Sci. 12, 6267–6280.*

fatal brain, and fetal liver and might be responsible for the observed pregnancy complications in the pregnant mice including smaller uteri and smaller fetuses than untreated control groups. However, larger size (from 300 to 1000 nm) of silica NPs did not generate these pregnant complications. Notably, these toxic effects disappeared when the surface of the silica NPs was modified with carboxyl and amine groups. All these observations demonstrated that silica NPs have a severe adverse effect on pregnancy, fetal growth and development, and parturition (Gabbe et al., 1991; Feldt-Rasmussen and Mathiesen, 2011).

The foremost step in the evaluation of the neurotoxicity of SiO_2 NPs is to explore their local biodistribution and accumulation in the brain. One study found that intranasal instillation of SiO_2 NPs for 1 and 7 days resulted in a significant increase in SiO_2 NPs content in the brain. The results also showed that SiO_2 NPs are present in subbrain regions in descending order as olfactory bulb, striatum, hippocampus, brain stem, cerebellum, and frontal cortex. The olfactory bulb is the most enriched in silica NPs, likely through the olfactory nerve pathway bypassing the BBB, as discussed earlier. Besides the olfactory bulb, the striatum seemed to be the other major distribution site of intranasal instillation of SiO_2 NPs. The striatum, a crucial component of the brain, plays essential roles in the regulation of movement pathways and is closely associated with neurodegenerative diseases such as Parkinson's disease and Huntington's disease (Huot et al., 2007). Moreover, studies also found that the SiO_2 NPs content in the hippocampus was behind that of the striatum. The hippocampus is responsible for learning and long-term memory, which is one of the first regions of the brain to suffer from damage in patients with AD (Laakso et al., 2000). Therefore it is essential to determine whether the presence of SiO_2 NPs in the striatum and hippocampus could induce potential injuries and increase the risk of developing neurodegenerative disorders. Xie et al. (2010) utilized a stable-isotopic tracing approach based on their published method to detect the distribution, localization, and accumulation of SiO_2 NPs in the brain after intranasal instillation and consequently to estimate the potential particle-induced injuries and function alterations in the brain. Whether silica NPs deposited in brain or not has been observed from repeated dermal and oral administrations for 90 days (Shim et al., 2014). After 90 days of repeated dermal and oral administration of SiO_2 NPs, there is no deposition of NPs detected in hippocampus, striatum, and cerebellum regions using transmission electron microscope analyses. These observations suggest that the BBB was possibly not passed through and was able to block penetration of SiO_2 NPs. However, professional workers exposed to NPs for 5–13 months exhibited symptoms of pulmonary fibrosis, with NPs present in the cytoplasm of pulmonary epithelial and mesothelial cells (Song et al., 2009). It is possible that 90 days may not be long enough to assess full neurotoxicity of NPs in vivo. Unfortunately, relatively few researches have focused on the distribution, localization, and accumulation studies of silica NPs through other pathways than intranasal instillation, so there is an urgent need to address this discrepancy in future research.

3.2 Factors Affecting Neurotoxicity of Silica Nanoparticles

Just as cytotoxicity of silica NPs to various cells such as human peripheral blood mononuclear cells, human embryonic kidney (HEK293) cells, and mouse keratinocytes depends on size, dose, and cell types, the neurotoxicity of silica NPs is also associated with these factors. Wang et al. (2011) reported that 20- and 50-nm SiO_2 induced free radical generation and GSH depletion in a dose-dependent manner, resulting in an increased level of ROS and a lowered GSH level. A dose- and time-dependent decrease in cell viability were also observed, which are associated with elevation of ROS level, especially after 24 h of nano-SiO_2 exposure, suggesting the role of oxidative stress on nano-SiO_2-induced cell death. Compared with 50-nm silica NPs, 20-nm silica NPs seemed to do more harm in terms of the level of increased intracellular oxidative stress and thiobarbituric acid reactive species (TBARS), as well as depleted GSH. Furthermore, silica NPs significantly reduced cell viability, induced cellular apoptosis, and elevated the levels of intracellular ROS in a dose-dependent manner (Xifei et al., 2014). Considering the complicated factors leading to silica NPs neurotoxicity, more and rigid in vitro and in vivo studies are needed.

3.3 Possible Mechanism of Neurotoxicity by Silica Nanoparticles

It is essential to explore the mechanism of silica NP neurotoxicity from different experiments. The widely used in vitro studies are summarized in Table 10.2. Several studies focused on pheochromocytoma cells (PC12 cells), a neuroendocrine cell line that has the capability to produce the neurotransmitter dopamine (DA) and contains functional DA metabolism pathways. Normal PC12 cells are around 25–30 mm, and thus they could extend themselves to several 100 micrometers through cell division (Caruthers et al., 2007). Thus PC12 cells have been widely used in both neurobiological and neurotoxicological studies as a model of neuronal differentiation (Das et al., 2004; Pisanic et al., 2007; Qiao et al., 2005). An important feature of PC12 cells is that they respond to nerve growth factor (NGF) with a dramatic change in phenotype with a number of properties resembling sympathetic neurons (Greene and Tischler, 1982). Wang et al. (2011) exposed cultured PC12 cells to SiO_2 NPs and evaluated the ability of these cells to respond to NGF and then compared levels of ROS, GSH, as well as TBARS in SiO_2 NPs treated groups with those of controls. The results show elevation of intracellular ROS and TBARS, as well as depletion of GSH in PC12 cells exposed to SiO_2 NPs. Apoptosis induced by SiO_2 NPs in PC12 cells was also examined. Besides, a dose-dependent increase in sub-G1 population in SiO_2-exposed cells was observed. Cell shrinkage and nuclear condensation were also obvious from morphological examination. These observations

TABLE 10.2 In Vivo Studies on Nerve Cells

Cell	Material	Model	References
SK-N-SH, neuro-2a	Silica nanoparticles	Alzheimer's disease	Song et al. (2009)
PC-12	Mesoporous silica nanoparticles	Alzheimer's disease	Xifei et al. (2014)
PC-12	Silica nanoparticles	Neuronal cell	Geng et al. (2012)
SH-SY5Y	Silica nanoparticles	Brain cell	Wang et al. (2011)
SH-SY5Y	Colloidal silica nanoparticle	Neuronal cell	Franziska et al. (2014)
Microglia	Silica nanoparticles	Central nervous system	Yang et al. (2013)

strongly support nano-SiO_2-induced cell apoptosis. Furthermore, nano-SiO_2 exposure diminished the ability of neurite extension in response to NGF in treated PC12 cells. The nervous system is especially vulnerable to oxidative stress (Gitto et al., 2002; Piga et al., 2007) as the brain contains much higher concentrations of polyunsaturated fatty acids susceptible to LPO, whereas it consumes much more oxygen with relatively less antioxidant glutathione peroxidase (GSH-Px) and catalase compared with other organs (Kishido et al., 2007). Concomitant cellular oxidative stress was characterized by elevated ROS levels, reduced GSH levels, and increased TABRS. The inverse linear relationship between ROS and GSH level seen in NPs-treated PC12 cells indicated that free radical species were generated by SiO_2 NPs exposure with reduction of intracellular antioxidant levels. In addition, free radicals also resulted in the production of malondialdehyde (MDA), an indicator of LPO. The elevated ROS level also resulted in an increase in the production of TABRS revealing an elevated LPO. Cell viability appeared to be strongly relevant to ROS level after 24 h exposure in our study (Wang et al., 2011), which suggested that oxidative stress is the main cause of SiO_2 NPs-induced cell death. The inverse linear relationship between cell viability and TBARS level after 24 h exposure demonstrated that the lower viability was likely due to membrane damage by LPO as cell membrane damage can trigger the release of intracellular enzymes such as lactate dehydrogenase (LDH), which induce a chain reaction further leading to cell death (Balduzzi et al., 2004; Sayes et al., 2005). Many studies from other research groups have found that the elevated level of LDH in cell culture medium after the cells are exposed to NPs correlates with cell death (Hussain et al., 2005; Lin et al., 2006). Furthermore, in our study, most of the cells exposed to SiO_2 NPs showed cell shrinkage and nuclear condensation, both of which are important markers for apoptosis. Therefore the low cell viability of PC12 may also involve mechanisms of apoptosis. Flow cytometric analysis indicated that SiO_2 NPs can cause phase arrest in a dose-dependent manner, and subsequently lead to the apoptotic sub-G1 phase after 24 h exposure. The observation suggested the sequential events of cell cycle arrest happened before apoptosis. We also explored the effects of SiO_2 NPs on human embryonic kidney cells (Wang et al., 2009). Based on the results of cytotoxicity studies from PC12 and human embryonic kidney cells, we found excessive intracellular ROS and oxidative stress contributed to silica NPs-induced cytotoxicity. Our discoveries are consistent with the predominant theory on NPs (Hamilton et al., 2008; Li et al., 2001; Simon et al., 2000).

Beside PC12 cells, microglia cells also play an important role against the disruption of brain homeostasis and prevent the entry of foreign particles or infectious agents as the sentinels of the CNS. Once activated, microglia can undergo morphological and functional changes, including proliferation, migration to the site of injury, and phagocytosis of cellular debris. Activated microglia can also generate ROS and reactive nitrogen species (RNS) and induce inflammatory responses (Streit et al., 2005, 1999; Tambuyzer et al., 2009). Thus microglia is among the brain cells most vulnerable to the entry of NPs, which has been suggested in several studies (Kircher et al., 2003; Ribot et al., 2007; Voisin et al., 2007). Choi et al. (2009) found that a significant increase of ROS and RNS was induced by very low levels of silica NPs. ROS generated by microglia are capable of damaging neurons in the vicinity. Silica NP-induced microglial dysfunction can lead to their inability to maintain neuronal health. Xifei et al. (2014) investigated the cellular uptake of silica NPs in human SK-N-SH and mouse neuro2a (N2a) neuroblastoma cells. These cells were treated with 10.0 g/mL of 15-nm silica NPs for 24 h by transmission electron microscopy. They found that silica NPs were mainly localized in the cytoplasm of the treated cells. After exposure to silica NPs, the morphology of SK-N-SH and N2a cells were undermined and the amount of round cells increased, whereas dendrite-like processes disappeared and cell density decreased. Obviously silica NPs reduced cell viability, elicited cellular apoptosis, and induced elevated levels of intracellular ROS in a dose-dependent manner in both cell lines. Because oxidative stress and inflammation are generally recognized to be the main pathogenic mechanisms of AD, silica NPs may have potential effects on the pathological development of AD. Human SH-SY5Y neuroblastoma cell line, which has the capability of undergoing neuronal maturation and differentiation into neuron-like cells with neuritis by all-trans-retinoic acid stimulation (Nostrandt and Ehrich, 1992), has been used as an in vitro cell model for studying potential mechanisms of neurotoxicity as inhibition of neurite formation is a valid indicator of neurotoxicity (Tosetti et al., 1998; Uberti et al., 2002; Sayes et al., 2007). To explore the underlying mechanism for colloidal silica NPs-induced neurotoxicity, Kim and Yang (2011) assessed the ability of colloidal silica NPs to interfere with the outgrowth of neurites in human SH-SY5Y neuroblastoma cells. Moreover, they evaluated the molecular mechanism underlying the induced cellular effects observed for colloidal silica NPs, including DNA damage, intracellular ROS, cell cycle arrest, and apoptosis in both differentiating and undifferentiated SH-SY5Y cells.

The results of in vitro experiments of various NPs did not correlate well with those of in vivo studies. To fully understand the potential effects of silica NPs on the brain, in vivo studies based on animal models must be carried out (Nostrandt and Ehrich, 1992). In particular, many nanotoxicological studies on the brain were performed on fish. Li et al. (2014) used adult zebrafish as an animal model to evaluate the effect of size (15 and 50 nm) and concentration (300 and 1000 mg/mL) of SiO_2 NPs on their neurotoxicity via behavioral and physiological analyses. They found that zebrafish treated with 15-nm SiO_2 NPs exhibited more apparent variation on advanced cognitive neurobehavioral patterns (color preference), possibly resembling the behavior in patients with Parkinson's disease. Based on the analysis at the tissue, cell, and molecular levels together with the correlated behavioral observations, it was concluded that silica NPs acted on the retina and dopaminergic neurons and finally changed color preference of zebrafish resulting in behavior change resembling Parkinson's disease. Based on the above-mentioned results, a neurodegenerative disease behavioral model based on zebrafish was produced. Considering the extensive application of silica NPs, especially in food, elucidation of the toxic effect and underlying mechanisms is the prerequisite for reducing the potential risk toward various diseases, such as pulmonary and neurodegenerative disorders. Previously evidence of NPs-related disease in humans was identified in seven Chinese workers from a print factory where a polyacrylic ester paste containing NPs was used. These workers suffered from an unusual and progressive lung disease and two workers died of the disease (Fruijtier-Polloth, 2012). Owing to the distinctive discrepancy between in vitro and in vivo studies, more in vivo studies are necessary to further assess the human health risks of ambient NPs and to aid in the establishment of regulations regarding exposure to silica NPs.

4. CYTOTOXICITY OF SILICA NANOPARTICLES

With the popularity of silica NPs in biomedical applications, such as drug release (Trewyn et al., 2007), injectable nanovectors, biosensor (Qi sheng et al., 2006; Slowing et al., 2007), and DNA delivery (Fuller et al., 2008), concerns about the safety issue of SiO_2 NPs in terms of their potential release into human tissues and the environment are also arising. Current limited evidences have suggested that silica NPs exposure may induce various autoimmune diseases, such as systemic sclerosis, rheumatoid arthritis, lupus, and chronic renal disease, as reviewed by Steenland and Goldsmith (1995). Another study by Haustein et al. (1990) found silica NPs may induce scleroderma by investigation of patients with long-term exposure to silica dust. Although data are limited, the potential risk of silica NPs toward chronic obstructive bronchitis disease, chronic obstructive pulmonary disease, or emphysema must be included (Merget et al., 2002). More researches have focused on the cytotoxicity of silica NPs to ensure their safety. In vitro cell-based cytotoxicity studies include the use of various human tissue cells such as human lung cancer cells, human liver cell line HepG2, HUVECs, Beas-2B, THP-1, A549, and MCF-7, and numerous animal cells, such as mouse embryonic stem cells, RAW 264.7 cell line, and a rodent fibroblast cell line (Geng et al., 2012; Lin et al., 2006; Eom and Choi, 2009; Mohamed et al., 2011; Ji et al., 2015; Park et al., 2009; Eun-Jung and Kwangsik, 2009; Chung et al., 2007; Fisichella et al., 2009). Among them, HeLa cells derived from the cervix carcinoma are most widely used. For example, Giri et al. (2005) showed that silica NPs at two different concentrations (10 and 100 mg/mL) show little in vitro cytotoxicity toward HeLa cells. In another study, Lu et al. (2009) observed that 20-nm silica NPs induced cytotoxicity, apoptosis, and oxidative stress in HeLa cells in both dose- and time-dependent manner.

4.1 Factors Contributing to Cytotoxicity of Silica Nanoparticles

There are various complex factors contributing to silica NP toxicity. Size and dose are just two factors among them. It is generally accepted that smaller size and higher dose come with increased cytotoxicity (Jie et al., 2007). Slowing et al. (2006) compared the adverse effects of 10-nm silica NPs with those of 100-nm ones on human peripheral blood mononuclear cells. They found that 10-nm silica NPs induced more serious oxidative stress than 100-nm silica NPs and were significantly more cytotoxic. Various cytokines including tumor necrosis factor α, interleukin (IL)-1Ra, IL-6, IL-8, IL-1β, and interferon γ were identified and proved to be expressed and released after exposure. For example, IL-1β production was enhanced by both 10- and 100-nm silica NPs. It seemed that the cytotoxicity of the NPs was associated with an increased ratio of IL-1β/IL-6. Yu et al. (2009) found amorphous silica NPs below 100 nm induced cytotoxicity in mouse keratinocytes. Another study on silica NPs with sizes 20 and 50 nm indicated a dose-dependent cytotoxicity on cultured human embryonic kidney (HEK293) cells. Furthermore, Park et al. (2013) found no apparent cytotoxicity in terms of cell proliferation when the mice were treated with silica NPs in a relatively low dose of 50 mg/kg. Meanwhile, it is necessary to point out that a few studies indicate valid cellular uptake of MSNs could be achieved at concentrations below 50 μg/mL, and the dose of 100 μg/mL in different mammalian cells may not cause cytotoxic effects (Chung et al., 2007; Fisichella et al., 2009; Giri et al., 2005; Lu et al., 2009; Jie et al., 2007; Slowing et al., 2006).

Besides the factors of size and dose, physical–chemical properties of NPs, surface area, and cell type in study also contributed to the cytotoxicity observed. One study explored RAW 264.7 murine macrophages treated with three kinds of 50-nm fluorescent silica-based NPs (Yu et al., 2009) and showed that the highly positively charged NPs were mostly adsorbed on the cell surface, which triggered more cytotoxicity than other NPs. Another study compared negatively charged (NC) and weakly negatively charged (WNC) silica NPs in their toxicity against keratinocytes and skin fibroblast cells, and showed that the NC silica NPs were more toxic than WNC silica NPs with the same size (Park et al., 2013). In addition, researchers also found that the cellular uptake of silica NPs and the observed cytotoxicity are cell type dependent. A comparison of HUVEC cells and HeLa cells found that the uptake of silica NPs into the cytoplasm is up to 10 times faster in HUVEC cells than in HeLa cells. Furthermore, silica NPs led to the death of HUVEC cells, whereas HeLa cells survived, demonstrating that HUVEC are more sensitive to silica NP exposure than HeLa cells (Mohd Javed et al., 2010; Julia et al., 2013). Tian et al. (2011) have observed that cancer epithelial cells were highly enduring to NP treatment, whereas macrophages were easier to be influenced by the toxicity probably due to different physiological functions between these two types of cells. The effects of other physical–chemical properties of silica NPs on health, such as porosity, chemical purity, surface chemistry, and solubility, are less well studied; therefore no definite conclusions can be drawn.

The morphology of silica NPs and exposure time to tissue cells are also essential factors that need to be verified (Ye et al., 2010). HepG2 treated with 43-nm silica NPs at various concentrations for exposure time of 3 and 24 h revealed mitochondrial membrane damage and exposure time–dependent cell apoptosis (Mohamed et al., 2011). The structural morphology of silica nanomaterials dramatically affects interaction with cells; a study result indicated that silica NPs have much less cytotoxicity than nanowires, even at the highest concentrations tested (Ji et al., 2015). Winter et al. (Meike et al., 2011) compared the in vitro influences between amorphous silica NPs and fine crystalline silica, and observed that both activated the inflammatory pathway and led to apoptosis, although some researchers classified crystalline silica as a group 1 carcinogen (Zuckerman, 1995) while synthetic amorphous silica as wet process, thermal or fumed, and chemically or physically modified silica. Moreover, there is remarkable diversity between natural amorphous silica and synthetic amorphous silica. The adverse health effects of nonsynthetic amorphous silica are often attributed to contamination with small amount of crystalline silica (Merget et al., 2002). Synthetic amorphous silica, such as colloidal silica, fumed silica, and precipitated silica, is not involved in progressive lung fibrosis (Lee and Kelly, 1992; Reuzel et al., 1991). However, high doses of amorphous silica may result in acute pulmonary inflammatory responses (Rosenbruch, 1992). The cytotoxicity of amorphous silica and MSNs were studied in depth as developmental materials for drug delivery and sensors (Slowing et al., 2008; Lin et al., 2005; Vallet-Regi, 2010).

The toxicity of silica NPs does not depend on their physical and chemical properties alone. It is also associated with their administration route (Hudson et al., 2008). Warheit (2001) concluded that some forms of amorphous silica are more potent in producing adverse pulmonary effects than others based on the inhalation toxicity data analysis. Moreover, in vivo inhalation studies involving synthetic amorphous silica showed at least partially reversible inflammation (Arts et al., 2007; Johnston et al., 2000). Besides inhalation, ingestion as a major route of uptake of NPs into the human body should also be considered (Kim et al., 2015). All in all, cytotoxicity of silica NPs depends on the size, dose, cell type, time, surface area, and structural difference in morphology (Kim et al., 2015; Sohaebuddin et al., 2010).

4.2 Possible Mechanism of Cytotoxicity of Silica Nanoparticles

Many studies have demonstrated that the cytotoxicity of silica NPs is closely linked to oxidative stress, a common index evaluated for the toxicities of artificial NPs (Xia et al., 2006, 2008; Yang et al., 2009). Some researchers argued that the cytotoxic and genotoxic effects induced by NPs might be primarily the consequence of oxidative stress (Kim et al., 2009; Sharma et al., 2009; Mroz et al., 2008). Indeed, previous studies had found oxidative stress as the toxic mechanism of silica NPs (Lin et al., 2006; Eom and Choi, 2009; Kaewamatawong et al., 2006). It is worthy to be noted that ROS generation should be one of the major damaging pathways but may not be the only factor for the toxicity induced by silica NPs (Geng et al., 2012).

ROS such as HO•, O_2^-, and H_2O_2 trigger various physiological and cellular reactions including apoptosis, inflammation, and DNA damages (Ahamed et al., 2011; Asharani et al., 2009; Nel et al., 2006). Ahmad et al. (2012) explored the ROS-mediated apoptosis induced by well-modified 14-nm silica NPs in HepG2, and found that oxidative stress is activated by formation of ROS and results in LPO and GSH depletion in a dose-dependent manner. Lei et al. (2011) found silica NPs-treated HepG2 cells led to increased cellular ROS production, which indicated that the observed apoptosis was caused by oxidative stress through the mitochondrial pathway. Therefore silica NPs could initiate oxidative stress and thus activate the apoptotic signaling pathway and induce cell cycle arrest (Wang and Youle, 2009). Studies on human lung epithelial (A549) cells and human skin epithelial (A431) cells in terms of the toxicity of silica NPs also observed oxidative stress–induced ROS, LPO, and depletion of antioxidants (eg, GSH), which also suggested that SiO_2 NPs possibly induce cytotoxicity and apoptosis mediated by ROS generation and oxidative stress (Maqusood, 2013). Duan et al. (2013a) aimed to study the cytotoxicity and mechanisms of silica NPs toward endothelial cells using HUVECs as the model cell line. After 24 h exposure, ROS-generated oxidative damage, increased MDA, and inhibited superoxide dismutase (SOD) and GSH-Px activities were observed. Besides, apoptosis and necrosis levels were increased obviously, whereas the mitochondrial membrane potential decreased significantly.

Silica NPs might increase cell membrane permeability and induce the release of intracellular substances, ROS, cytokines, and chemokines leading to inflammatory responses (Wang et al., 2011; Nabeshi et al., 2011; Passagne et al., 2012; Sergent et al., 2012; Yoshida et al., 2012). Several previous in vitro and in vivo studies revealed that silica NPs induce inflammatory response through nuclear factor kappa B activation and oxidative stress responses (Fruijtier-Polloth, 2012; Lin et al., 2006; Kaewamatawong et al., 2006). Silica NPs also caused DNA damage as they potentially enter the cell nucleus and bind to the DNA phosphate backbone (Chen and von Mikecz, 2005), and may generate ROS, particularly the hydroxyl radical, a highly reactive species that likely induces DNA strand break and base oxidation (Valko et al., 2006). However, NPs with sizes bigger than 200 nm fail to penetrate into the nucleus and interfere with nuclear structure and gene expression.

Silica NPs also promote the risk of cardiovascular diseases based on epidemiological evidences from persons with extensive exposure to environmental NPs (Kang et al., 2012; Zhao et al., 2011; Arden et al., 2004). It is generally accepted that silica NPs can cause an inflammatory response in the lung (Mccarthy et al., 2012; Chen et al., 2013). In a study, 70-nm silica particles may possess a worse effect than 300- or 1000-nm ones on inducing liver damage and inflammatory cytokines (Nishimori et al., 2009). However, little is known about the interaction between silica NPs with platelets. A new in vitro study identified that silica NPs with concentrations of 1, 5, and 25 μg/mL on mouse platelets induced platelet aggregation, oxidative stress, and upregulated intracellular calcium (Nemmar et al., 2015).

Results from in vitro cell-based cytotoxicity studies of various NPs may not necessarily reflect the toxicity observed in in vivo experiments. It is essential to use animal models to further evaluate the in vivo toxicity of SiO_2 for more accurate judgments. There are various types of animal models for studying toxicity, ranging from zebrafish to mice (Adams et al., 2006; Fent et al., 2010; Hoecke et al., 2008). Duan et al. (2013b) have investigated cardiovascular toxicity of silica NPs based on the zebrafish model. The results showed that silica NPs inhibit angiogenesis and hinder heart formation. Xie et al. (2010) explored tissue distribution, subcellular distribution, and potential toxicity of intravenously administered silica NPs in mice for a long-term exposure up to 30 days. The results indicated that silica NPs accumulate mainly in lungs, liver, and spleen and can be retained for over 30 days in the tissue due to endocytosis by macrophages, and then might possibly undermine normal functions of liver when intravenously injected. In short, many studies have explored and evaluated the toxicity of silica NPs in terms of cytotoxicity, genotoxicity, and phototoxicity (Kim and Yang, 2011; Fruijtier-Polloth, 2012; Kumar and Dhawan, 2013; Manke et al., 2013; Park et al., 2013; Musa et al., 2012; Choi et al., 2011).

5. SUMMARY

In conclusion, with the increased popularity of nanotechnology, nanomaterials have been increasingly applied in diverse fields, such as drug delivery, bioimaging, gene delivery, and engineering. However, there are relatively few available studies focused on the toxicity of silica NPs, especially compared with those of titanium dioxide and carbon nanotubes. Besides, the comparable lack of information on the safety or hazards of silica NPs occasionally resulted in contradicting conclusions in different documented reports on the same issue. For all researches, it is essential to exactly describe the characteristics of nanomaterials including size, morphology, composition, surface area, and surface features by the aid of numerous assay methods for more valid comparison of different studies (Sayes and Warheit, 2009). Moreover, careful selection of assay methods and conditions is also vital to characterize and evaluate the physical and chemical performance (Geys et al., 2010; Monteiro-Riviere et al., 2009). Only with the complete description of the NPs and assay conditions being documented can the results of reported data be directly compared among those studies on the similar nanomaterials (Park and Park, 2009; Murdock et al., 2008). According to the existing studies on silica NPs, silica exposure could probably induce various autoimmune diseases including systemic sclerosis, rheumatoid arthritis, lupus, and chronic renal disease. There seems to be various complicating factors other than their sizes and doses, which attribute to the toxicity of silica NPs. Cytotoxicity is also associated with physical and chemical properties of NPs, surface area, and cell type used in the study. The morphology of silica NPs and time of treatment of tissue cells are also essential factors. Besides, the toxicity of silica NPs in vivo is also associated with the route of administration.

Overall, cytotoxicity of silica NPs is tightly linked to but not limited to their sizes, doses, cell type in the study, treatment time, surface areas, and structural discrimination.

Many studies have linked silica NPs-associated cytotoxicity with oxidative stress, a common index for the evaluation of the toxicities of artificial NPs. Cell under oxidative stress is characterized by the increase of ROS, LPO, and MDA and the suppression of SOD and GSH-Px levels. ROS would trigger various physiological and cellular reactions including apoptosis, inflammation, and DNA damage. Existing studies to explore and assess the toxicity of silica NPs include three aspects: cytotoxicity, genotoxicity, and phototoxicity. Based on in vitro toxicity studies on PC12cells, microglia, human SK-N-SH and N2a neuroblastoma cells, and human SH-SY5Y neuroblastoma cells, silica NPs-related neurotoxicity is likely associated with oxidative stress, resulting in apoptosis, inflammation, and DNA damage, with an increasing level of ROS and LPO and depleting level of GSH. In vivo studies on animals such as mice and fish observed the adverse role of SiO_2 NPs including disturbance of neural behaviors, and a possible physiological rationalization for this phenomenon was given. Depending on the size and chemical composition, NPs may bypass the BBB and gain direct entry into the brain (Lockman et al., 2004; Kashiwada, 2006). Overall, in vitro and in vivo studies all suggest the underlying neurotoxicity of SiO_2 NPs and they present a potential hazard for neurodegenerative dysfunctions.

The astonishingly rapid development of nanotechnology has both advantages and disadvantages. On the one hand, it definitely generates a beneficial impact on human life; on the other hand, it also consequently brings concerns about the risks and potential toxicity and harm to human health and environment. Nanomaterials are more prone to induce toxicity due to their acute conductivity, reactivity, and optical sensitivity than other materials with the same chemical characteristics. The complexity of these properties might further impose toxic risks on biological tissues and hamper research to elucidate the underlying mechanisms. Therefore advanced techniques and modern instruments are needed to help the researches to better understand the potential toxic effects of nanomaterials for keeping up with the ever increasing use and more rigid safety requirement of silica NPs. The first and foremost step is to achieve reliable statistics for better understanding the physical and chemical characteristics, particularly molecular mechanisms behind the toxicity of silica NPs. The requirements on health and environmental factors would have a positive influence in commercialization of the nanomaterials ultimately. Considering the extensive applications of silica NPs, particularly in food industries, clarification of the toxic effects and underlying mechanisms is essential for us to better control the risk of various possible diseases, such as pulmonary health and neurodegenerative disorders. Validated data involving silica NP applications, potential toxic effects, and underlying mechanisms would provide researchers with better understanding for this relatively deficient research area of the adverse effect of silica NPs on human health.

REFERENCES

Adams, L.K., Lyon, D.Y., Alvarez, P.J.J., 2006. Comparative eco-toxicity of nanoscale TiO_2, SiO_2, and ZnO water suspensions. Water Res. 40 (19), 3527–3532.

Ahamed, M., et al., 2011. ZnO nanorod-induced apoptosis in human alveolar adenocarcinoma cells via p53, survivin and bax/bcl-2 pathways: role of oxidative stress. Nanomed. Nanotechnol. 7 (6), 904–913.

Ahmad, J., et al., 2012. Apoptosis induction by silica nanoparticles mediated through reactive oxygen species in human liver cell line HepG2. Toxicol. Appl. Pharm. 259 (2), 160–168.

Alric, C., et al., 2008. Gadolinium chelate coated gold nanoparticles as contrast agents for both X-ray computed tomography and magnetic resonance imaging. J. Am. Chem. Soc. 130 (18), 5908–5915.

Alshatwi, A.A., Athinarayanan, J., Periasamy, V.S., 2014. Biocompatibility assessment of rice husk-derived biogenic silica nanoparticles for biomedical applications. Mater. Sci. Eng. C 47, 8–16.

Ametamey, S.M., Honer, M., Schubiger, P.A., 2008. Molecular imaging with PET. Chem. Rev. 108 (5), 1501–1516.

Arden, P., et al., 2004. Cardiovascular mortality and long-term exposure to particulate air pollution epidemiological evidence of general pathophysiological pathways of disease. Circulation 109 (1), 71–77.

Arts, J.H., et al., 2007. Five-day inhalation toxicity study of three types of synthetic amorphous silicas in Wistar rats and post-exposure evaluations for up to 3 months. Food Chem. Toxicol. 45, 1856–1867.

Asharani, P.V., et al., 2009. Cytotoxicity and genotoxicity of silver nanoparticles in human cells. ACS Nano 3, 279–290.

Balduzzi, M., Diociaiuti, M., Berardis, B.D., 2004. In vitro effects on macrophages induced by noncytotoxic doses of silica particles possibly relevant to ambient exposure. Environ. Res. 96, 62–71.

Bhakta, G., et al., 2011. Multifunctional silica nanoparticles with potentials of imaging and gene delivery. Nanomed. Nanotechnol. 7 (4), 472–479.

Bulte, J.W.M., Kraitchman, D.L., 2004. Iron oxide MR contrast agents for molecular and cellular imaging. NMR Biomed. 17 (7), 484–499.

Calderón-Garcidueñas, L., et al., 2002. Air pollution and brain damage. Toxicol. Pathol. 30 (3), 373–389.

Caruthers, S.D., Wickline, S.A., Lanza, G.M., 2007. Nanotechnological applications in medicine. Curr. Opin. Biotechnol. 18, 26–30.

Chen, M., von Mikecz, A., 2005. Formation of nucleoplasmic protein aggregates impairs nuclear function in response to SiO_2 nanoparticles. Exp. Cell Res. 305 (1), 51–62.

Chen, J.M., Tan, M.G., Nemmar, A., 2006. Quantification of extrapulmonary translocation of intratracheal-instilled particles in vivo in rats: effect of lipopolysaccharide. Toxicology 222, 195–201.

Chen, Q., Xue, Y., Sun, J., 2013. Kupffer cell-mediated hepatic injury induced by silica nanoparticles in vitro and in vivo. Int. J. Nanomed. 8 (10), 1129–1140.

Chen, L., et al., 2015a. Intracellular pH-operated mechanized mesoporous silica nanoparticles as potential drug carries. Micropor. Mesopor. Mat. 201, 169–175.

Chen, X., et al., 2015b. Mesoporous silica nanoparticles capped with fluorescence- conjugated cyclodextrin for pH-activated controlled drug delivery and imaging. Micropor. Mesopor. Mat. 217, 46–53.

Choi, J., et al., 2009. Silica-based nanoparticle uptake and cellular response by primary microglia. Environ. Health Perspect. 118, 589–595.

Choi, J.E., Park, Y.-H., Young Lee, E., 2011. A safety assessment of phototoxicity and sensitization of SiO_2 nanoparticles. Mol. Cell Toxicol. 7 (2), 171–176.

Chu, M., et al., 2010. Transfer of quantum dots from pregnant mice to pups across the placental barrier. Small 6 (5), 670–678.

Chung, T.H., et al., 2007. The effect of surface charge on the uptake and biological function of mesoporous silica nanoparticles in 3T3-L1 cells and human mesenchymal stem cells. Biomaterial 28, 2959–2966.

Dai, L., et al., 2015. Dendrimerlike mesoporous silica nanoparticles as pH-responsive nanocontainers for targeted drug delivery and bioimaging. ACS Appl. Mater. Inter. 7 (13), 7357–7372.

Das, K.P., Freudenrich, T.M., Mundy, W.R., 2004. Assessment of PC12 cell differentiation and neurite growth: a comparison of morphological and neurochemical measures. Neurotoxicol. Teratol. 26, 397–406.

David, H., 2013. The next big things are tiny. Lancet Neurol. 12 (1), 31–32.

Dengler, E.C., et al., 2013. Mesoporous silica-supported lipid bilayers (protocells) for DNA cargo delivery to the spinal cord. J. Control Release 168 (2), 209–224.

Duan, J., et al., 2013a. Cardiovascular toxicity evaluation of silica nanoparticles in endothelial cells and zebrafish model. Biomaterial 34 (23), 5853–5862.

Duan, J.C., et al., 2013b. Toxic effect of silica nanoparticles on endothelial cells through DNA damage response via Chk1-dependent G2/M checkpoint. PLoS One 8 (4), e62087.

Elder, A., et al., 2006. Translocation of inhaled ultrafine manganese oxide particles to the central nervous system. Environ. Health Persp. 8, 1172–1178.

Eom, H.J., Choi, J., 2009. Oxidative stress of silica nanoparticles in human bronchial epithelial cell, Beas-2B. Toxicol. In Vitro 23 (7), 1326–1332.

Farokhzad, O.C., Lansar, R., 2006. Nanomedicine: developing smarter therapeutic and diagnostic modalities. Adv. Drug Deliv. Rev. 58, 1456–1459.

Feldt-Rasmussen, U., Mathiesen, E.R., 2011. Endocrine disorders in pregnancy: physiological and hormonal aspects of pregnancy. Best Pract. Res. Clin. Endocrinol. Metab. 25, 875–884.

Fent, K., Weisbrod, C.J., Wirth-Heller, A., Pieles, U., 2010. Assessment of uptake and toxicity of fluorescent silica nanoparticles in zebrafish (*Danio rerio*) early life stages. Aquat Toxicol. 100, 218–228.

Fisichella, M., et al., 2009. Mesoporous silica nanoparticles enhance MTT formazan exocytosis in HeLa cells and astrocytes. Toxicol. In Vitro 23, 697–703.

Franziska, K., et al., 2014. An in vitro toxicity evaluation of gold-, PLLA- and PCL-coated silica nanoparticles in neuronal cells for nanoparticle-assisted laser-tissue soldering. Toxicol. In Vitro 28 (5), 990–998.

Fruijtier-Polloth, C., 2012. The toxicological mode of action and the safety of synthetic amorphous silica-a nanostructured material. Toxicology 294 (2–3), 61–79.

Fuller, J.E., et al., 2008. Intracellular delivery of core-shell fluorescent silica nanoparticles. Biomaterial 29 (10), 1526–1532.

Gabbe, S.G., et al., 1991. Obstetrics: Normal and Problem Pregnancies. Churchill Livingstone.

Gambhir, S.S., 2002. Molecular imaging of cancer with positron emission tomography. Nat. Rev. Cancer 2 (9), 683–693.

Gao, X., et al., 2004. In vivo cancer targeting and imaging with semiconductor quantum dots. Nat. Biotechnol. 22 (8), 969–976.

Gary-Bobo, M., et al., 2012. Cancer therapy improvement with mesoporous silica nanoparticles combining targeting, drug delivery and PDT. Int. J. Pharmaceut. 423 (2), 509–515.

Ge, K., et al., 2015. Defect-Related luminescent mesoporous silica nanoparticles employed for novel detectable nanocarrier. Acs Appl. Mater. Inter 7, 1–10.

Geng, J., et al., 2012. Mesoporous silica nanoparticle-based H_2O_2 responsive controlled-release system used for Alzheimer's disease treatment. Adv. Healthc. Mater. 1 (3), 332–336.

Geys, J., Nemery, B., Hoet, P.H., 2010. Assay conditions can influence the outcome of cytotoxicity tests of nanomaterials: better assay characterization is needed to compare studies. Toxicol. In Vitro 24, 620–629.

Giri, S., et al., 2005. Stimuli-responsive controlled release delivery system based on mesoporous silica nanorods capped with magnetic nanoparticles. Angew. Chem. Int. Ed. Engl. 44, 5038–5044.

Gitto, E., Reiter, R.J., Karbownik, M., 2002. Causes of oxidative stress in the pre- and perinatal period. Biol. Neonate 81, 146–157.

Gong, C., et al., 2012. Methylation of PARP-1 promoter involved in the regulation of nano-SiO_2-induced decrease of PARP-1 mRNA expression. Toxicol. Lett. 209, 264–269.

Gonzalez, L., et al., 2014. Co-assessment of cell cycle and micronucleus frequencies demonstrates the influence of serum on the in vitro genotoxic response to amorphous monodisperse silica nanoparticles of varying sizes. Nanotoxicology 8 (8), 876–884.

Greene, L.A., Tischler, A.S., 1982. PC12 pheochromocytoma cells in neurobiological research. Adv. Cell Neurobiol. 3, 373–414.

Hamilton Jr., R.F., Thakur, S.A., Holian, A., 2008. Silica binding and toxicity in alveolar macrophages. Free Radic. Biol. Med. 44, 1246–1258.

Hanafi-Bojd, M.Y., et al., 2015. Surface functionalized mesoporous silica nanoparticles as an effective carrier for epirubicin delivery to cancer cells. Eur. J. Pharm. Biopharm. 89, 248–258.

Haustein, U.F., et al., 1990. Silica-induced scleroderma. J. Am. Acad. Dermatol. 22, 444–448.

Hemadi, A., et al., 2015. Bioconjugated fluorescent silica nanoparticles for the rapid detection of Entamoeba histolytica. Acta Trop. 145, 26–30.

Hoecke, K.V., et al., 2008. Ecotoxicity of silica nanoparticles to the green alga *Pseudokirchneriella subcapitata*: importance of surface area. Environ. Toxicol. Chem. 27, 1948–1957.

Hudson, S.P., et al., 2008. The biocompatibility of mesoporous silicates. Biomaterial 29, 4045–4055.

Huot, P., Levesque, M., Parent, A., 2007. The fate of striatal dopaminergic neurons in Parkinson's disease and Huntington's Chorea. Brain 130, 222–232.

Hussain, S.M., Hess, K.L., Gearhart, J.M., 2005. In vitro toxicity of nanoparticles in BRL 3A rat liver cells. Toxicol. In Vitro 19, 975–983.

Jain, T.K., et al., 1998. Nanometer silica particles encapsulating active compounds: a novel ceramic drug carrier. J. Am. Chem. Soc. 120 (43), 11092–11095.

Ji, H.S., et al., 2015. Short-term effects of ultrahigh concentration cationic silica nanoparticles on cell internalization, cytotoxicity, and cell integrity with human breast cancer cell line (MCF-7). J. Nanopart Res. 17 (1), 1–9.

Jia, L., et al., 2015. Multifunctional mesoporous silica nanoparticles mediated co-delivery of paclitaxel and tetrandrine for overcoming multidrug resistance. Int. J. Pharmaceut. 489 (1–2), 318–330.

Jie, L., et al., 2007. Mesoporous silica nanoparticles as a delivery system for hydrophobic anticancer drugs. Small 3 (8), 1341–1346.

Jin, Z., Hildebrandt, N., 2012. Semiconductor quantum dots for in vitro diagnostics and cellular imaging. Trends Biotechnol. 30 (7), 394–403.

Johnston, C.J., et al., 2000. Pulmonary chemokine and mutagenic responses in rats after subchronic inhalation of amorphous and crystalline silica. Toxicol. Sci. 56, 405–413.

Jonakait, G.M., 2007. The effects of maternal inflammation on neuronal development: possible mechanisms. Int. J. Dev. Neurosci. 25, 415–425.

Julia, B., et al., 2013. Cellular uptake: uptake kinetics and nanotoxicity of silica nanoparticles are cell type dependent. Small 9 (23), 3970–3980.

Kaewamatawong, T., et al., 2006. Acute and subacute pulmonary toxicity of low dose of ultrafine colloidal silica particles in mice after intratracheal instillation. Toxicol. Pathol. 34 (7), 958–965.

Kang, J.H., et al., 2012. Asian dust storm events are associated with an acute increase in pneumonia hospitalization. Ann. Epidemiol. 22, 257–263.

Kashiwada, S., 2006. Distribution of nanoparticles in the see-through medaka (*Oryzias latipes*). Environ. Health Persp 114, 1697–1702.

Kempen, P.J., et al., 2015. Theranostic mesoporous silica nanoparticles biodegrade after pro-survival drug delivery and ultrasound/magnetic resonance imaging of stem cells. Theranostics 5 (6), 631–642.

Kilpeläinen, M., et al., 2009. In vivo delivery of a peptide, ghrelin antagonist, with mesoporous silicon microparticles. J. Control Release 137, 166–170.

Kim, Y.J., Yang, S.I., 2011. Neurotoxic effects by silica TM nanoparticle is independent of differentiation of SH-SY5Y cells. Mol. Cell Toxicol. 7 (4), 381–388.

Kim, D., et al., 2007. Antibiofouling polymer-coated gold nanoparticles as a contrast agent for in vivo X-ray computed tomography imaging. J. Am. Chem. Soc. 129 (24), 7661–7665.

Kim, S., et al., 2009. Oxidative stress-dependent toxicity of silver nanoparticles in human hepatoma cells. Toxicol. In Vitro 23, 1076–1084.

Kim, I.Y., et al., 2015. Toxicity of silica nanoparticles depends on size, dose, and cell type. Nanomed. Nanotechnol. 11 (6), 1407–1416.

Kircher, M.F., et al., 2003. A multimodal nanoparticle for preoperative magnetic resonance imaging and intraoperative optical brain tumor delineation. Cancer Res. 63, 8122–8125.

Kishido, T., Unno, K., Yoshida, H., 2007. Decline in glutathione peroxidase activity is a reason for brain senescence. consumption of green tea catechin prevents the decline in its activity and protein oxidative damage in ageing mouse brain. Biogerontology 8, 423–430.

Kleinman, M.T., Araujo, J.A., Nel, A., 2008. Inhaled ultrafine particulate matter affects CNS inflammatory processes and may act via MAP kinase signaling pathways. Toxicol. Lett. 178, 127–130.

Kulvietis, V., et al., 2011. Distribution of nanoparticles in the pregnant rat: the morphologic and spectroscopic study. Pap. Anthropol. 20, 218–228.

Kumar, A., Dhawan, A., 2013. Genotoxic and carcinogenic potential of engineered nanoparticles: an update. Arch. Toxicol. 87 (11), 1883–1900.

Kuzminska, M., et al., 2015. Magnetic nanoparticles: improving chemical stability via silica coating and organic grafting with silanes for acidic media catalytic reactions. Appl. Catal. A Gen. 505, 200–212.

Laakso, M.P., et al., 2000. Hippocampus in Alzheimer's Disease: a 3-year follow-up MRI study. Biol. Psychiat. 6, 557–561.

Lanone, S., Boczkowski, J., 2006. Biomedical applications and potential health risks of nanomaterials: molecular mechanisms. Curr. Mol. Med. 6 (6), 651–663.

Lee, N., Hyeon, T., 2012. Designed synthesis of uniformly sized iron oxide nanoparticles for efficient magnetic resonance imaging contrast agents. Chem. Soc. Rev. 41 (7), 2575–2589.

Lee, K.P., Kelly, D.P., 1992. The pulmonary response and clearance of Ludox colloidal silica after a 4-week inhalation exposure in rats. Fundam. Appl. Toxicol. 19, 399–410.

Lei, S., et al., 2011. Cytotoxicity and mitochondrial damage caused by silica nanoparticles. Toxicol. In Vitro 25 (8), 1619–1629.

Li, L., Shou, Y., Borowitz, J.L., 2001. Reactive oxygen species mediate pyridostigmine-induced neuronal apoptosis: involvement of muscarinic and NMDA receptors. Toxicol. Appl. Pharmacol. 177, 17–25.

Li, X., et al., 2014. SiO_2 nanoparticles change colour preference and cause Parkinson's-like behaviour in zebrafish. Sci. Rep. 4 (4), 3810–3818.

Li, X., et al., 2015. Aerosol droplet delivery of mesoporous silica nanoparticles: a strategy for respiratory-based therapeutics. Nanomed. Nanotechnol. 11 (6), 1377–1385.

Lin, Y.S., et al., 2005. Well-ordered mesoporous silica nanoparticles as cell markers. Chem. Mater. 17, 4570–4573.

Lin, W., et al., 2006. In vitro toxicity of silica nanoparticles in human lung cancer cells. Toxicol. Appl. Pharm. 217 (3), 252–259.

Lin, X., et al., 2015. The shape and size effects of polycation functionalized silica nanoparticles on gene transfection. Acta Biomater. 11, 381–392.

Lockman, P.R., et al., 2004. Nanoparticle surface charges alter blood–brain barrier integrity and permeability. J. Drug Target. 12 (9–10), 635–641.

Lu, F., et al., 2009. Size effect on cell uptake in well suspended, uniform mesoporous silica nanoparticles. Small 5, 1408–1413.

Manke, A., Wang, L., Rojanasakul, Y., 2013. Mechanisms of nanoparticle-induced oxidative stress and toxicity. Biomed. Res. Int. 8, 616–617.

Mankoff, D.A., 2007. A definition of molecular imaging. J. Nucl. Med. 48 (6), 18–21.

Maqusood, A., 2013. Silica nanoparticles-induced cytotoxicity, oxidative stress and apoptosis in cultured A431 and A549 cells. Hum. Exp. Toxicol. 32 (2), 186–195.

Margolis, D.J., et al., 2007. Molecular imaging techniques in body imaging. Radiology 245 (2), 333–356.

Masserini, M., 2013. Nanoparticles for brain drug delivery. ISRN Biochem. 2013, 1–18.

Mccarthy, J., et al., 2012. Mechanisms of toxicity of amorphous silica nanoparticles on human lung submucosal cells in vitro: protective effects of Fisetin. Chem. Res. Toxicol. 25 (10), 2227–2235.

Medintz, I.L., et al., 2005. Quantum dot bioconjugates for imaging, labelling and sensing. Nat. Mater. 4 (6), 435–446.

Mehrasa, M., et al., 2015. Electrospun aligned PLGA and PLGA/gelatin nanofibers embedded with silica nanoparticles for tissue engineering. Int. J. Biol. Macromol. 79, 687–695.

Meike, W., et al., 2011. Activation of the inflammasome by amorphous silica and TiO_2 nanoparticles in murine dendritic cells. Nanotoxicology 5 (3), 326–340.

Merget, R., et al., 2002. Health hazards due to the inhalation of amorphous silica. Arch. Toxicol. 75, 625–634.

Meyer, U., Feldon, J., Fatemi, S.H., 2009. In-vivo rodent models for the experimental investigation of prenatal immune activation effects in neurodevelopmental brain disorders. Neurosci. Biobehav. Rev. 33, 1061–1079.

Michalet, X., et al., 2005. Quantum dots for live cells, in vivo imaging, and diagnostics. Science 307 (5709), 538–544.

Mirshafiee, V., et al., 2015. The importance of selecting a proper biological milieu for protein corona analysis in vitro: human plasma versus human serum. Int. J. Biochem. Cell Biol. 1–8.

Mirshafiee, V., et al., 2016. Impact of protein pre-coating on the protein corona composition and nanoparticle cellular uptake. Biomaterials 75, 295–304.

Mishra, A.K., et al., 2014. Nanoengineered mesoporous silica nanoparticles for smart delivery of doxorubicin. J. Nanopart. Res. 16 (8), 1–10.

Mistry, A., Stolnik, S., Illum, L., 2009. Nanoparticles for direct nose-to-brain delivery of drugs. Int. J. Pharm. 379, 146–157.

Mohamed, B.M., et al., 2011. Activation of stress-related signalling pathway in human cells upon SiO_2 nanoparticles exposure as an early indicator of cytotoxicity. J. Nanobiotechnol. 9 (1), 1–14.

Mohd Javed, A., et al., 2010. Nanotoxicity of pure silica mediated through oxidant generation rather than glutathione depletion in human lung epithelial cells. Toxicology 276 (2), 95–102.

Monteiro-Riviere, N.A., Inman, A.O., Zhang, L.W., 2009. Limitations and relative utility of screening assays to assess engineered nanoparticle toxicity in a human cell line. Toxicol. Appl. Pharmacol. 234, 222–235.

Mroz, R.M., et al., 2008. Nanoparticle-driven DNA damage mimics irradiation related carcinogenesis pathways. Eur. Respir. J. 31, 241–251.

Mulder, W.J., et al., 2009. Nanoparticulate assemblies of amphiphiles and diagnostically active materials for multimodality imaging. Acc. Chem. Res. 42, 904–914.

Murdock, R.C., et al., 2008. Characterization of nanomaterial dispersion in solution prior to in vitro exposure using dynamic light scattering technique. Toxicol. Sci. 101, 239–253.

Murphy, C.J., et al., 2005. Anisotropic metal nanoparticles: synthesis, assembly, and optical applications. J. Phys. Chem. B 109 (29), 13857–13870.

Musa, M., et al., 2012. Assessment of DNA damage caused by locally produced hydroxyapatite-silica nanocomposite using comet assay on human lung fibroblast cell line. Mol. Cell Toxicol. 8 (1), 53–60.

Nabeshi, H., et al., 2011. Amorphous nanosilica induce endocytosis-dependent ROS generation and DNA damage in human keratinocytes. Part. Fibre Toxicol. 8, 1.

Nel, A., et al., 2006. Toxic potential of materials at the nano level. Science 311, 622–627.

Nemmar, A., et al., 2004. Possible mechanisms of the cardiovascular effects of inhaled particles: systemic translocation and prothrombotic effects. Toxicol. Lett. 149, 243–253.

Nemmar, A., et al., 2015. In vitro platelet aggregation and oxidative stress caused by amorphous silica nanoparticles. Int. J. Physiol. 7 (1), 27–33.

Ngamcherdtrakul, W., et al., 2015. Cationic polymer modified mesoporous silica nanoparticles for targeted siRNA delivery to HER_2^+ breast cancer. Adv. Funct. Mater. 25 (18), 2646–2659.

Nishimori, H., et al., 2009. Silica nanoparticles as hepatotoxicants. Eur. J. Pharm. Biopharm. 72, 496–501.

Nooney, R.I., et al., 2015. Investigating the colloidal stability of fluorescent silica nanoparticles under isotonic conditions for biomedical applications. J. Colloid Interf. Sci. 456, 50–58.

Nostrandt, A.C., Ehrich, M., 1992. Development of a model cell culture system in which to study early effects of neuropathy-inducing organophosphorus esters. Toxicol. Lett. 60, 107–114.

Oberdorster, G., Sharp, Z., Atudorei, V., 2004. Translocation of inhaled ultrafine particles to the brain. Inhal Toxicol. 16, 437–445.

Palanikumar, L., et al., 2015. Noncovalent polymer-gatekeeper in mesoporous silica nanoparticles as a targeted drug delivery platform. Adv. Funct. Mater. 25 (6), 957–965.

Park, E.J., Park, K., 2009. Oxidative stress and pro-inflammatory responses induced by silica nanoparticles in vivo and in vitro. Toxicol. Lett. 184 (1), 18–25.

Park, M.V., et al., 2009. In vitro developmental toxicity test detects inhibition of stem cell differentiation by silica nanoparticles. Toxicol. Appl. Pharm. 240 (1), 108–116.

Park, Y.H., Bae, C.H., Jang, Y., 2013. Effect of the size and surface charge of silica nanoparticles on cutaneous toxicity. Mol. Cell Toxicol. 9 (1), 67–74.

Passagne, I., et al., 2012. Implication of oxidative stress in size-dependent toxicity of silica nanoparticles in kidney cells. Toxicology 299 (2–3), 112–124.

Peng, C., et al., 2012. PEGylated dendrimer-entrapped gold nanoparticles for in vivo blood pool and tumor imaging by computed tomography. Biomaterial 33 (4), 1107–1119.

Petkar, K.C., et al., 2011. Nanostructured materials in drug and gene delivery: a review of the state of the art. Crit. Rev. Ther. Drug Carrier Syst. 28 (2), 101–164.

Piga, R., et al., 2007. Cytotoxic effects of various stressors on PC12 cells: involvement of oxidative stress and effect of antioxidants. Neuro Toxicol. 28, 67–75.
Pisanic II, T.R., Blackwella, J.D., Shubayev, V.I., 2007. Nanotoxicity of iron oxide nanoparticle internalization in growing neurons. Biomaterial 28, 2572–2581.
Popovtzer, R., et al., 2008. Targeted gold nanoparticles enable molecular CT imaging of cancer. Nano Lett. 8 (12), 4593–4596.
Prasad, P.N., Masters, B.R., 2004. Nanophotonics. J. Biomed. Opt. 1 (1), 432.
Pysz, M.A., Gambhir, S.S., Willmann Jr., K., 2010. Molecular imaging: current status and emerging strategies. Clin. Radiol. 65 (7), 500–516.
Qi sheng, H., et al., 2006. A new class of silica cross-linked micellar core-shell nanoparticles. J. Am. Chem. Soc. 128 (128), 6447–6453.
Qian, J., et al., 2010. Using some nanoparticles as contrast agents for optical bioimaging. IEEE J. Sel. Top. Quant. 16 (3), 672–684.
Qiao, D., Seidler, F.J., Slotkin, T.A., 2005. Oxidative mechanisms contributing to the developmental neurotoxicity of nicotine and chlorpyrifos. Toxicol. Appl. Pharm. 206, 17–26.
Quan, G., et al., 2015. Lactosaminated mesoporous silica nanoparticles for asialoglycoprotein receptor targeted anticancer drug delivery. J. Nanobiotechnol. 13 (1), 1–12.
Re, F., Gregori, M., Masserini, M., 2012. Nanotechnology for neurodegenerative disorders. Maturitas 73 (1), 45–51.
Refuerzo, J.S., et al., 2011. Size of the nanovectors determines the transplacental passage in pregnancy: study in rats. Am. J. Obstet. Gynecol. 204 (6), 546.
Reuzel, P.G., et al., 1991. Subchronic inhalation toxicity of amorphous silicas and quartz dust in rats. Food Chem. Toxicol. 29, 341–354.
Ribot, E., et al., 2007. Microglia used as vehicles for both inducible thymidine kinase gene therapy and MRI contrast agents for glioma therapy. Cancer Gene Ther. 14, 724–737.
Rojas, S., et al., 2015. Novel methodology for labelling mesoporous silica nanoparticles using the 18F isotope and their in vivo biodistribution by positron emission tomography. J. Nanopart. Res. 17 (3), 1–9.
Rosenblum, L.T., et al., 2010. In vivo molecular imaging using nanomaterials: general in vivo characteristics of nano-sized reagents and applications for cancer diagnosis. (Review). Mol. Membr. Biol 27 (7), 274–285.
Rosenbruch, M., 1992. Inhalation of amorphous silica: morphological and morphometric evaluation of lung associated lymph nodes in rats. Exp. Toxicol. Pathol. 44, 10–14.
Sahoo, S.K., Labhasetwar, V., 2003. Nanotech approaches to drug delivery and imaging. Drug Discov. Today 8, 1112–1120.
Sayes, C.M., Warheit, D.B., 2009. Characterization of nanomaterials for toxicity assessment. Wiley Interdiscip. Rev. Nanomed. Nanobiotechnol. 1, 660–670.
Sayes, C.M., Gobin, A.M., Ausman, K.D., 2005. Nano-C_{60} cytotoxicity is due to lipid peroxidation. Biomaterial 26, 7587–7595.
Sayes, C.M., Reed, K.L., Warheit, D.B., 2007. Assessing toxicity of fine and nanoparticles: comparing in vitro measurements to in vivo pulmonary toxicity profiles. Toxicol. Sci. 97, 163–180.
Sergent, J.A., Paget, V., Chevillard, S., 2012. Toxicity and genotoxicity of nano-SiO_2 on human epithelial intestinal HT-29 cell line. Ann. Occup. Hyg. 56 (5), 622–630.
Sharma, P., et al., 2006. Nanoparticles for bioimaging. Adv. Colloid Interf. Sci. 123, 471–485.
Sharma, V., et al., 2009. DNA damaging potential of zinc oxide nanoparticles in human epidermal cells. Toxicol. Lett. 185, 211–218.
Shi, J., Votruba, A.R., Farokhzad, O.C., Langer, R., 2010. Nanotechnology in drug delivery and tissue engineering: from discovery to applications. Nano Lett. 10, 3223–3230.

Shim, K.H., et al., 2014. Assessment of ZnO and SiO_2 nanoparticle permeability through and toxicity to the blood–brain barrier using Evans blue and TEM. Int. J. Nanomed. 9 (2), 225–233.
Show, S., et al., 2015. Bacterial (BKH1) assisted silica nanoparticles from silica rich substrates: a facile and green approach for biotechnological applications. Colloid Surf. B 126, 245–250.
Shraboni, D., Tapan Kumar, J., Amarnath, M., 2002. Inorganic-organic hybrid nanoparticles from n-octyl triethoxy silane. J. Colloid Interf. Sci. 252 (1), 82–88.
Simon, H.U., Haj-Yehia, A., Levi-Schaffer, F., 2000. Role of reactive oxygen species (ROS) in apoptosis induction. Apoptosis 5, 415–418.
Slowing, I., Trewyn, B.G., Lin, V.S., 2006. Effect of surface functionalization of MCM- 1-type mesoporous silica nanoparticles on the endocytosis by human cancer cells. J. Am. Chem. Soc. 128, 14792–14793.
Slowing, I.I., et al., 2007. Mesoporous silica nanoparticles for drug delivery and biosensing applications. Adv. Funct. Mater. 17 (8), 1225–1236.
Slowing, I.I., et al., 2008. Mesoporous silica nanoparticles as controlled release drug delivery and gene transfection carriers. Adv. Drug Deliv. Rev. 60, 1278–1288.
Smith, A.M., et al., 2008. Bioconjugated quantum dots for in vivo molecular and cellular imaging. Adv. Drug Deliv. Rev. 60 (11), 1226–1240.
Sohaebuddin, S.K., et al., 2010. Nanomaterial cytotoxicity is composition, size, and cell type dependent. Part. Fibre Toxicol. 7 (2), 1–17.
Song, Y., Li, X., Du, X., 2009. Exposure to nanoparticles is related to pleural effusion, pulmonary fibrosis and granuloma. Eur. Respir. J. 34, 559–567.
Steenland, K., Goldsmith, D.F., 1995. Silica exposure and autoimmune diseases. Am. J. Ind. Med. 28, 603–608.
Streit, W.J., Walter, S.A., Pennell, N.A., 1999. Reactive microgliosis. Prog. Neurobiol. 57, 563–581.
Streit, W.J., et al., 2005. Role of microglia in the central nervous system's immune response. Neurol. Res. 27, 685–691.
Tambuyzer, B.R., Ponsaerts, P., Nouwen, E.J., 2009. Microglia: gatekeepers of central nervous system immunology. J. Leukoc. Biol. 85, 1–19.
Tan, W., et al., 2004. Bionanotechnology based on silica nanoparticles. Med. Res. Rev. 24 (5), 621–638.
Tao, C., et al., 2014. Binding of CpG oligodeoxynucleotides to mesoporous silica nanoparticles for enhancing delivery efficiency. Micropor. Mesopor. Mat. 204, 91–98.
Thorek, D.L., et al., 2006. Superparamagnetic iron oxide nanoparticle probes for molecular imaging. Ann. Biomed. Eng. 34 (1), 23–38.
Thorne, R.G., et al., 2004. Delivery of insulin-like growth factor-I to the rat brain and spinal cord along olfactory and trigeminal pathways following intranasal administration. Neuroscience 127 (2), 481–496.
Tian, Y., Alexander, M., Hamidreza, G., 2011. Impact of silica nanoparticle design on cellular toxicity and hemolytic activity. ACS Nano 5 (7), 5717–5728.
Tosetti, P., Taglietti, V., Toselli, M., 1998. Functional changes in potassium conductances of the human neuroblastoma cell line SH-SY5Y during in vitro differentiation. J. Neurophysiol. 79, 648–658.
Trewyn, B.G., et al., 2007. Mesoporous silica nanoparticle based controlled release, drug delivery, and biosensor systems. Chem. Inform. 38 (31), 3236–3245.
Uberti, D., et al., 2002. Pergolide protects SH-SY5Y cells against neurodegeneration induced by H_2O_2. Eur. J. Pharmacol. 434, 17–20.
Valko, M., et al., 2006. Free radicals, metals and antioxidants in oxidative stress-induced cancer. Chem. Biol. Interact. 160 (1), 1–40.

Vallet-Regi, M., 2010. Nanostructured mesoporous silica matrices in nanomedicine. J. Int. Med. 267, 22–43.
Voisin, P., et al., 2007. Use of lanthanide-grafted inorganic nanoparticles as effective contrast agents for cellular uptake imaging. Bioconjug. Chem. 18, 1053–1063.
Wahl, M., et al., 1988. Mediators of blood–brain barrier dysfunction and formation of vasogenic brain edema. J. Cerebr. Blood Flow Metab. 8 (5), 621–634.
Wang, C., Youle, R.J., 2009. The role of mitochondria in apoptosis. Annu. Rev. Genet. 43, 95–118.
Wang, Y.-X.J., Hussain, S.M., Krestin, G.P., 2001. Superparamagnetic iron oxide contrast agents: physicochemical characteristics and applications in MR imaging. Eur. Radiol. 11 (11), 2319–2331.
Wang, J.X., et al., 2008. Time-dependent translocation and potential impairment on central nervous system by intranasally instilled TiO_2 nanoparticles. Toxicology 12, 82–90.
Wang, F., Gao, F., Lan, M.B., 2009. Oxidative stress contributes to silica nanoparticle-induced cytotoxicity in human embryonic kidney cells. Toxicol. In Vitro 23, 808–815.
Wang, F., et al., 2011. Oxidative mechanisms contribute to nanosize silicon dioxide-induced developmental neurotoxicity in PC12 cells. Toxicol. In Vitro 25 (8), 1548–1556.
Wang, Y., et al., 2013. Ordered nanoporous silica as carriers for improved delivery of water insoluble drugs: a comparative study between three dimensional and two dimensional macroporous silica. Int. J. Nanomed. 8, 4015–4031.
Warheit, D.B., 2001. Inhaled amorphous silica particulates: what do we know about their toxicological profiles? J. Environ. Pathol. Toxicol. Oncol. 20 (1), 133–141.
Wick, P., et al., 2010. Barrier capacity of human placenta for nanosized materials. Environ. Health Persp. 118, 432.
Xavier, M., et al., 2006. Multivalent effects of RGD peptides obtained by nanoparticle display. J. Med. Chem. 49 (20), 6087–6093.
Xia, T., et al., 2006. Comparison of the abilities of ambient and manufactured nanoparticles to induce cellular toxicity according to an oxidative stress paradigm. Nano Lett. 6, 1794–1807.
Xia, T., et al., 2008. Comparison of the mechanism of toxicity of zinc oxide and cerium oxide nanoparticles based on dissolution and oxidative stress properties. ACS Nano 2, 2121–2134.
Xie, G., et al., 2010. Biodistribution and toxicity of intravenously administered silica nanoparticles in mice. Arch. Toxicol. 3, 183–190.
Xifei, Y., et al., 2014. Uptake of silica nanoparticles: neurotoxicity and Alzheimer-like pathology in human SK-N-SH and mouse neuro2a neuroblastoma cells. Toxicol. Lett. 229 (1), 240–249.
Yamashita, K., et al., 2011. Silica and titanium dioxide nanoparticles cause pregnancy complications in mice. Nat. Nanotechnol. 6, 321–328.
Yan, Y.L., Yang, N., Xue-Fan, G.U., 2013. Progress in research work on surface hydrophobic modification of silica nanoparticles and its effects of emulsification. China Surfactant Deterg. Cosmet. 43 (3), 224–231.
Yang, R.H., et al., 2007. Persistent tissue kinetics and redistribution of nanoparticles, quantum dot 705, in mice: ICP-MS quantitative assessment. Environ. Health Persp. 9, 1339–1343.
Yang, H., et al., 2009. Comparative study of cytotoxicity, oxidative stress and genotoxicity induced by four typical nanomaterials: the role of particle size, shape and composition. J. Appl. Toxicol. 29, 69–78.
Yang, X., et al., 2010. SiO_2 nanoparticles induce cytotoxicity and protein expression alteration in HaCa Tcells. Part. Fibre Toxicol. 196 (11), 273.
Yang, L., et al., 2013. The role of potassium channel in silica nanoparticle-induced inflammatory effect in human vascular endothelial cells in vitro. Toxicol. Lett. 223, 16–24.

Yang, L., Yi, Z., Bing, Y., 2014. Nanotoxicity overview: nano-threat to susceptible populations. Int. J. Mol. Sci. 15 (3), 3671–3697.

Ye, Y., et al., 2010. In vitro toxicity of silica nanoparticles in myocardial cells. Environ. Toxicol. Pharm. 29 (2), 131–137.

Yoshida, T., et al., 2012. Surface modification of amorphous nanosilica particles suppresses nanosilica-induced cytotoxicity, ROS generation, and DNA damage in various mammalian cells. Biochem. Biophys. Res. Commun. 427 (4), 748–752.

Youns, M., Hoheisel, J.D., Efferth, T., 2011. Therapeutic and diagnostic applications of nanoparticles. Curr. Drug Targets 12 (3), 357–365.

Yu, K.O., et al., 2009. Toxicity of amorphous silica nanoparticles in mouse keratinocytes. J. Nanopart. Res. 11 (1), 15–24.

Yun, X., et al., 2007. Bioconjugated quantum dots for multiplexed and quantitative immunohistochemistry. Nat. Protoc. 2 (5), 1152–1165.

Zhang, P., Kong, J., 2015. Doxorubicin-tethered fluorescent silica nanoparticles for pH-responsive anticancer drug delivery. Talanta 134, 501–507.

Zhang, Q., et al., 2012. Multifunctional mesoporous silica nanoparticles for cancer-targeted and controlled drug delivery. Adv. Funct. Mater. 22 (24), 5144–5156.

Zhao, Y., et al., 2011. Kinetics and mechanisms of heterogeneous reaction of gaseous hydrogen peroxide on mineral oxide particles. Environ. Sci. Technol. 45, 3317–3324.

Zhu, R., et al., 2010. Molecularly imprinted layer-coated silica nanoparticles for selective solid-phase extraction of bisphenol A from chemical cleansing and cosmetics samples. Anal. Chim. Acta 658 (2), 209–216.

Zuckerman, A.J., 1995. IARC Monographs on the evaluation of carcinogenic risks to humans. J. Clin. Pathol. 48 (1), 229–230.

Chapter 11

The Synthesis, Application, and Related Neurotoxicity of Carbon Nanotubes

D. Shi[1], G. Mi[1], T.J. Webster[1,2]
[1]*Northeastern University, Boston, MA, United States;* [2]*King Abdulaziz University, Jeddah, Saudi Arabia*

Chapter Outline

1. INTRODUCTION

Although the real interest in carbon nanotubes (CNTs) was revealed in 1991 by an NEC employee in Japan, Sumio Iijima, the original discovery of the CNT can be chased back to as early as the mid-20th century (Радушкевич, 1952). Then in 1976, by using a chemical vapor-growth technique, Morinobu Endo from the University of Orleans in France successfully observed hollow carbon tubes rolled up with a layer of graphite sheet, which later became known as single-walled carbon nanotubes (SWNTs) (Oberlin et al., 1976). In addition, in his review of vapor-phase-grown carbon fibers (VPGCFs) in 1988, Morinobu Endo also indicated a "hollow tube linearly extended with a parallel carbon layer

Neurotoxicity of Nanomaterials and Nanomedicine. http://dx.doi.org/10.1016/B978-0-12-804598-5.00011-8

near the fiber core," which later appeared to be multiwalled carbon nanotubes (MWNTs) at the center of the fiber (Endo, 1988). Owing to Endo's devotion of VPGCFs, the discovery of MWNTs followed right after SWNTs and therefore the process is called "Endo-process" (Endo, 1988). After CNTs were formally introduced by Sumio Iijima in 1991, this promising field quickly thrived and the first polymer resin–nanotube composite was reported by Ajayan et al. (1994) (Monthioux and Kuznetsov, 2006). Since then, all forms of CNTs have attracted attention in almost every avenue of science from electronics to medical applications, with no end in sight regarding their applications.

2. STRUCTURE OF CNTs

Similar to that of diamond and fullerenes, CNTs are formed by rolling up sheets of graphite into a cylindrical shape (Chico et al., 1996). Distinguished by a different number of layers, CNTs can be classified into two main categories with one important subsegment.

SWNTs are tubes that have only one cylindrical wall. Similar to graphene that is formed by a single-atom-thick layer of graphite, the structure of an SWNT looks like a seamless cylinder that is normally capped at the ends. The majority of SWNTs usually are 1 nm in diameter but are thousands or even millions of times longer in length. Compared with MWNTs, SWNTs are more elastic yet difficult to produce. They can be twisted and bent into small circles or be flattened without breaking their structure (Dervishi et al., 2009; Eatemadi et al., 2014; Sinnott and Andrews, 2001). Based on their unique electronic and mechanical properties, SWNTs have specific applications such as nanosensors, nanocomposite materials, tissue engineering, and cancer imaging despite their high cost.

MWNTs, on the other hand, consist of multiple rolled layers of graphene. There are basically two models describing the structure of MWNTs. One is called the "Russian Doll model," which means that they are in the form of a coaxial assembly of SWNTs, and the other is called the "Parchment model," in which a single sheet of graphite is rolled and resembles that of a scroll of parchment, having an interlayer distance close to the one in graphite (which ranges from 0.34 to 0.41 nm) (Ajayan and Ebbesen, 1997; Ren et al., 2002; Vander Wal et al., 2003).

Double-walled carbon nanotubes (DWNTs) are an important subsegment of MWNTs. These nanomaterials have obtained special interest because of their similar morphology and properties to SWNTs with significantly improved chemical resistance, which is especially important while functionalizing those nanotubes. In addition, because the intrinsic properties of DWNTs are preserved when modifying their outer wall, DWNTs tend to be more immune to breakage than SWNTs while maintaining their attractive properties (Vander Wal et al., 2003). As DWNTs exhibit both the flexibility of SWNTs and electrical/thermal stability of MWNTs, they have been investigated for a variety of highly specific applications (such as gas sensors and field-emission displays).

CNTs can also be classified by the orientation of the hexagon with respect to the axis of the nanotube (Fig. 11.1). There are three possible structures for the

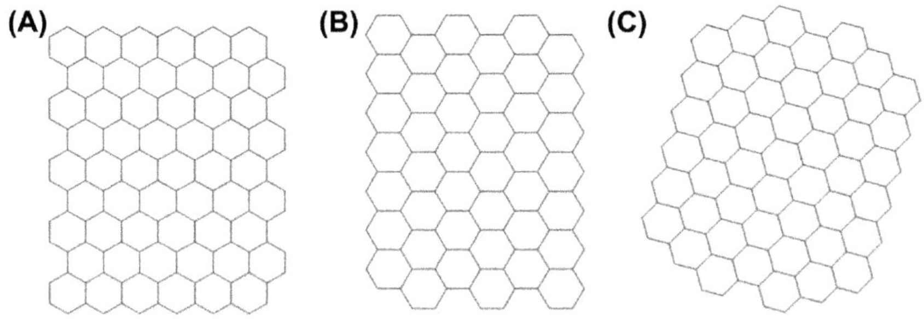

FIGURE 11.1 Various configurations of CNTs. (A) Armchair, (B) zigzag, and (C) chiral. *Adapted from Eatemadi, A., Daraee, H., Karimkhanloo, H., Kouhi, M., Zarghami, N., Akbarzadeh, A., Abasi, M., Hanifehpour, Y., Joo, S., 2014. Carbon nanotubes: properties, synthesis, purification, and medical applications. Nanoscale Res. Lett. 9 (1), 393.*

nanotube: achiral structures including armchair and zigzag that have identical mirror images, and a chiral structure which means their mirror image is asymmetric or nonsuperimposable to the original object (Eatemadi et al., 2014).

3. SYNTHESIS

There are mainly three conventional methods (the chemical vapor deposition technique (Abbasi et al., 2014; José-Yacamán et al., 1993), carbon arc discharge technique (Hirlekar et al., 2009; Hou et al., 2002; Thess et al., 1996), and laser ablation technique (Ajayan and Ebbesen, 1997; Chico et al., 1996)) that have been developed to fabricate CNTs and all of them involve gas phase processes (Table 11.1). The most common technique to produce nanotubes is called chemical vapor deposition (CVD), which is also believed to be the most practical method for manufacturing (Abbasi et al., 2014; José-Yacamán et al., 1993).

Generally, most CNT synthesis techniques can be classified into two types, physical methods and chemical methods. Among various chemical processes, CVD is one of the most applicable techniques in synthesizing CNTs, especially when talking about controllable large-scale production (Li et al., 1996; Sinnott and Andrews, 2001). Taking catalytic CVD as an example, the process involves decomposition of hydrocarbon gas flow (such as ethylene, methane, or ethanol) over transition metal surfaces (usually Ni, Co, and/or Fe) (Rafique and Iqbal, 2011; Journet and Bernier, Landi et al., 2005), under inert gas at ~700°C at atmospheric pressure. Carbon will be precipitated due to the saturation of carbon atoms on the metal particle site, which then forms CNTs as a result (Choudhary et al., 2014; Sayangdev and Ishwar, 2008). Depending on the different types and sizes of the metal catalyst, the diameter of the CNTs can be controlled, whereas the length of the nanotubes can be controlled by reaction time (Journet and Bernier). Because the high purity of the products and the use of metal catalysts reduce the requirement for high-temperature operations, CVD is considered to be an economical and practical technique for commercial manufacturing (Dervishi et al., 2009; Patole et al., 2008; Sinnott and Andrews, 2001)

TABLE 11.1 Comparisons of Three Major CNT Synthesis Techniques

Technique	Purity	Yield (%)	SWNT/ MWNT	Require Catalytic Metal	Temperature/ Energy	Pressure	Scalability
Arc discharge	Low purity but high quality	~60	Both	Yes	~4000°C High energy cost	50–700 mBar	Laboratory-scale production
Laser ablation	Relatively high	~90	SWNT > MWNT	Yes	~1200°C High energy cost	500–700 mBar	Laboratory-scale production
CVD	High	~90	MWNT > SWNT	Yes	500–700°C Low energy cost	Atmosphere	Large-scale production

CNT, carbon nanotube; *CVD*, chemical vapor deposition; *MWNT*, multiwalled carbon nanotube; *SWNT*, single-walled carbon nanotube.
Revised from Eatemadi, A., Daraee, H., Karimkhanloo, H., Kouhi, M., Zarghami, N., Akbarzadeh, A., Abasi, M., Hanifehpour, Y., Joo, S., 2014. Carbon nanotubes: properties, synthesis, purification, and medical applications. Nanoscale Res. Lett. 9 (1), 393.

Besides chemical processes, such as CVD, physical processes such as arc discharge and laser ablation methods also play important roles in producing CNTs (Fig. 11.2) (Sane, 2011). The arc discharge deposition synthesis of CNTs was the first technique used for the production of CNTs by Iijima in 1991. It has a very

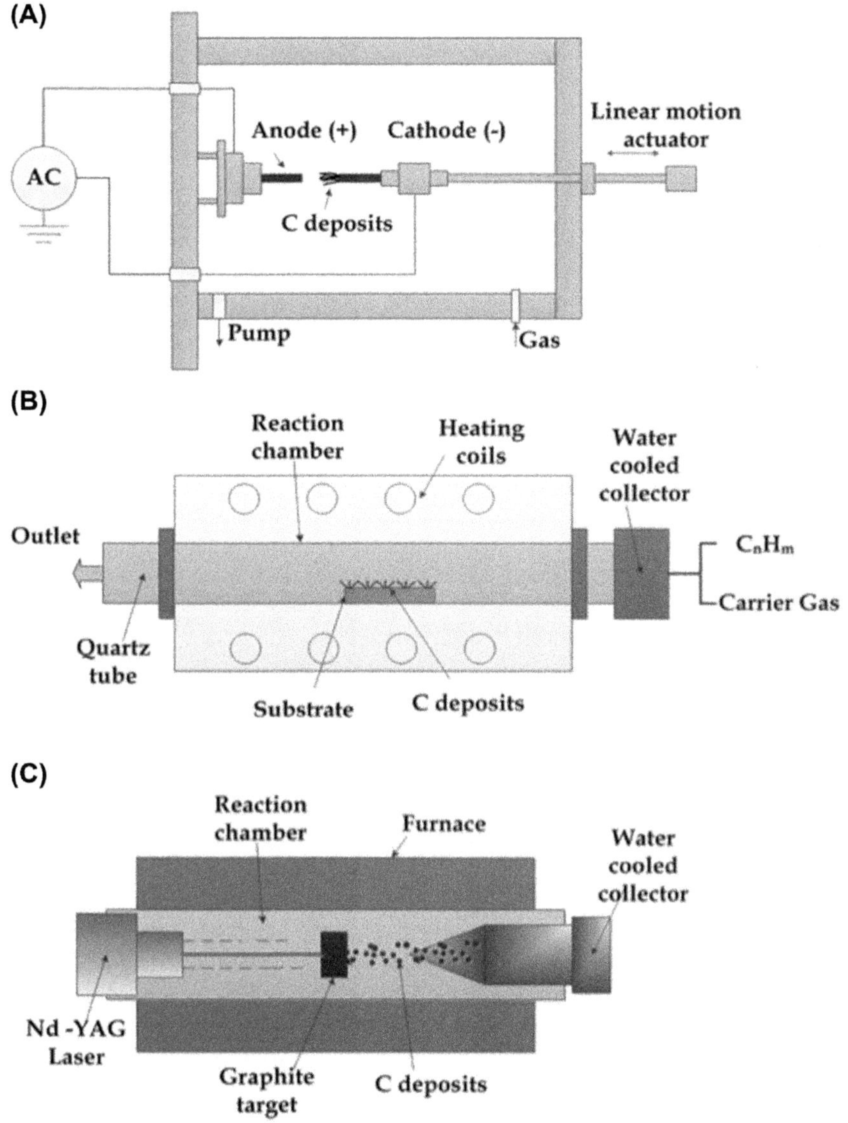

FIGURE 11.2 Schematic representation of methods used for CNT synthesis. (A) Arc discharge, (B) chemical vapor deposition, and (C) laser ablation *(© 2011, Gore, J., Sane, A. Published in Sane, A., Gore, J., 2011. Flame synthesis of carbon nanotubes. In: Yellampalli, S. (Eds), Carbon Nanotubes – Synthesis, Characterization, Applications, InTech, under CC BY-NC-SA 3.0 license. Available from: http://dx.doi.org/10.5772/21012).*

high processing temperature (3000–4000°C) (Eftekhari, 2011), which can lead to a CNT with fewer structural defects, and generally contains two different processing conditions: the synthesis of SWNTs, which involves different catalyst precursors, and the synthesis of MWNTs, which could be accomplished without any catalyst precursors. To be more specific, pure graphite is used for MWNT generation and metal catalyst precursors such as Fe, Ni, and/or Co are used with the pure graphite cathode for SWNT synthesis (Askeland et al., 2010; Hutagalung, 2012). Although this arc discharge method has the advantage of producing relatively good yield (~60%) (Isaacs et al., 2009) and CNTs with fewer defects, the cost and energy consumption plus by-product contamination (such as the mixture of chiral and achiral products as well as purification problem caused by metallic catalysts) keep it from wide application (Feng, 2015; Jones et al., 1996).

In laser ablation or pulsed laser deposition method, material from a solid surface is removed by irradiating the surface with a pulsed laser beam (Fig. 11.2). Similar to the electric-arc discharge technique that is used for CNT formation, pure graphite blocks are used to produce MWNTs and graphite mixed with catalytic metals (such as Co, Ni, Pt, and/or Cu) are introduced for SWNTs synthesis (Guo et al., 1995a,b). The mixture of the graphite/metal block is made with graphite powder, carbon cement, and the catalytic metals, and the block is ready to use after the paste is placed in a mold and backed for hours (Nguyen, 2006).

4. MODIFICATION/FUNCTIONALIZATION

4.1 Covalent Functionalization

Despite the wide use of CNTs, there are several limitations that hinder CNTs from further applications. First of all, owing to the significantly large surface area, the surface energy of CNTs differs from that of the host matrices, such as organic solvents. Hence, these nanotubes are known to aggregate during fabrication and the dispersion of CNTs remains a big obstacle in practice (Gupta and Choudary, 2011; In-Yup Jeon, 2011; Song and Youn, 2005). In addition, no physical interaction with the host matrices can be provided because of the seamless, defect free, surface of the CNTs (Ajayan et al., 2000). Other concerns about conventional CNT synthesis techniques (such as arc discharge, laser ablation, and CVD) are that they all are hindered by high-temperature processing (>700°C) and metal contaminations (Ebbesen and Ajayan, 1992; Guo et al., 1995a,b, Kumar and Ando, 2010).

Several surface modification and functionalization methods have been addressed in recent years to improve their dispersion abilities and optimize their interactions with polymeric matrices, and those methods can be mainly classified into two categories, covalent functionalization strategies (eg, functionalize defect sites/sidewalls) and noncovalent functionalization strategies (eg, surfactant or polymer wrapping) as described later.

Owing to the high aspect ratio (usually >1000) and tangled structure of conventionally synthesized CNTs, dispersion of those nanotubes remains a big problem. Compared with the highly reactive, fullerene-like end caps, the sidewall of a CNT usually contains several defect sites, such as Stone-Wales defects, *sp3*-hybridized defects, and vacancy defects (Hirsch, 2002, Sharma et al., 2012). Covalent functionalization can be accomplished by modifying defect sites of the CNTs, such as introducing the oxidizing group (eg, strong acid groups or strong oxidants) onto the defect sites to form carboxylic acid (—COOH), alcohol, or hydroxyl group (—OH) on the CNT surface (Coleman et al., 2000; Singh et al., 2009).

This oxidation method has been reported to enhance dispersion of CNTs and improve their stability in polar solvents and, therefore, can be introduced to water-soluble polymers to perform homogeneous dispersions (Feng et al., 2008; Zhao et al., 2008). In addition, those acid functional groups can also function as useful precursors for other chemical reactions including grafting polymer chains on the surface of nanotube side walls and improve the interfacial interactions between CNTs and polymeric matrices, which can lead to various further modifications with solution-based chemistry (Kong et al., 2004; Regev et al., 2004; Sano et al., 2001).

Another approach to covalently functionalize CNTs is known as the direct covalent sidewall functionalization technique, which engages the switch of hybridization of carbon atoms from sp^2 to sp^3 with the loss of conjugation on the graphene layer. The process of direct covalent functionalization involves reaction with some molecules, for example, fluorine or chlorine, which has a high chemical reactivity. Compared with the two-step modification of CNTs by introducing oxidizing groups, direct covalent functionalization directly coupled the functional groups onto the framework, which not only has a strong effect on mechanical properties, but also enhances the solubility of the CNTs while providing additional functional sites for CNTs (Touhara et al., 2004). Based on different reaction species, direct covalent functionalization can be divided into several derivatives such as fluorination and derivative reactions, hydrogenation, cycloaddition, and radical attachment (Kim et al., 2002; Ma et al., 2010).

4.2 Noncovalent Functionalization

Rather than chemical functionalization approaches, physical approaches, also known as noncovalent functionalization approaches, have the advantages of not only preserving the intrinsic properties of the sidewalls but also improving CNT solubility. One of the most common noncovalent functionalization methods is to apply surfactant or employ polymer wrapping by van der Waals interactions and π–π stacking (Ma et al., 2010), which can insulate each individual nanotube by either lowering the surface tension or forming supramolecular complexes and, thus, lead to homogeneous dispersion in polar or organic solvents (Matarredona et al., 2003; Regev et al., 2004; Vigolo et al., 2000). Several variations of surfactants (anionic, cationic, and nonionic surfactants) and polymers have been studied and results indicated that the capability of these polymers to disperse CNTs is closely related to their hydrophilic groups (Moore et al., 2003; O'Connell et al., 2001). For instance, O'Connell reported that wrapping SWNTs with polymers like poly(vinyl pyrrolidone) or poly(styrene sulfonate) that contains polar side chains would present CNT/polymer complexes and form a stable solution (O'Connell et al., 2001). In addition, a study shows that employing porphyrin derivatives or FITC-terminated polyethylene glycol (PEG) chains onto the CNT surface can also lead to enhanced biocompatibility and reduced toxicity (Yan et al., 2011).

5. CARBON NANOTUBE-RELATED APPLICATIONS

5.1 Tissue Engineering

Conventional bone regeneration therapies (like autografts, allografts, or metal prostheses) function as artificial materials and, although they allow for bone growth juxtaposed to the biomaterial, they do not improve regenerating bone tissue itself. As a result, one of the central concerns in modern biomedical engineering is to develop materials that possess similar mechanical and biological properties to bone, and hence stimulate bone tissue regeneration in trauma or disease. Because the structure and function of bone tissue is intimately related to its micro- to nanoscale cellular arrangement, a good scaffold for bone tissue engineering is required to mimic the bone surface roughness to the maximize bone growth, in addition to having a large porosity, inhibiting infection, reducing inflammation, and having overall good biocompatibility (Karageorgiou and Kaplan, 2005). Clearly, CNTs possess dimensions that closely resemble that of bone constituents (such as collagen), which has lead them for extensive investigation in orthopedics.

Although the specific mechanism of whether CNTs are internalized by either endocytosis or penetration has yet to be decided (Kam et al., 2006; Kostarelos et al., 2007), CNTs have shown potential and promise in the tissue engineering field based on their unique physical and chemical properties (such as large surface area and accessibility for multiple derivatives) (Kostarelos et al., 2009; Lacerda et al., 2008a,b; Singh et al., 2006; Zhang and Webster, 2009). As discussed earlier, the sidewall of the CNTs can be either covalently or noncovalently modified with various biologically activated molecules, and behave as therapeutic agents or cargo carriers to transport and deliver biochemical and/or pharmaceutical stimulus (Klingeler and Sim, 2011). Current tissue engineering applications of CNTs include delivering antitumor agents (such as antibodies or peptides) for immunotherapy (Chakravarty et al., 2008; Fan et al., 2012; Madani et al., 2011), and being employed as nonviral molecules to deliver short interfering RNA (siRNA) into human T cells and primary cells (Fig. 11.3) (Liu et al., 2007).

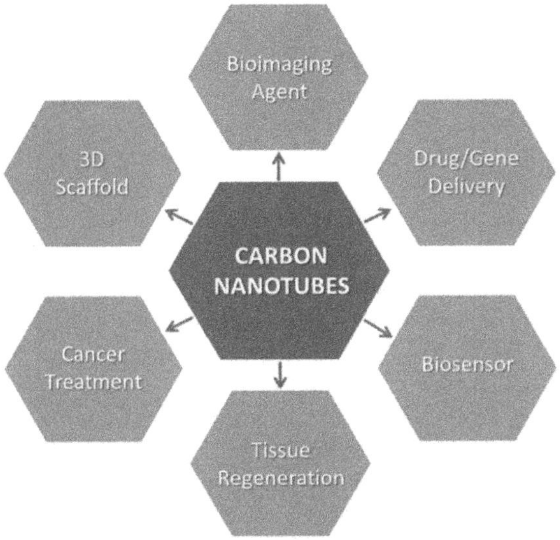

FIGURE 11.3 CNT-related applications in medicine.

5.1.1 Bone Tissue Engineering

Previous studies have indicated that carbon-based nanomaterials improved bone tissue implant integration (Hartgerink et al., 2001; Martin and Kohli, 2003; Thomas et al., 2004; Webster, 2001). As postadministration rejection of artificial joints remains a large problem in bone regeneration, both SWNTs and MWNTs have been used to diminish host rejection by attaching CNTs to proteins and/or amino acids. More interestingly, with the ability to offer higher tensile strength with lighter weight than steel, CNTs can also behave as bone itself if combined with calcium and, thus, mimic the shape/roughness of the natural bone structure (Bian et al., 2004; Marquis, 2003). Other studies showed that the nanofibrous geometry of hydroxyapatite (HA)-incorporated SWNT chitosan nanocomposites is able to simulate extracellular matrix production and trigger osteoblast adhesion and proliferation during bone repair (Im et al., 2012; Webster, 2001).

In various studies, CNTs have been applied either alone or in combination with other biomaterials to enhance their mechanical properties. For instance, Shokuhfar et al. (2008) mentioned that the combination of HA/polymethyl methacrylate (PMMA) with CNTs significantly increased the strength and toughness of HA and overall had improved mechanical properties than traditionally synthesized HA/PMMA nanocomposites. Moreover, porous nanotubes like ultrashort SWNTs and functionalized ultrashort SWNTs have also been considered as injectable scaffolds for bone tissue regeneration strategies based on their ability to act like an inert matrix and promote cells to deposit new living materials (Shi et al., 2007; Sitharaman et al., 2007).

5.1.2 Neural Tissue Engineering

The complexity of neural systems makes neural tissue engineering one of the most challenging as well as most investigated areas in biomedical research. Generally, neurodegeneration or neuron loss can be caused by various factors including external trauma, anoxia, diabetes, and viral infection. Diseases like Parkinson, Alzheimer, and Huntington usually occur as a result of neurodegeneration (Hague et al., 2005). Although nerve axons exhibit some ability to regenerate themselves, the length is limited to about 5 mm and surgical co-optation is usually required (Sedaghati et al., 2011). Other than co-optation, traditional treatments such as nerve grafts (autografts, allografts, and xenografts) and nerve conduits have revealed several disadvantages such as tissue rejection, incomplete function recovery, prolonged immunosuppression, and instability (Marcaccio and Paolucci, 2014). Some of the key configurations for neural tissue engineering design, addressed by Sedaghati et al. and Arslantunali et al., are to mimic the scaffold nanotopography to support cell distribution, promote axonal growth in three dimensions, appropriate tensile strength, and develop material flexibility (Arslantunali et al., 2014; Bradbury and McMahon, 2006; Sedaghati et al., 2011). Therefore designing a biocompatible and biostable material that meets all of these requirements for neural regeneration and engineering becomes necessary.

Several experiments have been carried out and one of the most promising candidates for modern neural tissue regeneration are CNTs, not only because they have a structure that can possibly mimic neural fibers but also because of their relatively high conductivity essential for neuron cells to communicate and function. It has been indicated that different types of neuronal cells exhibit high viability on various functionalized CNT surfaces with different nanotopographies for neural interaction and maturation. For instance, Hu et al. (2004) showed that 4-hydroxynonenal-coated CNTs improved hippocampal neurons long-term viability and cell growth for embryonic Sprague-Dawley rats. In addition, based on the ability to increase the tensile strength and electrical conductivity for polymer scaffolds (such as poly-L-lactic acid) or biomaterials such as collagen, CNTs have been used to combine with these materials to reinforce their properties. Experiment carried out by Cho et al. demonstrated that CNT-collagen scaffolds significantly improved neuronal cell growth and differentiation in pheochromocytoma rats, even at concentrations as low as 5% (Cho and Borgens, 2010). More importantly, they also indicated that CNT-collagen nanocomposites could facilitate neurite spreading when electrical stimulation was applied. Other studies revealed that aligned CNTs (like MWNT-chitosan scaffolds) have the potential to be designed to direct neuron cell migration and create neural networks (Bian et al., 2004; Gabay et al., 2005; Gupta et al., 2015). Furthermore, it has also been demonstrated that carboxylated MWNTs were able to induce human bone marrow mesenchymal stem cell differentiation without any external differentiating cues (Chen and Hsiue, 2013). The exact mechanisms with which CNTs affect neural function are still under investigation. One possible route proposed for this conductive CNT-induced neuron stimulation and directive growth is called "electrotonic hypothesis." It suggested that CNT-generated electrical shortcuts between axons and dendrites by tightly attaching to the cell membranes, which lead to dendritic interaction directing calcium electrogenesis. As a result, the increased calcium concentration causes a prolonged potential, which facilitates synaptic formation and excites neural networks (Cellot et al., 2009).

5.2 Cancer Diagnostics and Treatment

For the past few decades, research groups across the world have been focusing on the development of safer and more efficient techniques to overcome carcinoma, and have thoroughly investigated a variety of nanomaterials including quantum dots, gold nanoparticles, CNTs, dendrimers, liposomes, and micelles for cancer treatment.

CNTs have attracted tremendous attention as one of the most promising nanomaterials for multimodality cancer therapy and imaging because of their unique physiochemical, electromagnetic, and optical properties, which cannot be concurrently found in any other nanomaterial. Ultrahigh surface area and high internal volume allows for a variety of small molecules, such as chemotherapeutic agents, to be loaded onto or inside the nanotubes. The high reactivity of the CNT surfaces enables easy functionalization with multiple modalities to target cancer cells. The unique physical properties of CNTs permit efficient electromagnetic stimulation for thermal ablation and highly sensitive detection using various imaging modalities. These unique properties of CNTs make them suitable candidates for drug delivery, thermal ablation, and early diagnosis in cancer.

5.2.1 Target Recognition

The first question that always arises when investigating any drug delivery system is how these systems can recognize their site of action accurately and efficiently. Owing to the characteristics of the tumor microenvironment, such as changes in vascular pathophysiology, targeted delivery involves two main mechanisms: passive targeting by the enhanced permeability and retention (EPR) effect and active targeting using disease-specific ligands. The EPR effect is a unique phenomenon of solid tumors, where angiogenesis leads large gaps between endothelial cells in tumor blood vessels (Fang et al., 2011). Macromolecular drugs and delivery systems within a certain size range (1–1000 nm) (Petros and DeSimone, 2010) can therefore escape through these abnormally leaky vasculatures and be selectively retained in tumors.

In addition, active targeting using targeting ligands (such as antibodies, proteins, peptides, nucleic acid–based ligands, and small molecules [Bertrand et al., 2014]) has been widely explored to achieve higher selectively and, thus, reducing systematic side effects. The major type of targeting ligands used in the context with CNTs are antibodies that have a high affinity for the target cancer cell. For example, McDevitt et al. found that CNTs can be functionalized with rituximab, an antibody against CD 20, to achieve improved targeting efficiency (McDevitt et al., 2007). In another experiment, a HER2 IgY-SWNT complex was designed to specifically target HER2-expressing SK-BR-3 cells but not receptor-negative cells in breast cancer, and it was also shown in vitro that the resulting complex was indeed able to detect and selectively destruct HER2-expressing cells via photothermal ablation (Xiao et al., 2009).

5.2.2 Drug Loading and Release

Another important aspect of CNT drug delivery systems is the method by which the drug can be loaded into the system and the mode of drug release on reaching the intended target. Three main methodologies of drug attachment have been commonly used in the literature: covalent attachment, noncovalent attachment, and drug filling. Among different chemistries used for covalent attachment, the so-called smart systems have attracted much attention due to their ability to deliver drugs on demand. Drug molecules are typically grafted directly onto the backbone of CNTs through a stimuli-responsive linker, such as a disulfide bond, which is susceptible to cleavage on exposure to various external stimuli such as pH, temperature, and reducing agents. Incorporation of these "smart" linkers imparts to the CNTs the ability to release therapeutically relevant molecules in a controlled manner, thereby facilitating intracellular transport of these biological cargos and reducing possible side effects caused by burst release. In an experiment carried out by Kam et al. (2005), a novel nanotube-biomolecule platform with an incorporated disulfide bond was created, which showed promising efficiency in transporting and releasing both DNA oligonucleotides and small interfering RNA (siRNA) into the cells. It was hypothesized based on previous work (Arunachalam et al., 2000) that a disulfide bond in the system can be cleaved by thiol-reducing enzymes aided by the acidic pH in the lysosomes. Further investigations found that disulfide bridges are susceptible to endogenous thiols in the blood circulation, and therefore suffer from poor stability in vivo (Saito et al., 2003). To improve in vivo efficiency, a better system was designed and characterized by Samori et al. (2010), where an enzyme-sensitive peptide linker was used to achieve high cytotoxicity. Other examples of covalent linkage of CNTs to anticancer agents include delivery of paclitaxel (Liu et al., 2008a,b) and 10-hydroxy camptothecin (Wu et al., 2009), both of which achieved higher antitumor activity in vivo than drug molecules alone.

6. TOXICITY

6.1 Neurotoxicity

The novel properties of CNTs allow them to be a promising material for biological applications, especially in the delivery of therapeutic cargos in a less invasive manner. However, one major hurdle that needs to be addressed is the concern over their toxicity due to conflicting results presented in the literature (Saito et al., 2014). Also, because of similar properties (such as extremely high aspect ratios, low solubility, and needle-like fiber shapes) with asbestos, their toxicology profiles are often linked together (Pacurari et al., 2010). It has been shown in a pilot study that mice can develop asbestos-like pathogenic behavior on exposure of the mesothelial lining to long MWCTs (Poland et al., 2008). Therefore potential risks of CNTs used in biological applications still need to be carefully assessed and extensive toxicology studies, especially neurotoxicity, should be undertaken to further clarify their fate both in vivo and in vitro. This chapter provides a summary of neurotoxicity studies on CNTs as well as methods to improve their pharmacological and toxicological profiles with the hope to provide some insights into the future development of CNT-based products.

6.1.1 In Vivo Studies

Neurotoxicity remains one of the largest obstacles when talking about CNT applications in the central nervous system (CNS), and because the neurological discovery of CNTs is still at an early stage, more work is expected to be completed to establish a way to enhance the short-term and long-term biocompatibility of CNTs. Previous research has addressed that some of the nanomaterials, including CNTs, can pass through the biological membrane barriers and get into the CNS either by olfactory or systemic administration (Kafa et al., 2015; Oberdörster et al., 2004). In an olfactory administration study, Oberdörster et al. (2004) claimed that inhalation of ^{13}C for more than 6 h could lead to CNT accumulation in the olfactory bulb. Specifically, reports showed that some of the carbon nanoparticles are able to pass olfactory neurons into the olfactory bulb, which will then induce an inflammatory response by activating microglial cells (Tin Tin Win et al., 2006). In addition, impurity is another leading cause that contributes to the intrinsic toxicity of CNTs in vivo. When implanted subcutaneously, extremely pure CNTs, ca. 98% as determined by SEM observations, showed good biocompatibility as evidenced by no skin hair loss and lower levels of cytokines, whereas contaminated tubes clearly induced immunological toxicity and localized alopecia (Koyama et al., 2009).

Many studies have indicated that one of the factors that cause various neurotoxic effects of CNTs is their different diameters and/or lengths; however, the conclusions made by different research group vary. In a study by Poland et al., mesothelial lining of the body cavity of mice were exposed to 50 μg of nonfunctionalized MWNTs of different lengths dispersed in saline using bovine serum albumin. Results indicated that rigid MWNTs longer than 20 μm accumulated significantly in the diaphragmatic mesothelium causing subsequent granuloma lesion formation possibly as a result of frustrated phagocytosis. On the other hand, short MWNTs did not induce any inflammatory responses (Poland et al., 2008). A complementary study has also shown that 825-nm-long CNTs, when implanted subcutaneously, were found to have a higher degree of inflammatory responses as compared with CNTs 220 nm in length, although no severe necrosis or neutrophil infiltration was observed (Sato et al., 2005). A similar mechanism was proposed that short CNTs can be taken up by macrophages more efficiently, whereas longer ones cannot. Other studies have indicated that short MWNTs potentially lead to the alternation of hippocampal CA1 neuron excitability and consequently induced abnormal neuronal action potential with high firing frequency and a spiked half-width, which is considered as an early diagnostic phenomenon of neuronal dysfunctional diseases (Chen et al., 2013; Liu et al., 2009; Meng et al., 2013; Shan et al., 2012). This contradiction underlines the importance of investigating the mechanisms and physical–chemical properties of CNTs, as different functionalization strategies may lead to altered cytotoxic results.

However, what should always be kept in mind when investigating toxicity in vivo is that toxicity findings from small animals might not be predictive for human responses. In cancer research, for example, the rate of successful translation from animal models to clinical trials is reported to be less than 8% mainly due to the extreme complexities of carcinogenesis and tumor progression that animal models fail to mimic (Mak et al., 2014). Although seemingly not predictive most of the time, animal models can still serve as an important source for in vivo toxicity on certain aspects until new approaches emerge to fill the gap between in vitro studies and clinical applications. In the case of CNTs, the migration of CNTs from the blood stream to tissue and CNT interactions with cells were deemed to be highly representative due to similar microvessel structure and same cell size (Saito et al., 2014). Learning how CNTs interact with the human body, however small, is a step further toward developing safer CNT-based products. Therefore more rigorous in vitro and in vivo testing is not only worth pursuing, but also very necessary. It is critical in such studies that not only toxicity studies , but also accurate CNT characterization (including chemistry, geometry, and size) and biodistribution studies be completed, parts that are often collectively missed in today's CNT toxicity studies.

6.1.2 In Vitro Studies

PC12 cells are involved in many in vitro neurobiological studies because they retain the characteristics of dopaminergic neurons. A previous study investigated the effect of SWNTs on mitochondrial membrane potentials in PC12 cells, as mitochondria is the primary site responsible for producing reactive oxygen species (ROS) and programmed cell apoptosis (Morin and Simon, 2006). The group assessed the toxicities of SWNTs on PC12 mitochondria and revealed that SWNTs, especially long SWNTs, inhibited the proliferation of PC12 cells and even induced cell apoptosis when the concentration of SWNTs increased from 100 to 400 μg/mL. To be more specific, SWNTs caused mitochondrial membrane potential reduction and consequently promoted ROS formation and lipid peroxidization in a dose-dependent manner (Wang et al., 2011). In addition, it has been reported that astrocyte (glial scar tissue forming cell) function can be minimized when increasing surface energy of CNTs (125–140 mJ/m^2), hence decreased glial scar tissue formation as well as increased neuronal implant efficacy was achieved as a result (McKenzie et al., 2004).

Another important underlying factor that influences the biosafety of CNTs is the impurities included in CNTs during the manufacturing process, namely iron that is used as a catalyst during CNT fabrication. Previous studies have shown that some metal and metal oxide nanoparticles as well as ultrafine CNTs have the ability to pass through either in vitro or in vivo blood–brain barrier (Di et al., 2014; Oberdörster et al., 2004). Researchers have indicated that catalysts, or impurities, present in CNTs were able to transit from a solid phase to metal ion. As a result, the metal residues will inhibit cell proliferation by facilitating ROS generation via the Fenton reaction (Kagan et al., 2006; Koyama et al., 2009; Pulskamp et al., 2007). Moreover, researchers also indicated that the accumulation of impurities in the long term (14 days) had significant cytotoxicity on neuron cells (Vittorio et al., 2009). Similar to what was claimed here, it has been shown by Pulskamp et al. (2007) that traces of metals within commercial nanotubes are responsible for the increased presence of intracellular ROS. Another observation made by Shvedova et al. (2009) was that CNTs have low cytotoxicity when metallic impurities are properly removed. Although it is unrealistic to expect all impurities to be removed completely, it has been reported that carbon purity of 99% or more is considered safe (Saito et al., 2014; Vittorio et al., 2009). Again, this is a critical point because many studies highlighting CNT toxicity are for CNTs with carbon purities less than 99%.

To reduce the cytotoxicity of CNTs, one of the approaches that have been applied is to have functionalized CNTs (Table 11.2). For instance, it has been proved by Zhang et al. that PEG-coated CNTs have a less toxic effect on mitochondrial function and membrane integrity than the pristine CNTs. Based on mechanical studies carried out by the same group, this phenomenon is probably because the PEG coating protected the PC12 cells from oxidative stress involved in the CNT toxic pathway (Zhang et al., 2011). Other groups have also confirmed this functionalization strategy to minimize neurotoxicity to promote neurite outgrowth (Roman et al., 2011). For instance, Dal Bosco et al. (2015) showed that compared with pristine SWNTs that induced oxidative stress and cognitive deficits in the hippocampus region (Wang et al., 2011), PEGylated SWNTs were able to activate antioxidant species to overcome the oxidative stress in the long term. Also, PEG-SWNTs were shown to enhance CpG uptake by tumor-related cells and increase its antiglioma immunopotency (Zhao et al., 2011). The group investigating the neurotoxicity of SWNTs on PC cells also indicated that the presence of vitamin E blocked ROS formation, prevented a decrease of mitochondrial membrane potential caused by SWNTs, and protected PC12 cells from oxidative stress and apoptosis as a result (Wang et al., 2012). In addition, researchers have indicated that surface wettability also plays an important role in promoting cell viability. It has been shown that CNT and/or carbon nanofibers coated with a hydrocarbon outer layer became hydrophobic when

TABLE 11.2 Different Functionalization of CNTs With Reduced Cytotoxicity (Dutta et al., 2007; Kam et al., 2005; Khang et al., 2007; Mattson et al., 2000; Podesta et al., 2009; Singh et al., 2006; Singh et al., 2005; Zhang et al., 2011; Moon et al., 2012)

Functional Group	SWNTs/MWNTs	Results	In Vivo/In Vitro	References
PEGylated	SWNTs	Decreased ROS-mediated toxicological response	In vitro	Zhang et al. (2011)
4-Hydroxynonenal	MWNTs	Induced multiple neurites than pristine MWNT	In vitro	Mattson et al. (2000)
Acid oxidized	SWNTs	Efficient delivery of siRNA with no significant cytotoxicity	In vitro	Kam et al. (2005)
Amino functionalized	MWNTs	Significant tumor growth inhibition when delivering siRNA sequence	In vivo	Podesta et al. (2009)
[^{111}In] DTPA	Both	No significant cytotoxicity and no accumulation in blood	In vivo	Singh et al. (2006)
Glucosamine	MWNTs	Improved biocompatibility	In vivo	Guo et al. (2007)
Albumin	SWNTs	No significant cytotoxicity	In vitro	Dutta et al. (2007)
Poly(carbonate) urethane	MWNTs	Enhanced fibronectin adsorption due to increased nanoroughness	In vitro	Khang et al. (2007)
DNA	Both	No toxic effect on lymphocytes	In vivo	Singh et al. (2005)
Hydrocarbon outer layer	SWNT/carbon nanofibers	Promote stem cell functions for neural tissue healing	In vivo	Moon et al. (2012)

CNT, carbon nanotube ; *DTPA*, diethylenetriaminopentaacetic acid; *MWNT*, multiwalled carbon nanotube; *PEG*, polyethylene glycol; *ROS*, reactive oxygen species; *SWNT*, single-walled carbon nanotube.

compared with the those without the hydrocarbon outer layer. The CNTs with hydrophobic outer layers exhibited improved abilities in promoting stem cells, in this case subventricular zone neural progenitor cells, adhesion and differentiation (Fig. 11.4), promoting damaged neural tissue healing as a result (Moon et al., 2012). In short, additional systematic and mechanical investigations for both in vivo and in vitro neurotoxicity are definitely required to further understand the impact of surface functionalization of CNTs on their biocompatibility.

6.2 Other Toxicity Studies

Despite the effort that has been made to explore the potential toxic effects of CNTs, toxicities often yield inconsistent or even contradictory findings in vitro.

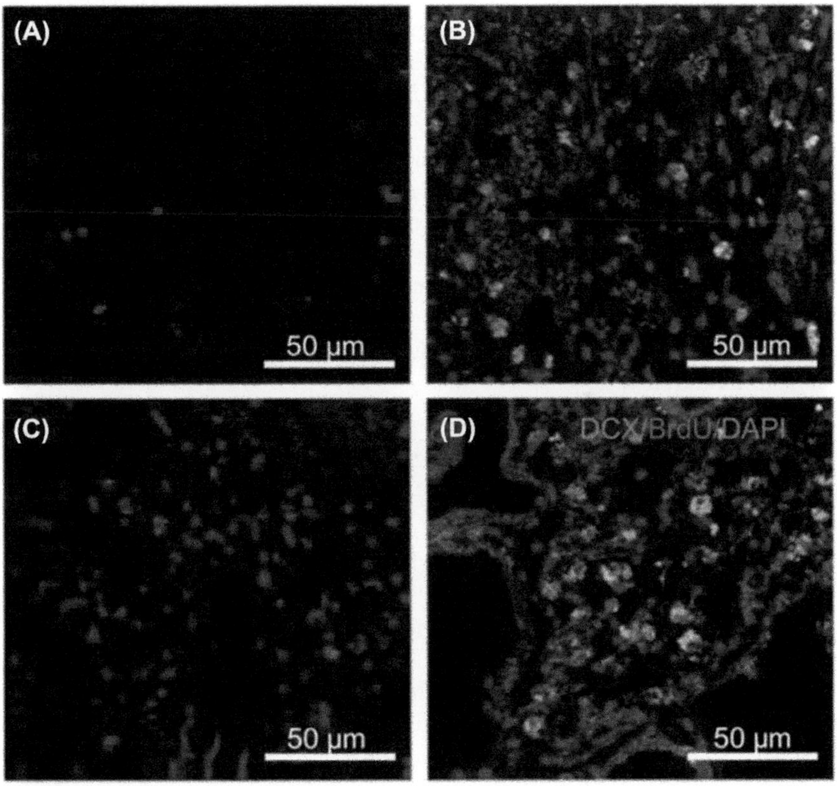

FIGURE 11.4 Cell staining for detecting the migration of exogenous subventricular zone neural progenitor cells 5 weeks after middle cerebral artery occlusion injury. The images show (A) control group without subventricular zone neural progenitor cell transplantation, (B) injury-subjected rats transplanted with subventricular zone neural progenitor cells, (C) injury-subjected rats transplanted with hydrophilic carbon nanotubes impregnated with subventricular zone neural progenitor cells, and (D) injury-subjected rats transplanted with hydrophobic carbon nanotubes impregnated with subventricular zone neural progenitor cells (Moon et al., 2012). *BrdU*, bromodeoxyuridine; *DAPI*, 4′,6-diamidino-2-phenylindole; *DCX*, doublecortin.

Some in vitro studies have reported CNT toxicity in a variety of relevant cell types, such as fibroblasts (Patlolla et al., 2010), macrophages (Di Giorgio et al., 2011), human embryonic kidney cells (Cui et al., 2005), and human keratinocytes (Manna et al., 2005), whereas other reports considered CNTs to be low in toxicity or even nontoxic (Pulskamp et al., 2007). Moreover, the lack of reference materials for safety evaluation makes it even harder to compare different experiments (Saito et al., 2014). Yet, close scrutiny of these results in the literature reveal that in vitro CNT toxicity might be attributed to several factors, such as impurities, fiber length, and degree of agglomeration. The phagocytosis of CNTs by macrophages, a process to prevent fiber-induced disease, is a length-dependent process (Boyles et al., 2015). The failure of macrophages to completely engulf longer CNTs, termed frustrated phagocytosis, can induce oxidative stress leading to chronic inflammation. One report stated that fibers that are longer than 20 μm are more slowly cleared by macrophages, resulting in fiber buildup, chronic inflammation, and eventually mesothelioma (Donaldson et al., 2010). One last factor that contributes to in vitro toxicity of CNTs involves the degree of agglomeration. When evaluating cytotoxicity in vitro using human MSTO-211H cells, it was shown that well-dispersed CNTs had much lower cytotoxicity effect as compared with purified agglomerated CNTs at the same concentrations (Wick et al., 2007).

Collectively, the above-mentioned studies among others have led to a conclusion that CNTs have minimal toxic effects in vitro provided that they are properly purified, have limited length, and are well dispersed. Most importantly, understanding the correlation between CNT properties and its in vitro/in vivo toxicity paves the way for designing and manufacturing safer CNT-based products in the future.

7. CONCLUSIONS AND FUTURE DIRECTIONS

Despite the remaining obstacles such as cytotoxicity issues, CNTs exhibit fantastic advantages and strong potentials in medicine including tissue engineering, biosensor, and cancer treatment. In addition, numerous researchers have demonstrated ways to limit the toxicity of CNTs. For instance, there here has been a consensus that functionalized CNTs represent a major improvement in biocompatibility over pristine CNTs. Among all techniques explored, functionalization of CNTs with hydrophilic polymers, which helps to improve their water solubility/dispersion and to shorten CNT length, has been considered as a few of the most promising ways to reduce toxicity. As one of the most commonly performed modification techniques for CNTs, PEGylated CNTs were shown to have relatively lower cytotoxicity in neural PC12 cells (Zhang et al., 2011). Moreover, when injected intravenously, PEGylated CNTs were found to have low uptake by the reticuloendothelial system and can be almost completely cleared out from organs in about 2 months with no toxic effects (Liu et al., 2008a,b). As a conclusion, with their excellent physical and chemical features, CNTs provide numerous possibilities to be investigated and optimized for further medical applications.

REFERENCES

Abbasi, E., Aval, S.F., Akbarzadeh, A., Milani, M., Nasrabadi, H.T., Joo, S.W., Hanifehpour, Y., Nejati-Koshki, K., Pashaei-Asl, R., 2014. Dendrimers: synthesis, applications, and properties. Nanoscale Res. Lett. 9 (1), 247.

Ajayan, P.M., Stephan, O., Colliex, C., Trauth, D., 1994. Aligned carbon nanotube arrays formed by cutting a polymer resin-nanotube composite. Science 265 (5176), 1212–1214.

Ajayan, P.M., Ebbesen, T.W., 1997. Nanometre-size tubes of carbon. Rep. Prog. Phys. 60 (10), 1025.

Ajayan, P.M., Schadler, L.S., Giannaris, C., Rubio, A., 2000. Single-walled carbon nanotube–polymer composites: strength and weakness. Adv. Mater. 12 (10), 750–753.

Arslantunali, D., Budak, G., Hasirci, V., 2014. Multiwalled CNT-phema composite conduit for peripheral nerve repair. J. Biomed. Mater. Res. A 102 (3), 828–841.

Arunachalam, B., Phan, U.T., Geuze, H.J., Cresswell, P., 2000. Enzymatic reduction of disulfide bonds in lysosomes: characterization of a gamma-interferon-inducible lysosomal thiol reductase (gilt). Proc. Natl. Acad. Sci. USA 97 (2), 745–750.

Askeland, D.R., Fulay, P.P., Wright, W.J., 2010. The Science and Engineering of Materials. Cengage Learning.

Bertrand, N., Wu, J., Xu, X., Kamaly, N., Farokhzad, O.C., 2014. Cancer nanotechnology: the impact of passive and active targeting in the era of modern cancer biology. Adv. Drug Deliv. Rev. 66, 2–25.

Bian, Z., Wang, R.J., Wang, W.H., Zhang, T., Inoue, A., 2004. Carbon-nanotube-reinforced Zr-based bulk metallic glass composites and their properties. Adv. Funct. Mater. 14 (1), 55–63.

Boyles, M.S.P., Young, L., Brown, D.M., MacCalman, L., Cowie, H., Moisala, A., Smail, F., Smith, P.J.W., Proudfoot, L., Windle, A.H., Stone, V., 2015. Multi-walled carbon nanotube induced frustrated phagocytosis, cytotoxicity and pro-inflammatory conditions in macrophages are length dependent and greater than that of asbestos. Toxicol. In Vitro 29 (7), 1513–1528.

Bradbury, E.J., McMahon, S.B., 2006. Spinal cord repair strategies: why do they work? Nat. Rev. Neurosci. 7 (8), 644–653.

Cellot, G., Cilia, E., Cipollone, S., Rancic, V., Sucapane, A., Giordani, S., Gambazzi, L., Markram, H., Grandolfo, M., Scaini, D., Gelain, F., Casalis, L., Prato, M., Giugliano, M., Ballerini, L., 2009. Carbon nanotubes might improve neuronal performance by favouring electrical shortcuts. Nat. Nano 4 (2), 126–133.

Chakravarty, P., Marches, R., Zimmerman, N.S., Swafford, A.D.-E., Bajaj, P., Musselman, I.H., Pantano, P., Draper, R.K., Vitetta, E.S., 2008. Thermal ablation of tumor cells with antibody-functionalized single-walled carbon nanotubes. Proc. Natl. Acad. Sci. U.S.A 105 (25), 8697–8702.

Chen, Y.-S., Hsiue, G.-H., 2013. Directing neural differentiation of mesenchymal stem cells by carboxylated multiwalled carbon nanotubes. Biomaterials 34 (21), 4936–4944.

Chen, T., Yang, J., Ren, G., Yang, Z., Zhang, T., 2013. Multi-walled carbon nanotube increases the excitability of hippocampal CA1 neurons through inhibition of potassium channels in rat's brain slices. Toxicol. Lett. 217 (2), 121–128.

Chico, L., Crespi, V.H., Benedict, L.X., Louie, S.G., Cohen, M.L., 1996. Pure carbon nanoscale devices: nanotube heterojunctions. Phys. Rev. Lett. 76 (6), 971–974.

Cho, Y., Borgens, R.B., 2010. The effect of an electrically conductive carbon nanotube/collagen composite on neurite outgrowth of PC12 cells. J. Biomed. Mater. Res. A 95A (2), 510–517.

Choudhary, N., Hwang, S., Choi, W., 2014. Carbon nanomaterials: a review. Handb. Nanomater. Prop. 709.

Coleman, J.N., Dalton, A.B., Curran, S., Rubio, A., Davey, A.P., Drury, A., McCarthy, B., Lahr, B., Ajayan, P.M., Roth, S., Barklie, R.C., Blau, W.J., 2000. Phase separation of carbon nanotubes and turbostratic graphite using a functional organic polymer. Adv. Mater. 12 (3), 213–216.

Cui, D., Tian, F., Ozkan, C.S., Wang, M., Gao, H., 2005. Effect of single wall carbon nanotubes on human HEK293 cells. Toxicol. Lett. 155 (1), 73–85.

Dal Bosco, L., Weber, G.E.B., Parfitt, G.M., Paese, K., Gonçalves, C.O.F., Serodre, T.M., Furtado, C.A., Santos, A.P., Monserrat, J.M., Barros, D.M., 2015. Pegylated carbon nanotubes impair retrieval of contextual fear memory and alter oxidative stress parameters in the rat hippocampus. BioMed Res. Int. 2015:104135.

Dervishi, E., Li, Z., Xu, Y., Saini, V., Biris, A.R., Lupu, D., Biris, A.S., 2009. Carbon nanotubes: synthesis, properties, and applications. Part. Sci. Technol. 27 (2), 107–125.

Di, S., Linlin, S., Gujie, M., Lubna, S., Soumya, B., Suprabha, N., Thomas, J.W., 2014. Controlling ferrofluid permeability across the blood–brain barrier model. Nanotechnology 25 (7), 075101.

Di Giorgio, M.L., Bucchianico, S.D., Ragnelli, A.M., Aimola, P., Santucci, S., Poma, A., 2011. Effects of single and multi walled carbon nanotubes on macrophages: cyto and genotoxicity and electron microscopy. Mutat. Res. 722 (1), 20–31.

Donaldson, K., Murphy, F.A., Duffin, R., Poland, C.A., 2010. Asbestos, carbon nanotubes and the pleural mesothelium: a review of the hypothesis regarding the role of long fibre retention in the parietal pleura, inflammation and mesothelioma. Part. Fibre Toxicol. 7, 5.

Dutta, D., Sundaram, S.K., Teeguarden, J.G., Riley, B.J., Fifield, L.S., Jacobs, J.M., Addleman, S.R., Kaysen, G.A., Moudgil, B.M., Weber, T.J., 2007. Adsorbed proteins influence the biological activity and molecular targeting of nanomaterials. Toxicol. Sci. 100 (1), 303–315.

Eatemadi, A., Daraee, H., Karimkhanloo, H., Kouhi, M., Zarghami, N., Akbarzadeh, A., Abasi, M., Hanifehpour, Y., Joo, S., 2014. Carbon nanotubes: properties, synthesis, purification, and medical applications. Nanoscale Res. Lett. 9 (1), 393.

Ebbesen, T.W., Ajayan, P.M., 1992. Large-scale synthesis of carbon nanotubes. Nature 358 (6383), 220–222.

Eftekhari, A., 2011. Nanostructured Conductive Polymers. Wiley.

Endo, M., 1988. Grow carbon fibers in the vapor phase. ChemTech 18 (9), 568–576.

Fan, H., Zhang, I., Chen, X., Zhang, L., Wang, H., Fonseca, A.D., Manuel, E.R., Diamond, D.J., Raubitschek, A., Badie, B., 2012. Intracerebral CpG immunotherapy with carbon nanotubes abrogates growth of subcutaneous melanomas in mice. Clin. Cancer Res. 18 (20), 5628–5638.

Fang, J., Nakamura, H., Maeda, H., 2011. The EPR effect: unique features of tumor blood vessels for drug delivery, factors involved, and limitations and augmentation of the effect. Adv. Drug Deliv. Rev. 63 (3), 136–151.

Feng, J., Sui, J., Cai, W., Gao, Z., 2008. Microstructure and mechanical properties of carboxylated carbon nanotubes/poly(l-lactic acid) composite. J. Compos. Mater. 42 (16), 1587–1595.

Feng, X., 2015. Nanocarbons for Advanced Energy Conversion. Wiley.

Gabay, T., Jakobs, E., Ben-Jacob, E., Hanein, Y., 2005. Engineered self-organization of neural networks using carbon nanotube clusters. Physica A 350 (2–4), 611–621.

Guo, T., Nikolaev, P., Rinzler, A.G., Tomanek, D., Colbert, D.T., Smalley, R.E., 1995a. Self-assembly of tubular fullerenes. J. Phys. Chem. 99 (27), 10694–10697.

Guo, T., Nikolaev, P., Thess, A., Colbert, D.T., Smalley, R.E., 1995b. Catalytic growth of single-walled nanotubes by laser vaporization. Chem. Phys. Lett. 243 (1–2), 49–54.

Guo, J.X., Zhang, X., Li, Q.N., Li, W.X., 2007. Biodistribution of functionalized multiwall carbon nanotubes in mice. Nucl. Med. Biol. 34 (5), 579–583.

Gupta, P., Sharan, S., Roy, P., Lahiri, D., 2015. Aligned carbon nanotube reinforced polymeric scaffolds with electrical cues for neural tissue regeneration. Carbon 95, 715–724.

Gupta, A., Choudhary, V., 2011. Carbon Nanotubes - Polymer Nanocomposites. InTech, US.

Hague, S., Klaffke, S., Bandmann, O., 2005. Neurodegenerative disorders: Parkinson's disease and Huntington's disease. J. Neurol. Neurosurg. Psychiatr. 76 (8), 1058–1063.

Hartgerink, J.D., Beniash, E., Stupp, S.I., 2001. Self-assembly and mineralization of peptide-amphiphile nanofibers. Science 294 (5547), 1684–1688.

Hirlekar, R., Yamagar, M., Garse, H., Vij, M., Kadam, V., 2009. Carbon nanotubes and its applications: a review. Asian J. Pharm. Clin. Res. 2 (4), 17–27.

Hirsch, A., 2002. Functionalization of single-walled carbon nanotubes. Angew. Chem. Int. Ed. 41 (11), 1853–1859.

Hou, P.X., Bai, S., Yang, Q.H., Liu, C., Cheng, H.M., 2002. Multi-step purification of carbon nanotubes. Carbon 40 (1), 81–85.

Hu, H., Ni, Y., Montana, V., Haddon, R.C., Parpura, V., 2004. Chemically functionalized carbon nanotubes as substrates for neuronal growth. Nano Lett. 4 (3), 507–511.

Hutagalung, S.D. (Ed.), 2012. Materials Science and Technology. InTech, Croatia.

Im, O., Li, J., Wang, M., Zhang, L.G., Keidar, M., 2012. Biomimetic three-dimensional nanocrystalline hydroxyapatite and magnetically synthesized single-walled carbon nanotube chitosan nanocomposite for bone regeneration. Int. J. Nanomed. 7, 2087–2099.

In-Yup Jeon, D.W.C., Kumar, N.A., Baek, J.-B., 2011. Carbon Nanotubes – Polymer Nanocomposites. InTech, US.

Isaacs, J.A., Tanwani, A., Healy, M.L., Dahlben, L.J., 2009. Economic assessment of single-walled carbon nanotube processes. J. Nanopart. Res. 12 (2), 551–562.

Jones, J.M., Malcolm, R.P., Thomas, K.M., Botrell, S.H., 1996. The anode deposit formed during the carbon-arc evaporation of graphite for the synthesis of fullerenes and carbon nanotubes. Carbon 34 (2), 231–237.

José-Yacamán, M., Miki-Yoshida, M., Rendón, L., Santiesteban, J.G., 1993. Catalytic growth of carbon microtubules with fullerene structure. Appl. Phys. Lett. 62 (2), 202–204.

Journet, C. and Bernier, P.: Production of carbon nanotubes. Appl. Phys. A 67(1):1–9.

Kafa, H., Wang, J.T.-W., Rubio, N., Venner, K., Anderson, G., Pach, E., Ballesteros, B., Preston, J.E., Abbott, N.J., Al-Jamal, K.T., 2015. The interaction of carbon nanotubes with an in vitro blood–brain barrier model and mouse brain in vivo. Biomaterials 53, 437–452.

Kagan, V.E., Tyurina, Y.Y., Tyurin, V.A., Konduru, N.V., Potapovich, A.I., Osipov, A.N., Kisin, E.R., Schwegler-Berry, D., Mercer, R., Castranova, V., Shvedova, A.A., 2006. Direct and indirect effects of single walled carbon nanotubes on raw 264.7 macrophages: role of iron. Toxicol. Lett. 165 (1), 88–100.

Kam, N.W.S., Liu, Z., Dai, H., 2005. Functionalization of carbon nanotubes via cleavable disulfide bonds for efficient intracellular delivery of siRNA and potent gene silencing. J. Am. Chem. Soc. 127 (36), 12492–12493.

Kam, N.W.S., Liu, Z., Dai, H., 2006. Carbon nanotubes as intracellular transporters for proteins and DNA: an investigation of the uptake mechanism and pathway. Angew. Chem. Int. Ed. 45 (4), 577–581.

Karageorgiou, V., Kaplan, D., 2005. Porosity of 3D biomaterial scaffolds and osteogenesis. Biomaterials 26 (27), 5474–5491.

Khang, D., Kim, S.Y., Liu-Snyder, P., Palmore, G.T.R., Durbin, S.M., Webster, T.J., 2007. Enhanced fibronectin adsorption on carbon nanotube/poly(carbonate) urethane: independent role of surface nano-roughness and associated surface energy. Biomaterials 28 (32), 4756–4768.

Kim, K.S., Bae, D.J., Kim, J.R., Park, K.A., Lim, S.C., Kim, J.J., Choi, W.B., Park, C.Y., Lee, Y.H., 2002. Modification of electronic structures of a carbon nanotube by hydrogen functionalization. Adv. Mater. 14 (24), 1818–1821.

Klingeler, R., Sim, R.B., 2011. Carbon Nanotubes for Biomedical Applications. Springer, Berlin, Heidelberg.

Kong, H., Gao, C., Yan, D., 2004. Controlled functionalization of multiwalled carbon nanotubes by in situ atom transfer radical polymerization. J. Am. Chem. Soc. 126 (2), 412–413.

Kostarelos, K., Lacerda, L., Pastorin, G., Wu, W., Wieckowski, S., Luangsivilay, J., Godefroy, S., Pantarotto, D., Briand, J.-P., Muller, S., Prato, M., Bianco, A., 2007. Cellular uptake of functionalized carbon nanotubes is independent of functional group and cell type. Nat. Nano 2 (2), 108–113.

Kostarelos, K., Bianco, A., Prato, M., 2009. Promises, facts and challenges for carbon nanotubes in imaging and therapeutics. Nat. Nano 4 (10), 627–633.

Koyama, S., Kim, Y.A., Hayashi, T., Takeuchi, K., Fujii, C., Kuroiwa, N., Koyama, H., Tsukahara, T., Endo, M., 2009. In vivo immunological toxicity in mice of carbon nanotubes with impurities. Carbon 47 (5), 1365–1372.

Kumar, M., Ando, Y., 2010. Chemical vapor deposition of carbon nanotubes: a review on growth mechanism and mass production. J. Nanosci. Nanotechnol. 10 (6), 3739–3758.

Lacerda, L., Ali-Boucetta, H., Herrero, M.A., Pastorin, G., Bianco, A., Prato, M., Kostarelos, K., 2008a. Tissue histology and physiology following intravenous administration of different types of functionalized multiwalled carbon nanotubes. Nanomedicine 3 (2), 149–161.

Lacerda, L., Herrero, M.A., Venner, K., Bianco, A., Prato, M., Kostarelos, K., 2008b. Carbon-nanotube shape and individualization critical for renal excretion. Small 4 (8), 1130–1132.

Landi, B.J., Raffaelle, R.P., Castro, S.L., Bailey, S.G., 2005. Single-wall carbon nanotube–polymer solar cells. Prog. Photovoltaics 13 (2), 165–172.

Li, W.Z., Xie, S.S., Qian, L.X., Chang, B.H., Zou, B.S., Zhou, W.Y., Zhao, R.A., Wang, G., 1996. Large-scale synthesis of aligned carbon nanotubes. Science 274 (5293), 1701–1703.

Liu, Z., Winters, M., Holodniy, M., Dai, H., 2007. SiRNA delivery into human T cells and primary cells with carbon-nanotube transporters. Angew. Chem. Int. Ed. 46 (12), 2023–2027.

Liu, Z., Chen, K., Davis, C., Sherlock, S., Cao, Q., Chen, X., Dai, H., 2008a. Drug delivery with carbon nanotubes for in vivo cancer treatment. Cancer Res. 68 (16), 6652–6660.

Liu, Z., Davis, C., Cai, W., He, L., Chen, X., Dai, H., 2008b. Circulation and long-term fate of functionalized, biocompatible single-walled carbon nanotubes in mice probed by Raman spectroscopy. Proc. Natl. Acad. Sci. USA 105 (5), 1410–1415.

Liu, Z., Ren, G., Zhang, T., Yang, Z., 2009. Action potential changes associated with the inhibitory effects on voltage-gated sodium current of hippocampal CA1 neurons by silver nanoparticles. Toxicology 264 (3), 179–184.

Ma, P.-C., Siddiqui, N.A., Marom, G., Kim, J.-K., 2010. Dispersion and functionalization of carbon nanotubes for polymer-based nanocomposites: a review. Compos. A 41 (10), 1345–1367.

Madani, S.Y., Naderi, N., Dissanayake, O., Tan, A., Seifalian, A.M., 2011. A new era of cancer treatment: carbon nanotubes as drug delivery tools. Int. J. Nanomed. 6, 2963–2979.

Mak, I.W.Y., Evaniew, N., Ghert, M., 2014. Lost in translation: animal models and clinical trials in cancer treatment. Am. J. Transl. Res. 6 (2), 114–118.

Manna, S.K., Sarkar, S., Barr, J., Wise, K., Barrera, E.V., Jejelowo, O., Rice-Ficht, A.C., Ramesh, G.T., 2005. Single-walled carbon nanotube induces oxidative stress and activates nuclear transcription factor-κb in human keratinocytes. Nano Lett. 5 (9), 1676–1684.

Marcaccio, M., Paolucci, F., 2014. Making and Exploiting Fullerenes, Graphene, and Carbon Nanotubes. Springer, Berlin, Heidelberg.

Marquis, F., 2003. Fully integrated hybrid polymeric carbon nanotube composites. Trans Tech Publ. 100, 85–88.

Martin, C.R., Kohli, P., 2003. The emerging field of nanotube biotechnology. Nat. Rev. Drug Discov. 2 (1), 29–37.

Matarredona, O., Rhoads, H., Li, Z., Harwell, J.H., Balzano, L., Resasco, D.E., 2003. Dispersion of single-walled carbon nanotubes in aqueous solutions of the anionic surfactant naddbs. J. Phys. Chem. B 107 (48), 13357–13367.

Mattson, M.P., Haddon, R.C., Rao, A.M., 2000. Molecular functionalization of carbon nanotubes and use as substrates for neuronal growth. J. Mol. Neurosci. 14 (3), 175–182.

McDevitt, M.R., Chattopadhyay, D., Kappel, B.J., Jaggi, J.S., Schiffman, S.R., Antczak, C., Njardarson, J.T., Brentjens, R., Scheinberg, D.A., 2007. Tumor targeting with antibody-functionalized, radiolabeled carbon nanotubes. J. Nucl. Med. 48 (7), 1180–1189.

McKenzie, J.L., Waid, M.C., Shi, R., Webster, T.J., 2004. Decreased functions of astrocytes on carbon nanofiber materials. Biomaterials 25 (7–8), 1309–1317.

Meng, L., Jiang, A., Chen, R., Li, C-z, Wang, L., Qu, Y., Wang, P., Zhao, Y., Chen, C., 2013. Inhibitory effects of multiwall carbon nanotubes with high iron impurity on viability and neuronal differentiation in cultured pc12 cells. Toxicology 313 (1), 49–58.

Monthioux, M., Kuznetsov, V.L., 2006. Who should be given the credit for the discovery of carbon nanotubes? Carbon 44 (9), 1621–1623.

Moon, S.U., Kim, J., Bokara, K.K., Kim, J.Y., Khang, D., Webster, T.J., Lee, J.E., 2012. Carbon nanotubes impregnated with subventricular zone neural progenitor cells promotes recovery from stroke. Int. J. Nanomed. 7, 2751–2765.

Moore, V.C., Strano, M.S., Haroz, E.H., Hauge, R.H., Smalley, R.E., Schmidt, J., Talmon, Y., 2003. Individually suspended single-walled carbon nanotubes in various surfactants. Nano Lett. 3 (10), 1379–1382.

Morin, C., Simon, N., 2006. Mitochondria: a target for neuroprotective interventions in cerebral ischemia-reperfusion. Curr. Pharm. Design 12 (6), 739–757.

Nguyen, K.C., 2006. Solid phase synthesized carbon nano fiber and tube, US Patent Number US 20100247419 A1.

Oberdörster, G., Sharp, Z., Atudorei, V., Elder, A., Gelein, R., Kreyling, W., Cox, C., 2004. Translocation of inhaled ultrafine particles to the brain. Inhal. Toxicol. 16 (6–7), 437–445.

Oberlin, A., Endo, M., Koyama, T., 1976. Filamentous growth of carbon through benzene decomposition. J. Cryst. Growth 32 (3), 335–349.

O'Connell, M.J., Boul, P., Ericson, L.M., Huffman, C., Wang, Y., Haroz, E., Kuper, C., Tour, J., Ausman, K.D., Smalley, R.E., 2001. Reversible water-solubilization of single-walled carbon nanotubes by polymer wrapping. Chem. Phys. Lett. 342 (3–4), 265–271.

Pacurari, M., Castranova, V., Vallyathan, V., 2010. Single- and multi-wall carbon nanotubes versus asbestos: are the carbon nanotubes a new health risk to humans? J. Toxicol. Environ. Health A 73 (5–6), 378–395.

Patlolla, A., Knighten, B., Tchounwou, P., 2010. Multi-walled carbon nanotubes induce cytotoxicity, genotoxicity and apoptosis in normal human dermal fibroblast cells. Ethnicity Dis. 20 (1 Suppl. 1), S1–S65 72.

Patole, S.P., Alegaonkar, P.S., Lee, H.-C., Yoo, J.-B., 2008. Optimization of water assisted chemical vapor deposition parameters for super growth of carbon nanotubes. Carbon 46 (14), 1987–1993.

Petros, R.A., DeSimone, J.M., 2010. Strategies in the design of nanoparticles for therapeutic applications. Nat. Rev. Drug Discov. 9 (8), 615–627.

Podesta, J.E., Al-Jamal, K.T., Herrero, M.A., Tian, B., Ali-Boucetta, H., Hegde, V., Bianco, A., Prato, M., Kostarelos, K., 2009. Antitumor activity and prolonged survival by carbon-nanotube-mediated therapeutic siRNA silencing in a human lung xenograft model. Small 5 (10), 1176–1185.

Poland, C.A., Duffin, R., Kinloch, I., Maynard, A., Wallace, W.A.H., Seaton, A., Stone, V., Brown, S., MacNee, W., Donaldson, K., 2008. Carbon nanotubes introduced into the abdominal cavity of mice show asbestos-like pathogenicity in a pilot study. Nat. Nano 3 (7), 423–428.

Pulskamp, K., Diabaté, S., Krug, H.F., 2007. Carbon nanotubes show no sign of acute toxicity but induce intracellular reactive oxygen species in dependence on contaminants. Toxicol. Lett. 168 (1), 58–74.

Rafique, M., Iqbal, J., 2011. Production of carbon nanotubes by different routes – a review. J. Encapsul. Adsorpt. Sci. 1 (2), 29–34.

Regev, O., ElKati, P.N.B., Loos, J., Koning, C.E., 2004. Preparation of conductive nanotube–polymer composites using latex technology. Adv. Mater. 16 (3), 248–251.

Ren, W., Li, F., Chen, J., Bai, S., Cheng, H.-M., 2002. Morphology, diameter distribution and Raman scattering measurements of double-walled carbon nanotubes synthesized by catalytic decomposition of methane. Chem. Phys. Lett. 359 (3–4), 196–202.

Roman, J.A., Niedzielko, T.L., Haddon, R.C., Parpura, V., Floyd, C.L., 2011. Single-walled carbon nanotubes chemically functionalized with polyethylene glycol promote tissue repair in a rat model of spinal cord injury. J. Neurotrauma 28 (11), 2349–2362.

Saito, G., Swanson, J.A., Lee, K.-D., 2003. Drug delivery strategy utilizing conjugation via reversible disulfide linkages: role and site of cellular reducing activities. Adv. Drug Deliv. Rev. 55 (2), 199–215.

Saito, N., Haniu, H., Usui, Y., Aoki, K., Hara, K., Takanashi, S., Shimizu, M., Narita, N., Okamoto, M., Kobayashi, S., Nomura, H., Kato, H., Nishimura, N., Taruta, S., Endo, M., 2014. Safe clinical use of carbon nanotubes as innovative biomaterials. Chem. Rev. 114 (11), 6040–6079.

Samori, C., Ali-Boucetta, H., Sainz, R., Guo, C., Toma, F.M., Fabbro, C., da Ros, T., Prato, M., Kostarelos, K., Bianco, A., 2010. Enhanced anticancer activity of multi-walled carbon nanotube-methotrexate conjugates using cleavable linkers. Chem. Commun. 46 (9), 1494–1496.

Sane, A., Gore, J., 2011. Flame synthesis of carbon nanotubes. In: Yellampalli, S. (Ed.), Carbon Nanotubes – Synthesis, Characterization, Applications. InTech.

Sano, M., Kamino, A., Okamura, J., Shinkai, S., 2001. Self-organization of peo-graft-single-walled carbon nanotubes in solutions and Langmuir–Blodgett films. Langmuir 17 (17), 5125–5128.

Sato, Y., Yokoyama, A., Shibata, K-i, Akimoto, Y., Ogino, S-i, Nodasaka, Y., Kohgo, T., Tamura, K., Akasaka, T., Uo, M., Motomiya, K., Jeyadevan, B., Ishiguro, M., Hatakeyama, R., Watari, F., Tohji, K., 2005. Influence of length on cytotoxicity of multi-walled carbon nanotubes against human acute monocytic leukemia cell line thp-1 in vitro and subcutaneous tissue of rats in vivo. Mol. Biosyst. 1 (2), 176–182.

Sayangdev, N., Ishwar, K.P., 2008. A model for catalytic growth of carbon nanotubes. J. Phys. D 41 (6), 065304.

Sedaghati, T., Yang, S.Y., Mosahebi, A., Alavijeh, M.S., Seifalian, A.M., 2011. Nerve regeneration with aid of nanotechnology and cellular engineering. Biotechnol. Appl. Biochem. 58 (5), 288–300.

Shan, D., Xie, Y., Ren, G., Yang, Z., 2012. Inhibitory effect of tungsten carbide nanoparticles on voltage-gated potassium currents of hippocampal CA1 neurons. Toxicol. Lett. 209 (2), 129–135.

Sharma, K., Saxena, K.K., Shukla, M., 2012. Effect of multiple stone-wales and vacancy defects on the mechanical behavior of carbon nanotubes using molecular dynamics. Procedia Eng. 38, 3373–3380.

Shi, X., Sitharaman, B., Pham, Q.P., Liang, F., Wu, K., Edward Billups, W., Wilson, L.J., Mikos, A.G., 2007. Fabrication of porous ultra-short single-walled carbon nanotube nanocomposite scaffolds for bone tissue engineering. Biomaterials 28 (28), 4078–4090.

Shokuhfar, T.M., Makradi, A., Titus, E., Cabral, G., Ahzi, S., Sousa, A.C.M., Belouettar, S., Gracio, J., 2008. Prediction of the mechanical properties of hydroxyapatite/polymethyl methacrylate/carbon nanotubes nanocomposite. J. Nanosci. Nanotechnol. 8 (8), 4279–4284.

Shvedova, A.A., Kisin, E.R., Porter, D., Schulte, P., Kagan, V.E., Fadeel, B., Castranova, V., 2009. Mechanisms of pulmonary toxicity and medical applications of carbon nanotubes: two faces of janus? Pharmacol. Ther. 121 (2), 192–204.

Singh, R., Pantarotto, D., McCarthy, D., Chaloin, O., Hoebeke, J., Partidos, C.D., Briand, J.-P., Prato, M., Bianco, A., Kostarelos, K., 2005. Binding and condensation of plasmid DNA onto functionalized carbon nanotubes: toward the construction of nanotube-based gene delivery vectors. J. Am. Chem. Soc. 127 (12), 4388–4396.

Singh, R., Pantarotto, D., Lacerda, L., Pastorin, G., Klumpp, C., Prato, M., Bianco, A., Kostarelos, K., 2006. Tissue biodistribution and blood clearance rates of intravenously administered carbon nanotube radiotracers. Proc. Natl. Acad. Sci. USA 103 (9), 3357–3362.

Singh, P., Campidelli, S., Giordani, S., Bonifazi, D., Bianco, A., Prato, M., 2009. Organic functionalisation and characterisation of single-walled carbon nanotubes. Chem. Soc. Rev. 38 (8), 2214–2230.

Sinnott, S.B., Andrews, R., 2001. Carbon nanotubes: synthesis, properties, and applications. Crit. Rev. Solid State Mater. Sci. 26 (3), 145–249.

Sitharaman, B., Shi, X., Tran, L.A., Spicer, P.P., Rusakova, I., Wilson, L.J., Mikos, A.G., 2007. Injectable in situ cross-linkable nanocomposites of biodegradable polymers and carbon nanostructures for bone tissue engineering. J. Biomater. Sci. Polym. Ed. 18 (6), 655–671.

Song, Y.S., Youn, J.R., 2005. Influence of dispersion states of carbon nanotubes on physical properties of epoxy nanocomposites. Carbon 43 (7), 1378–1385.

Thess, A., Lee, R., Nikolaev, P., Dai, H., Petit, P., Robert, J., Xu, C., Lee, Y.H., Kim, S.G., Rinzler, A.G., Colbert, D.T., Scuseria, G.E., Tománek, D., Fischer, J.E., Smalley, R.E., 1996. Crystalline ropes of metallic carbon nanotubes. Science 273 (5274), 483–487.

Thomas, J.W., Michael, C.W., Janice, L.M., Rachel, L.P., Jeremiah, U.E., 2004. Nano-biotechnology: carbon nanofibres as improved neural and orthopaedic implants. Nanotechnology 15 (1), 48.

Tin Tin Win, S., Yamamoto, S., Ahmed, S., Kakeyama, M., Kobayashi, T., Fujimaki, H., 2006. Brain cytokine and chemokine mRNA expression in mice induced by intranasal instillation with ultrafine carbon black. Toxicol. Lett. 163 (2), 153–160.

Touhara, H., Yonemoto, A., Yamamoto, K., Komiyama, S., Kawasaki, S., Okino, F., Yanagisawa, T., Endo, M., 2004. Fluorination of cup-stacked carbon nanotubes, structures and properties. MRS Online Proc. Libr. Arch. 858.

Vander Wal, R.L., Berger, G.M., Ticich, T.M., 2003. Carbon nanotube synthesis in a flame using laser ablation for in situ catalyst generation. Appl. Phys. A 77 (7), 885–889.

Vigolo, B., Pénicaud, A., Coulon, C., Sauder, C., Pailler, R., Journet, C., Bernier, P., Poulin, P., 2000. Macroscopic fibers and ribbons of oriented carbon nanotubes. Science 290 (5495), 1331–1334.

Vittorio, O., Raffa, V., Cuschieri, A., 2009. Influence of purity and surface oxidation on cytotoxicity of multiwalled carbon nanotubes with human neuroblastoma cells. Nanomedicine 5 (4), 424–431.

Wang, J., Sun, P., Bao, Y., Liu, J., An, L., 2011. Cytotoxicity of single-walled carbon nanotubes on PC12 cells. Toxicol. In Vitro 25 (1), 242–250.

Wang, J., Sun, P., Bao, Y., Dou, B., Song, D., Li, Y., 2012. Vitamin E renders protection to PC12 cells against oxidative damage and apoptosis induced by single-walled carbon nanotubes. Toxicol. In Vitro 26 (1), 32–41.

Webster, T.J., 2001. Nanophase ceramics: the future orthopedic and dental implant material. In: Advances in Chemical Engineering. Academic Press, pp. 125–166.

Wick, P., Manser, P., Limbach, L.K., Dettlaff-Weglikowska, U., Krumeich, F., Roth, S., Stark, W.J., Bruinink, A., 2007. The degree and kind of agglomeration affect carbon nanotube cytotoxicity. Toxicol. Lett. 168 (2), 121–131.

Wu, W., Li, R., Bian, X., Zhu, Z., Ding, D., Li, X., Jia, Z., Jiang, X., Hu, Y., 2009. Covalently combining carbon nanotubes with anticancer agent: preparation and antitumor activity. ACS Nano 3 (9), 2740–2750.

Xiao, Y., Gao, X., Taratula, O., Treado, S., Urbas, A., Holbrook, R.D., Cavicchi, R.E., Avedisian, C.T., Mitra, S., Savla, R., Wagner, P.D., Srivastava, S., He, H., 2009. Anti-her2 IgY antibody-functionalized single-walled carbon nanotubes for detection and selective destruction of breast cancer cells. BMC Cancer 9 (1), 1–11.

Yan, L., Zhao, F., Li, S., Hu, Z., Zhao, Y., 2011. Low-toxic and safe nanomaterials by surface-chemical design, carbon nanotubes, fullerenes, metallofullerenes, and graphenes. Nanoscale 3 (2), 362–382.

Zhang, L., Webster, T.J., 2009. Nanotechnology and nanomaterials: promises for improved tissue regeneration. Nano Today 4 (1), 66–80.

Zhang, Y., Xu, Y., Li, Z., Chen, T., Lantz, S.M., Howard, P.C., Paule, M.G., Slikker, W., Watanabe, F., Mustafa, T., Biris, A.S., Ali, S.F., 2011. Mechanistic toxicity evaluation of uncoated and pegylated single-walled carbon nanotubes in neuronal PC12 cells. ACS Nano 5 (9), 7020–7033.

Zhao, B., Wang, J., Li, Z., Liu, P., Chen, D., Zhang, Y., 2008. Mechanical strength improvement of polypropylene threads modified by PVA/CNT composite coatings. Mater. Lett. 62 (28), 4380–4382.

Zhao, D., Alizadeh, D., Zhang, L., Liu, W., Farrukh, O., Manuel, E., Diamond, D.J., Badie, B., 2011. Carbon nanotubes enhance CpG uptake and potentiate anti-glioma immunity. Clin. Cancer Res. 17 (4), 771–782.

Радушкевич, Л.В., 1952. О структуре углерода, образующегося при термическом разложении окиси углерода на железном контакте. Журнал Физической Химии 26, 88–95.

Chapter 12

The Application, Neurotoxicity, and Related Mechanism of Cationic Polymers*

Y. Li[1,2], D. Ju[1]

[1]Fudan University, Shanghai, China; [2]University of Pennsylvania, Philadelphia, PA, United States

Chapter Outline

---

*__Conflict of Interests__: All the Figures and Table in "The application, neurotoxicity, and related mechanism of cationic polymers" are original, unpublished materials designed and prepared by Yubin Li and Dianwen Ju. The authors declared that there's no conflict of interests.

Neurotoxicity of Nanomaterials and Nanomedicine. http://dx.doi.org/10.1016/B978-0-12-804598-5.00012-X

The application of cationic polymers in medical research has been widely expected to change the landscape of pharmaceutical industry in the foreseeable future, owing to their unique structure, monodispersity, effective cellular uptake property, and tunable properties (Li et al., 2013; Zhang et al., 2013). As ideal medicinal nanomaterials, cationic polymers are widely used for therapeutic and diagnostic purposes, and have been developed as antipathogen or antitumor agents (Kang et al., 2015; Nurhasni et al., 2015; Shim and Kwon, 2012). Owing to their special properties, cationic polymers have the capacity to enter the central nervous system (CNS) by passing the blood–brain barrier (Katare et al., 2015; Serramia et al., 2015; Zarebkohan et al., 2015). But the toxicity of cationic polymers, especially neurotoxicity, is still a serious challenge in their nanomedicinal applications, and related mechanisms of cationic polymers–induced neurotoxicity are being investigated (Win-Shwe et al., 2014; Yuan et al., 2015). This chapter introduces neurotoxicity and related mechanisms of cationic polymers. It starts by presenting classification of cationic polymers including chitosan, poly(L-lysine) (PLL), polyethyleneimine (PEI), polypropylenimine (PPI), poly(lactic-co-glycolic acid) (PLGA), poly(ester amine)s (PEAs), poly(2-(dimethylamino)ethyl methacrylate) (PDMAEMA), and poly(amidoamine) (PAMAM) dendrimers, and their related applications in nanomedicinal field: gene transfer, drug delivery, antimicrobial agents, and diagnostic imaging. It then presents neurotoxicity of different cationic polymers. The end of the chapter explains the related mechanisms of cationic polymers-induced neurotoxicity, including physicochemical and biochemical mechanisms. Physicochemical mechanisms mainly include charge, size, and shape of cationic polymers; biochemical mechanisms primarily focus on apoptosis, necrosis, reactive oxygen species (ROS), autophagy, and inflammation triggered by cationic polymers. This chapter mainly lays emphasis on the applications, neurotoxicity, and related mechanisms of synthetic cationic polymers through listing and summarizing representative examples.

1. CLASSIFICATION OF CATIONIC POLYMERS

Cationic polymers are defined as macromolecules bearing positive charges in the polymer backbone or side chains. The different positions of positive charges and different polymeric structures make cationic polymers of many different kinds. The common characters of cationic polymers are functional amine groups that can be protonated. According to the origin of cationic polymers, they can be divided into two categories: natural cationic polymers and synthetic cationic polymers. Among synthetic cationic polymers, according to the different components of origin, they can be mainly divided into several categories: polypeptides, polyamines, polymethacrylic acids, and PLGA (Fig. 12.1).

1.1 Chitosan

Chitosan is a polysaccharide composed of 2-amino-2-deoxy-beta-D-glucan combined with glycosidic linkages (Borchard, 2001). Being a cationic polysaccharide in neutral or basic pH conditions, chitosan contains free amino groups,

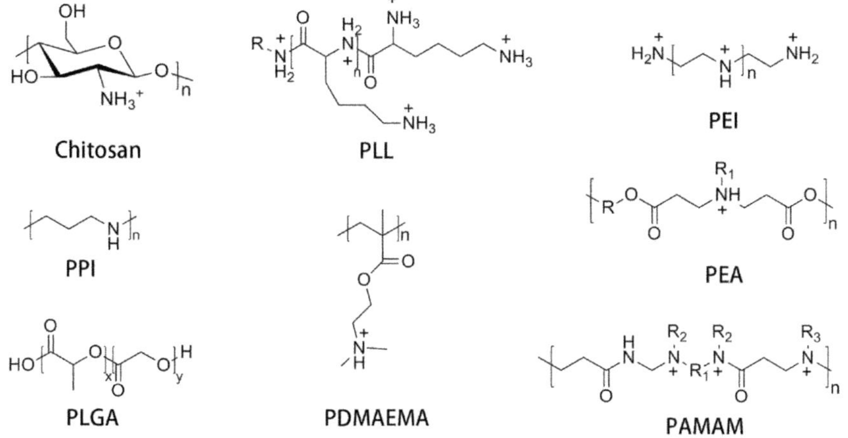

FIGURE 12.1 Chemical structure of cationic polymers. *PAMAM*, poly(amidoamine); *PDMAEMA*, poly(2-(dimethylamino)ethyl methacrylate); *PEA*, poly(ester amine); *PEI*, polyethyleneimine; *PLGA*, poly(lactic-co-glycolic acid); *PLL*, poly(L-lysine); *PPI*, polypropylenimine.

and is insoluble in water. The primary amine groups present special properties making chitosan very useful in pharmaceutical applications (Kato et al., 2003; Kumar et al., 2004).

1.2 Poly(L-lysine)

PLL polymers are linear polypeptides with the amino acid lysine for the repeat unit. As they consist of amino acid lysine, PLL polymers and their derivatives have well-defined architecture, excellent biocompatibility, and biodegradable properties. PLL polymers have sparked growing interest in biomedicine, and have become one of the first cationic polymers employed for gene transfer (Wagner et al., 1998). PLL polymers were also used for food preservation and as antimicrobial agents (Moschonas et al., 2012). But PLL polymers with high molecular weight showed potential cytotoxicity, and modified PLL polymers were designed to overcome these disadvantages (Kim et al., 2009).

1.3 Polyethylenimines

PEI is synthesized by acid-catalyzed polymerization of aziridine, and yields a highly branched network with a high cationic charge-density potential (von Harpe et al., 2000). PEI is one of the most typical polycationic polymers and lysosomal-escape agents at nano size. It can be divided into linear PEI and branched PEI based on different synthesized structures. The difference between linear PEI and branched PEI is also shown by the different amino groups: linear PEI just possesses primary and secondary amino groups, whereas branched PEI also possesses tertiary amino groups (Gao et al., 2011). Different shaped or sized PEI presents different transfection efficiency and cytotoxicity (Goyal et al., 2012).

1.4 Polypropylenimine

PPI dendrimers were initiated by the Michael addition of amine with acrylonitrile and followed by the reduction of nitrile groups to yield primary amine groups that provide new branching points (Shao et al., 2011). Since PPI dendrimers were first synthesized by Meijer in 1993, PPI dendrimers are widely used as effective delivery vectors for DNA and drugs (Mishra and Jain, 2014; Wang et al., 2012).

1.5 Poly(lactic-co-glycolic acid)

PLGA is a copolymer of poly lactic acid (PLA) and poly glycolic acid. PLA contains an asymmetric alpha-carbon that is typically described as the D or L form in classical stereochemical terms. PLGA is normally an acronym for poly D, L-lactic-co-glycolic acid where D- and L-lactic acid forms are in equal ratio (Makadia and Siegel, 2011). As PLGA polymer exhibits excellent biocompatibility, long-standing track record, and properties for continuous drug release, it is widely used in biomedical field (Acharya and Sahoo, 2011).

1.6 Poly(2-(dimethylamino)ethyl methacrylate)

PDMAEMA is a polymer that consists of covalently bound 2-(dimethylamino) ethyl methacrylate monomers. PDMAEMA possesses tertiary amino groups on its monomer molecule (Bromberg et al., 2005). As a mucoadhesive polymer, PDMAEMA is cationic in acidified media or when quaternized with an alkylating agent. Being investigated as a potential antibacterial drug and as a non-viral gene transfer vector, PDMAEMA was usually modified to overcome its disadvantages on cytotoxic and nonbiodegradable properties (Rawlinson et al., 2010).

1.7 Poly(ester amine)s

PEAs include linear PEAs and branched PEAs. Linear PEAs are synthesized by conjugate addition of low-molecular-weight linear amines over diacrylate monomers. Branched PEAs are synthesized by Michael addition based on hydrophobic polycaprolactone and low-molecular-weight PEI (Arote et al., 2009). Compared with linear PEAs, branched PEAs showed much higher gene transfer efficiency and lower toxicity.

1.8 Poly(amidoamine) Dendrimers

PAMAM dendrimers were firstly synthesized in 1984 by Donald Tomalia (Abbasi et al., 2014). They are a family of dendritic polymers based on an ethylenediamine core and repeated amidoamine branching. Up to now, PAMAM dendrimers grow to generations 10. From generation 1 to generation 10, there are more repeated amidoamine branching based on an ethylenediamine core, and the sizes of PAMAM dendrimers increases from 1.1 to 12.4 nm (Sun et al., 2015). As a family of highly branched polymers that possess valuable inherent properties, such as monodispersity, multifunctionality, and tenability, PAMAM dendrimers are potentially suitable and intensively investigated for biomedical applications in gene transfection, drug delivery, antimicrobial agent, and diagnostic imaging (Madaan et al., 2014).

2. APPLICATIONS OF CATIONIC POLYMERS

2.1 Gene Delivery

Gene therapy, as an adjuvant therapy, is one of the most promising strategies to treat abnormal diseases. It is the treatment of human disorders by the introduction of genetic material into specific target cells of patients, where the encoded protein is expressed, or targeted gene expression is silenced (Germano and

Binello, 2009). Cationic polymers, which are nonimmunogenic, have tightly controlled structural and chemical properties, and little acute toxicity, and they have several advantages in the application of gene delivery (Lv et al., 2006).

2.1.1 Chitosan

Chitosan is one of the most promising polymers for drug delivery because of its polycationic, biocompatible, and biodegradable properties (Cadete et al., 2012). Chitosan could interact with the negatively charged DNA and form polyelectrolyte complex (DNA-chitosan nanoparticles) for gene transfer (Luo and Wang, 2014).

Wan et al. (2015) used hydroxybutyl chitosan for small interfering RNA (siRNA) delivery targeting tissue factor inducing growth inhibition and apoptosis in human umbilical vein smooth muscle cells. In addition to chitosan-based nanoparticles, researchers also developed copolymer nanoparticles through combining chitosan, PEI, or other cationic polymers for gene delivery.

Wang et al. (2015c) developed a targeted iron oxide nanoparticle coated with chitosan-polyethylene glycol (PEG)-PEI copolymer and chlorotoxin and evaluated its effect in delivering human tumor necrosis factor (TNF)-alpha–related apoptosis-inducing ligand (TRAIL) in vitro and in vivo. Iron oxide nanoparticle coated with chitosan-PEG-PEI copolymer could successfully deliver TRAIL into human T98G glioblastoma cells and induce secretion of TRAIL in vitro. Systemic combined administration of iron oxide nanoparticle coated with chitosan-PEG-PEI copolymer and chlorotoxin to mice bearing T98G-derived flank xenografts resulted in near-zero tumor growth and induced apoptosis in tumor tissue. Nam et al. (2013) designed a novel nanocarrier using low-molecular-weight water-soluble O-carboxymethyl chitosan, branched low-molecular-weight PEI, and targeting ligand. The copolymer has low cytotoxicity, site-specific target function, and high transfection efficiency for gene therapy.

2.1.2 Poly(L-lysine)

PLL was able to form complex with plasmid DNA via ionic interactions, leading to compaction of DNA molecules. Also, PLL protected DNA from nuclease degradation and served as a platform for enhancing the cellular delivery of DNA (Shih et al., 2006).

Kodama et al. (2014) developed novel gene carriers composed of dendrigraft PLL. The gene transfer efficiency of the pDNA/dendrigraft PLL complexes was much higher than that of the control pDNA/PLL complex. However, the dendrigraft PLL complexes caused cytotoxicity and erythrocyte agglutination at high doses. Gomez-Valades et al. (2005) used copolymers of PLL with serine and tryptophan to form stable DNA vectors. All three copolymers are able to form compacted particles of plasmid DNA in the absence of salts and promote nonspecific gene transfer at similar levels into established cell lines in vitro.

Gao et al. (2015) used the PLL-modified PEI copolymer as a cationic polymer transfection material for nucleic acids through combining the high transfection efficiency of PEI and the excellent biodegradability of PLL. PEI-PLL copolymer was confirmed to transfect plasmid and oligonucleotide more effectively than PEI: Compared with negative control, PEI-PLL/microRNA-21 sponge plasmid DNA or PEI-PLL/antimicroRNA-21 oligonucleotide exhibited lower miR-21 expression and cell viability, and showed higher rate of cell cycle arrest and programmed cell death.

2.1.3 Polyethyleneimine

PEI is one of the most promising nonviral vectors that have been widely studied since first introduced in 1995 by Boussif (Huang et al., 2010). PEI, which strongly interacts with DNA, is widely used as a nucleic acid precipitant for nucleic acid purification, and considered as a golden standard in gene delivery (Mori et al., 2006).

Branched and linear PEIs were cationic polymers that have been used to deliver nucleic acids both in vitro and in vivo. The branched polymers exhibit high transfection efficiency, owing to their high cationic charge (Kafil and Omidi, 2011).

Bansal et al. (2015) synthesized two polymers derived from low-molecular-weight branched and linear PEI by the stepwise reactions with methylacrylate and amidation with tetraethylenepentamine. These two polymers not only showed abilities to bind and condense plasmid DNA into nanosized complexes, but also provided protection against nucleases in cellular milieu. Both polymers exhibited good buffering capacity and efficiently delivered nucleic acid (plasmid DNA or siRNA) across cells. Schafer et al. (2010) combined the constructive properties of PEI and liposomes for DNA and siRNA delivery by addition of liposomes to preformed polyplexes based on low-molecular-weight PEI.

2.1.4 Polypropylenimine

PPI dendrimers have been investigated as attractive cationic carriers for the delivery of nucleic acid (Wu et al., 2013). The lower generations of PPI/diaminobutane dendrimers have been reported to be effective gene delivery agents as well as being effective transfer systems for antisense oligonucleotides, siRNA and drugs (Aillon et al., 2009).

Russ et al. (2008) modified PPI dendrimers generation 2 or generation 3 via 1.6-hexandioldiacrylate with branched oligoethylenimine 800 Da as degradable and biocompatible synthetic carriers for gene delivery. All grafted dendrimers were able to efficiently compact DNA to nanosized polyplexes and exhibited an increased colloidal stability when compared with their unmodified counterparts. Transfection efficacy of oligoethylenimine-grafted PPI dendrimers was the highest, being similar or even higher as compared with standard linear or branched PEI.

Gu et al. (2015) explored the viability of using pluronic P123-conjugated PPI dendrimers as a vector for small hairpin RNA (shRNA) delivery to overcome tumor drug resistance, and found that the P123-conjugated PPI/plasmid DNA nanocomplexes with lower cytotoxicity revealed higher transfection efficiency when compared with the PPI/plasmid DNA nanocomplexes. Hollins et al. (2004) evaluated low-generation PPI dendrimers for the potential cellular delivery of antisense oligonucleotides targeting the epidermal growth factor receptor (EGFR). Anti-EGFR antisense oligonucleotides–dendrimers complexes treatment significantly inhibited cell growth, protein, and mRNA expression when compared with oligofectamine-mediated delivery.

2.1.5 Poly(lactic-co-glycolic acid)

Nanomaterials formulated with biodegradable polymers, for example, PLGA are being highly investigated as nonviral gene delivery systems because of their controlled release characteristics and biocompatibility (Marin et al., 2013). PLGA nanoparticles used for DNA delivery are mainly formulated by an emulsion-solvent evaporation technique using PVA as a stabilizer, which generates negatively charged particles and heterogeneous size distribution (Ravi Kumar et al., 2004).

Bivas-Benita et al. (2004) prepared PLGA nanomaterials bearing PEI on their surface for nonviral gene carriers targeting pulmonary epithelium, and found cytotoxicity of the nanomaterials mainly depended on the ratio of PEI and DNA. Ravi Kumar et al. (2004) prepared cationic PLGA nanoparticles as DNA carriers, and found these uniform cationic PLGA nanoparticles could bind DNA readily by electrostatic interaction. These cationic surface modified by PLGA nanoparticles could avoid plasmid usage during the particle preparation process.

2.1.6 Poly(2-(dimethylamino)ethyl methacrylate

PDMAEMA showed great potential as an ideal gene vector because of its relatively low toxicity, high transfection efficiency, and versatile synthesis strategies (Zhu et al., 2012).

Qian et al. (2013) prepared phage-displayed TGN peptide-decorated polymeric micelle-like polyplexes based on PEG-PDMAEMA for efficient brain-targeted gene delivery. The polyplexes maintained stable nanosize, excellent condensation capacity, lower toxicity, and higher cellular uptake when compared with the unmodified polyplexes. The transfection efficiency of TGN-modified polyplexes was also higher than that of unmodified polyplexes in vitro and in vivo. Cheng et al. (2011) synthesized PEHO-star-PDMAEMA with a hydrophobic hyperbranched PEHO (poly(3-ethyl-3-(hydroxymethyl)-oxetane)) core and cationic PDMAEMA arms, and explored their gene delivery capability and cytotoxicity: the gene delivery capability and the cytotoxicity of PEHO-star-PDMAEMA could be significantly improved when compared with PDMAEMA or PEI.

2.1.7 Poly(ester amine)s

Various biodegradable cationic polymers have been investigated as nonviral gene carriers, such as linear PEAs and PEA networks (Arote et al., 2007). Arote et al. (2009) supplied a novel biodegradable PEA based on polycaprolactone and PEI showing higher transfection efficiency and significant lower cytotoxicity when compared with PEI of 25 kDa molecular weight. Kim et al. (2013) developed a novel PEA by a Michael addition reaction between trimethylolpropane triacrylate and spermine as an efficient siRNA vector. The polyplex could effectively condense siRNA into a nanoscale structure and protect siRNA against RNase A enzyme. The complexes also showed excellent intracellular uptake and efficacious gene silencing effect with lower toxicity when compared with PEI of 25 kDa molecular weight. Jere et al. (2008) evaluated PEA carrier composed of low-molecular-weight PEI and PEG for si/shRNA delivery. PEA carriers could successfully deliver enhanced green fluorescence protein (EGFP) siRNA and silence the expression of EGFP. The silencing efficacy of PEA carrier was nearly 1.5 times higher than PEI 25 kDa.

2.1.8 PAMAM Dendrimers

PAMAM dendrimers are commercially available macromolecules that can be used as gene transfer vectors for plasmid DNA, antisense oligonucleotides, and siRNA (Eichman et al., 2000).

Choi et al. (2004) presented L-arginine-grafted-PAMAM dendrimers, a novel three-dimensional artificial protein and gene delivery vector. By introducing arginine residues to the dendritic surfaces, gene delivery potency of L-arginine-grafted PAMAM dendrimers was significantly increased by improving cell penetration and internalization mechanisms when compared with native PAMAM or PEI. Imamura et al. (2014) developed a ternary complex of plasmid DNA electrostatically assembled with PAMAM dendrimers and chondroitin sulfate for effective and secure gene delivery. Perez et al. (2009) screened the formation of complexes between ethylendiamine core PAMAM dendrimers and a short interfering RNA, and found that the complex combined with siEGFP and PAMAM dendrimers G7 produced highest suppression of EGFP expression in the cells, although it induced less protection of siRNA against RNase A degradation.

2.2 Drug Delivery

Cationic polymers could enhance the electrostatic interaction of nanoparticles with negatively charged site on cell surface and intestinal epithelial cells. Also, it could promote the uptake of drugs and intestinal absorption (Zou et al., 2007). Cationic polymers have played an integral role in the advancement of drug delivery technology by providing controlled release of therapeutic agents in constant doses over long periods, cyclic dosage, and tunable release of both hydrophilic and hydrophobic drugs because of their high level of control over the dendritic architecture (size, branching density, and surface functionality) (Liechty et al., 2010).

There are mainly three approaches for nanomaterials-mediated drug delivery: physical encapsulation, chemical conjunction, and physical encapsulation combined with chemical conjunction (Doane and Burda, 2013).

2.2.1 Physical Encapsulation

There are two methods for drug loading into nanomaterials by physical encapsulation: during the preparation of nanoparticles (incorporation) or after the formation of particles (incubation) (Zamani et al., 2013).

2.2.2 Chitosan

Wu et al. (2005) used chitosan nanoparticles as a novel delivery system for ammonium glycyrrhizinate encapsulation. They prepared the ammonium glycyrrhizinate-loaded chitosan nanoparticles by ionic gelation of chitosan with tripolyphosphate anions, and studied chitosan molecular weight, chitosan concentration, and ammonium glycyrrhizinate concentration on the physicochemical properties of the nanoparticles. The encapsulation efficiency decreased with the increase of ammonium glycyrrhizinate concentration and chitosan concentration, whereas the introduction of PEG decreased the positive charge of particle surface. Sarmento et al. (2007) found that alginate/chitosan nanoparticles were effective for oral insulin delivery. Insulin loaded into chitosan nanoparticles obviously enhanced intestinal absorption of insulin following oral administration. The hypoglycemic effect and insulinemia levels were much higher than that obtained from insulin solution or physical mixture of oral insulin and empty nanoparticles.

2.2.3 Poly(lactic-co-glycolic acid)

PLGA nanomaterials could protect the patient against toxic effects associated with high-concentration bolus doses, and release the drug in a controlled manner. Thus concentration of drug could be kept within therapeutic levels for longer periods of time (Acharya and Sahoo, 2011).

Hill et al. (2013) prepared PLGA polymers containing cinnamon bark extract using an emulsion-solvent evaporation method, and detected their antimicrobial efficacy: PLGA polymers loaded with cinnamon bark extract could significantly inhibit *Listeria monocytogenes* and *Salmonella typhimurium* at concentrations ranging from 224.42 to 549.23 μg/mL Betancourt et al. (2007) reported the preparation and in vitro evaluation of PLGA nanoparticles loaded with doxorubicin. Although cytotoxicity induced by doxorubicin encapsulated within the PLGA nanoparticles did not seem to increase compared with free drug, cellular intake with increased concentration and increased time of exposure were shown in doxorubicin-loaded nanoparticles when compared with free doxorubicin. Carboplatin was encapsulated in PLGA to develop a novel drug delivery system. Carboplatin/PLGA complex showed greater glioblastomas cytotoxicity and prolonged tissue half-life (Arshad et al., 2015).

Singhal et al. (2013) also evaluated catalase-loaded PLGA nanoparticles in human neuronal protection against oxidative damage. Although human neurons were highly sensitive to hydrogen peroxide, nanomaterials-mediated catalase delivery could successfully protect cultured human neurons against hydrogen peroxide–induced oxidative stress and cell death. Catalase-loaded nanoparticles could significantly impair hydrogen peroxide–induced protein oxidation, DNA damage, mitochondrial membrane potential depolarization, and loss of cell membrane integrity. It also restored neuronal morphology, neurite network, and microtubule-associated protein-2 levels.

2.2.4 Polypropylenimine

Murugan et al. (2013) developed hydroxylation and methylation of generation 2 and 3 PPI dendrimers for effective delivery of the drug norfloxacin. The modified PPI dendrimers showed excellent solubilizing potential and drug-loading abilities of norfloxacin in aqueous medium and could sustain delivery of norfloxacin. Surface-hydroxylated and internally quaternized PPI dendrimers might be chosen as potential biocompatible drug vectors for fluoroquinolines deliver under suitable conditions. Gupta et al. (2014) explored the targeting potential of sulfasalazine-loaded fucose-tethered PPI dendritic nanomaterials to Kupffer cells for effective treatment of cytokine-triggered liver injury. The study confirmed that surface-modified PPI dendrimers showed biocompatible framework, exhibited lower toxicity, and provided a sustained drug release activity in vitro and in vivo.

2.2.5 PAMAM Dendrimers

Winnicka et al. (2013) reported that PAMAM dendrimers could increase solubility of antifungal agents clotrimazole and ketoconazole, and significantly enhance the antifungal activity of clotrimazole and ketoconazole. Hayder et al. (2011) created azabisphosphonate-capped dendrimers that showed targeting human monocytes and directing them toward an antiinflammatory response and preventing osteoclastogenesis. Azabisphosphonate-capped dendrimers could double up against inflammation and bone erosion, and became more effective at treating rheumatoid arthritis when compared with existing antibody- and small molecule–based biologic drugs.

2.2.6 Chemical Conjugation

Ganguly et al. (2015) designed microsphere of chitosan cross-linked with PEG by emulsion cross-linking followed by solvent evaporation technique. 5-Fluorouracil (5-FU)-loaded microspheres could slowly release 5-FU up to 24 h to colonic region and improve its antitumor activity. Nascimento et al. (2015) applied the combinatorial design to chitosan, EGFR-binding peptide, and lipid derivatives with varying chain lengths to encapsulate hydrophobic drug cisplatin. The lipid-modified palatinate encapsulation into chitosan nanomaterials significantly enhanced cellular cytotoxicity of the cisplatin. Chen and Cooper (2002) synthesized quaternary ammonium-functionalized PPI dendrimers and demonstrated their antimicrobial activities as a function of dendrimer size and quaternary ammonium structure.

2.2.7 Physical Encapsulation Combined With Chemical Conjugation

Parveen and Sahoo (2011) developed paclitaxel-loaded chitosan and PEG-coated PLGA nanoparticles, and evaluated in vitro cytotoxicity in retinoblastoma, breast cancer, and pancreatic cancer cells. Superior antiproliferative effect and cell cycle inhibition were found in paclitaxel-loaded chitosan and PEG coated PLGA nanoparticles because of higher cellular uptake resulting in greater antiproliferative activity of nanomaterials. Xu et al. (2009) explored drug delivery mechanism of PLGA nanomaterials to determine their utility and limitations as an intracellular drug delivery system. PLGA nanomaterials physically encapsulated Nile red (hydrophobic fluorescent dye) and chemically conjugated to a fluorescent dye. Xu et al. observed their uptake by the cells using confocal microscopy. Although cells did not readily take up PLGA nanoparticles, extracellular drug release successfully delivered the payload to cells. Cheng et al. (2008) investigated the potential of PAMAM dendrimers as drug carriers of quinolones through exploring aqueous solubility and antibacterial activity of PAMAM dendrimers-coated quinolones. Chen found that aqueous solubility of quinolones was significantly enhanced by PAMAM dendrimers, whereas PAMAM dendrimers did not influence their strong antimicrobial activities.

2.3 Gene Transfer Combined With Drug Delivery

The appropriate combined treatment of chemotherapeutic agents and therapeutic genes have been proved as an effective strategy for enhancing chemosensitivity, improving the therapeutic outcome and patient compliance due to reduced dose, and decreased development of drug resistance (Prados et al., 2012). Rational drug combinations aim to exploit either additive or synergistic effects arising from the action of several species with the final goal of maximizing therapeutic efficacy (Kang et al., 2015). Thus, researchers designed a series of copolymers containing gene transfer and drug delivery properties.

Wang et al. (2015a) developed a new cationic polymer eprosartan-graft-PEI containing eprosartan and PEI 1.8K as a gene vector to assemble a drug (eprosartan) and gene codelivery complex. Chondroitin sulfate was used as a coating polymer to shield the surface charge of the complexes. The coated complexes exhibited amidase-responsive drug release and chondroitin sulfate–mediated endocytosis in vitro. Li et al. (2015a) designed captopril-PEI conjugate modified gold nanoparticles for codelivery of drug and gene in antiangiogenesis breast cancer therapy. Captopril-PEI containing low-molecular-weight PEI and antiangiogenesis drug captopril was fabricated to modify gold nanoparticles and complex with siRNA to construct the complexes for the codelivery of drug and siRNA in antiangiogenesis breast cancer therapy. Costa et al. (2015) synthesized the polyamine plasmid DNA hydrogels that conjugates plasmid DNA condensation by polyamines and cross-linking reaction with ethylene glycol diglycidyl ether. The cell-binding ligand transferrin was incorporated into PEI formulations. The systems were photodegradable, which allowed for the controlled release of different plasmids and anticancer drugs (doxorubicin, epirubicin, or paclitaxel), and showed stronger potential to suppress the development of cancer cells.

Cationic PLGA/folate-coated PEGylated polymeric liposome core–shell nanoparticles were self-assembled from a hydrophobic PLGA core and a hydrophilic folate-coated PEGylated lipid shell for targeting codelivery of drug and gene (Wang et al., 2010b).

Qian et al. (2014) synthesized three amphiphilic star-branched copolymers comprising polylactic acid and PDMAEMA with molecular architectures to codeliver doxorubicin and miR-21 inhibitor into LN229 glioma cells. The micelles could mediate miR-21 inhibitor escaping from lysosome degradation and the release of doxorubicin to the nucleus.

PAMAM dendrimers possess empty internal cavities and much higher density of surface functional group, and dendrimers could interact with drug agents and gene molecules by simple encapsulations, electrostatic interactions, and covalent conjugations (Patel et al., 2011). Han et al. (2011) employed T7 peptide-conjugated PEG-modified PAMAM dendrimers for the codelivery of plasmid DNA and doxorubicin. Compared with single doxorubicin or plasmid DNA delivery system, the codelivery system induced more apoptosis of tumor cells in vitro and in vivo.

2.4 Diagnostic Imaging

Diagnostic imaging required certain intensity of signal from an area of interest to differentiate structures under observation from surrounding tissues (Torchilin, 1996). Some nanomaterials have innate imaging abilities; others could be used as vectors to deliver various fluorescent dyes or photosensitizers for diagnostic imaging (Lim et al., 2015). Thus, it might be possible to improve the efficacy of cationic polymers as contrast medium by increasing the quantity of imaging reporter agents and enhancing the signal intensity (Alhaddad et al., 2012).

2.4.1 Chitosan

Mansur et al. (2015) synthesized novel multifunctional nanomaterials composed of fluorescent inorganic semiconductor quantum dot cores and tripeptide-modified polysaccharide organic shells. Chitosan was bound with the RGD peptide using a cross-linker to form bioconjugates. The quantum dot-RGD-chitosan systems were effective at specifically targeting integrin when assayed in vitro using fluorescence microscopy.

2.4.2 Poly(L-lysine)

Niidome et al. (2014) constructed a contrast agent based on 2,2,5,5-tetramehylpyrrolidine-1-oxyl groups (having nitroxyl radicals) and PEG-modified dendritic PLL, which could accumulate in the cells by enhanced absorptivity and maintenance of copolymers, and confirmed that nitroxyl radicals–modified dendritic PLL could be used as a contrast agent for Overhauser-enhanced magnetic resonance imaging.

2.4.3 Polyethyleneimine

Liu et al. (2012) designed a nanovector for gene delivery and bioimaging through combining carbon dots with PEI-passivation-enhanced fluorescence. The use of PEI as the surface passivation agent facilitated generating the carbon dots with both enhanced fluorescent properties and gene transfer efficiency. Han et al. (2012) also modified fluorescent carbon dots using PEI to improve fluorescence quality. After PEI modification, fluorescence intensity enhanced 300-folds and a red shift in emission wavelength occurred. Yang et al. (2015) created a multifunctional theranostic platform based on photosensitizer-loaded branched PEI-PEGylated ceria nanoparticles for effective cancer treatments using imaging-guided synchronous photochemotherapy. Photosensitizer-loaded branched PEI-PEGylated ceria nanoparticles with high chlorin e6 photosensitizer loading significantly enhanced the delivery of chlorin e6 into cells. Chlorin e6 mainly accumulated in lysosomes and obviously enhanced photodynamic therapeutic efficacy levels compared with free chlorin e6 treatment at ultralow drug doses.

2.4.4 Polypropylenimine

Shukla et al. (2014) designed a selective urea electrochemical biosensor based on electro-co-deposited zirconia-PPI dendrimers nanocomposites modified screen printed carbon electrode. The synergistic effect of high active surface area of PPI dendrimers and zirconia nanomaterials improved the electrocatalytic properties of the biosensor and encouraged the immobilization of the urease.

2.4.5 Poly(lactic-co-glycolic acid)

Wang et al. (2015b) created perfluoropropane (C3F8)-filled PLGA nanoparticles and explored the possibility of using PLGA nanoparticles as an ultrasound contrast agent. Ultrasound imaging of hollow PLGA nanoparticles showed that ultrasound signal was obviously enhanced in vitro and in vivo. Mieszawska et al. (2012) presented a complex nanoparticle system with dual drug-loading and diagnostic properties. Chemically modified PLGA polymers were devised into a polymer-lipid nanoparticle that consisted of doxorubicin in the polymeric core and sorafenib in the lipidic corona. The nanoparticle core also contained golden nanoparticles for imaging purposes. Pandey et al. (2013) prepared cationic PLGA nanomaterials encapsulating the iron oxide nanoparticles, and electrophoretically deposited onto indium tin oxide–coated glass substrate used as cathode and parallel platinum. The platform was successfully employed to invent a DNA biosensor by immobilizing a probe sequence specific to *Escherichia coli*.

2.4.6 Poly(2-(dimethylamino)ethyl methacrylate)

You et al. (2014) synthesized two star polycations, poly(2-aminoethyl methacrylate) (PAEMA) and PEMAEMA with perylene diimide for the central fluorophore. The star polymers presented excellent fluorescence properties in aqueous solution, which provided fluorescent tracing and imaging during gene delivery. Both polymers could successfully condense DNA into stable nanoparticles, and showed higher gene transfer efficiency and lower cytotoxicity when compared with PEI of 25 kDa molecular weight, but PDMAEMA presented higher capacity for gene transfer than PAEMA because of its better buffering property and faster rate of cellular internalization.

2.4.7 PAMAM Dendrimers

Wang et al. (2013b) reported a new usage of folic acid–modified dendrimers-entrapped gold nanoparticles as nanoprobes for in vitro and in vivo targeted computed tomography imaging of human lung adenocarcinoma. Lesniak et al. (2013) designed conjugates by combining an ethylenediamine-core generation 4 hydroxyl-terminated PAMAM dendrimers with fluorescent photostable, near-infrared cyanine dye and detected quantitative and qualitative biodistribution of the PAMAM dendrimers–cyanine dye conjugates. The quantitative biodistribution is as excellent as previous studies based on radiolabeling. The methods for dendrimers quantification were easier, provided excellent sensitivity, and allowed for quantification of dendrimers in different organs over longer time periods. Kojima et al. (2011) reviewed nanosized magnetic resonance imaging contrast agents consisting of dendrimer cores of various sizes and properties providing sufficient contrast enhancement for various applications, and found that smaller-sized PAMAM dendrimers-based contrast agents could easily cross the vascular wall inducing rapid perfusion throughout the body, quickly excretion through the kidney, and persistence in circulation. Smaller-sized PAMAM dendrimers-based contrast agents showed suitable potential as functional renal contrast agents or blood pool contrast agents.

2.5 Antimicrobial Agents

Cationic polymers, which possess a number of positive charges and membrane-disrupting activities to negative charged microbial surface phospholipids, showed potent antimicrobial activity against bacteria, fungi, and viruses (Thoma et al., 2014) (Table 12.1).

TABLE 12.1 Antibiotic Spectrum of Cationic Polymers

	Chitosan	PLL	PEI	PPI	PLGA	PDMAEMA	PEAs	PAMAM
Staphylococcus aureus	✔		✔	✔	✔	✔	✔	✔
Staphylococcus epidermidis			✔		✔	✔		✔
Bacillus subtilis							✔	
Listeria monocytogenes		✔						
Escherichia coli		✔		✔	✔			✔
Pseudomonas aeruginosa	✔			✔	✔	✔		✔
Salmonella typhimurium		✔						
Acinetobacter baumannii			✔					
Trichophyton rubrum			✔					
Candida albicans	✔		✔					
Aspergillus niger							✔	
Leishmania infantum	✔							
Fusarium oxysporium	✔						✔	
Alternaria solani	✔							
Valsa mali	✔							
Coleotricum capcii							✔	

PAMAM, Poly-amidoamine; *PDMAEMA*, poly(2-(dimethylamino)ethyl methacrylate); *PEA*, poly(ester amine); *PEI*, polyethyleneimine; *PLGA*, poly(lactic-co-glycolic acid); *PLL*, poly(l-lysine); *PPI*, polypropylenimine.

2.5.1 Chitosan

Chitosan and its derivatives are toxic to several bacteria, fungi, and parasites (El-Sharif and Hussain, 2011). Researchers investigated that an 87-kDa chitosan was more effective than a 532-kDa chitosan against both *Pseudomonas aeruginosa* and *Staphylococcus aureus* (Kean and Thanou, 2010). A lipid emulsion of the same chitosans was found to have antimycotic effect against *Candida albicans* and *Aspergillus niger* (Jumaa et al., 2002). It was also found that the chitosan excipient had antiparasitic properties against *Leishmania infantum*. Guo et al. (2006) prepared three kinds of Schiff bases of carboxymethyl chitosan and detected their antifungal activities using Jasso de Rodriguez's method; Schiff bases of carboxymethyl chitosan showed much better inhibitory effects against *Fusarium oxysporium*, *Alternaria solani*, and *Valsa mali*, when compared with carboxymethyl chitosan.

2.5.2 Poly(L-lysine)

Ye et al. (2013) found that epsilon PLL had antibacterial activity against *E. coli* O157:H7 through membrane lysis, ROS induction, and DNA damage response. The proposed mechanism of the inhibitory effect of epsilon PLL on microbial growth mainly depends on its electrostatic adsorption onto the cell surface of microorganisms, based on the cationic properties of PLL, leading to stripping of the outer membrane and abnormal distribution of cytoplasm. The difference in minimum inhibitory concentrations for bacteria, fungi, and yeasts might derive from their cell surface conditions. Geornaras et al. (2007) also confirmed antimicrobial activity of epsilon-PLL against *E. coli* O157:H7, *S. typhimurium*, and *L. monocytogenes* in various food extracts.

2.5.3 Polyethyleneimine

Azevedo et al. (2014) reported that PEI and PEI-based nanoparticles had antimicrobial and antibiofilm metabolic activity against *S. aureus*, *Staphylococcus epidermidis*, *Acinetobacter baumannii*, and *C. albicans*. Vignesh et al. (2012) evaluated the interaction, antibacterial, and antifungal properties of two water-soluble branched PEI–copper complexes containing bipyridine/phenanthroline with bovine serum albumin. The antimicrobial activity of antibiotic disk increased significantly in the presence of test sample. The maximum antimicrobial activity was observed against *Trichophyton rubrum* MTCC 3272 in both water-soluble branched PEI–copper complexes.

2.5.4 Polypropylenimine

Chen et al. (2000) synthesized quaternary ammonium functionalized PPI dendrimers and evaluated their antibacterial properties using a bioluminescence method. The antibacterial properties mainly depended on the size of the dendrimers, the length of hydrophobic chains in the quaternary ammonium groups, and the counteranion. Sun et al. (2012) designed the nitric oxide donor-functionalized PPI dendrimers with exterior functionality in dendrimers molecular weight. The synthesis of structurally diverse N-diazeniumdiolate-functionalized PPI dendrimers was capable of tunable nitric oxide storage and release, and influenced the bactericidal efficacy against gram-positive bacteria (*S. aureus*) and gram-negative bacteria (*P. aeruginosa*).

2.5.5 Poly(lactic-co-glycolic acid)

Aslan et al. (2010) investigated the antimicrobial nature of single-walled carbon nanotubes incorporated within the biomedical polymer PLGA: the viability and metabolic activity of *E. coli* and *S. epidermidis* were significantly decreased in the presence of single-walled carbon nanotubes–PLGA. Nurhasni et al. (2015) explored nitric oxide–releasing PLGA–PEI nanomaterials for prolonged nitric oxide release and antibacterial efficacy. The nanoparticles exhibited potent bactericidal efficacy against methicillin-resistant *S. aureus* and *P. aeruginosa* in a dose-dependent manner, and showed the ability to bind on the surface of bacteria.

2.5.6 Poly(2-(dimethylamino)ethyl methacrylate)

Quaternized poly(2-(dimethylamino)ethyl methacrylate) (PDMAEMAQ), polymers with quaternary ammonium groups, have been used as antimicrobial agents because of their good antimicrobial activities against huge numbers and types of microbes, such as *S. aureus*, *S. epidermidis*, and *P. aeruginosa* (Alvarez-Paino et al., 2015). Mei et al. (2014) synthesized PDMAEMA-silver nanoparticles, detected their antibacterial activity by evaluating growth inhibition of *P. aeruginosa* and *S. aureus*, and explored their related mechanism: the novel nanomaterials increased bacterial cytoplasmic membrane permeability, caused membrane disruption, subsequently penetrated cells, and strongly inhibited intracellular enzyme activity, leading to cell death.

Cationic antimicrobial agents may prevent device-associated infections caused by *S. epidermidis* and *S. aureus*. Rawlinson et al. reported that the cationic antimicrobial polymer PDMAEMA was more effective at antagonizing growth of clinical isolates of *S. epidermidis* than of *S. aureus*, and was influenced by cell-surface charge and hydrophobicity (Rawlinson et al., 2011).

2.5.7 Poly(ester amine)s

Pramanik et al. (2014) reported the fabrication of biobased hyperbranched PEA/polyaniline nanofiber–modified montmorillonite nanocomposites, and found the prepared polymers exhibited potent efficacy against gram-positive bacteria such as *Bacillus subtilis* and *S. aureus* as compared with gram-negative bacteria such as *P. aeruginosa* and *E. coli*. The nanocomposites also showed significant antifungal activity against *A. niger*, *Fusarium oxysporum*, and *Coleotricum capcii* and antialgal activity against algal consortium comprising *Chlorella*, *Hormidium*, and *Cladophorella* species.

2.5.8 PAMAM Dendrimers

Labena et al. (2016) investigated the antimicrobial activity and antibiofilms evaluation of hyperbranched PAMAM dendrimers, and found that hyperbranched PAMAM dendrimers with amine termination exhibited high activity against anti-methicillin-resistant *S. aureus*. Wang et al. (2010a) showed that topical cervical application of PAMAM dendrimers had potential to treat ascending uterine infection induced by *E. coli* in guinea pig model of chorioamnionitis. Besides, PAMAM dendrimers also showed good inhibitory effects on *S. epidermidis*, *E. coli*, and *P. aeruginosa* (Lopez et al., 2009; Navath et al., 2011).

3. NEUROTOXICITY OF CATIONIC POLYMERS

3.1 Neurotoxicity of Chitosan

Yuan et al. (2015) explored brain localization and evaluated neurotoxicity of polysorbate 80–modified chitosan nanoparticles in rats. Chitosan nanoparticles could enter into the brain and specially deposit in the frontal cortex and cerebellum after systemic injection. After exposure to chitosan nanoparticles for 7 days, the body weight of rats was remarkably decreased in a dose-dependent manner. Apoptosis and necrosis of neurons, slight inflammatory response in frontal cortex, and decrease of glial fibrillary acidic protein (GFAP) (a hallmark of the activation of astrocytes) expression in cerebellum also confirmed the neurotoxicity of chitosan nanoparticles. Hu et al. (2011) also evaluated the toxicity of biodegradable chitosan nanoparticles in a zebrafish embryo model, as zebrafish embryo model is usually used to evaluate neurotoxicity. Embryo exposure to chitosan nanoparticles resulted in a decreased hatching rate and increased mortality in a dose-dependent manner. The exposure showed an increased rate of cell death, high expression of ROS, as well as overexpression of heat shock protein 70, indicating that chitosan nanoparticles caused physiological stress and toxicity in zebrafish.

3.2 Neurotoxicity of PEI

Newland et al. (2013) reported that PEI could induce acute brain toxicity in adult Sprague-Dawley rats through detecting the size of gross tissue loss and measuring host response including reactive astrocytes and microglial cell recruitment by GFAP (marker of the activation of astrocytes) and OX-42 (microglia marker) immunohistochemical analysis. Compared with negative control, PEI induced the volume of the gross tissue loss on brain tissues and triggered the accumulation of astrocytes and microglia during its induced neurotoxicity. Besides, PEI polymers reduced the intracellular ATP activity, changed cell membrane integrity, and triggered direct necrosis and apoptosis during inhibiting cell viability in primary astrocytes.

3.3 Neurotoxicity of PLGA

Halamoda Kenzaoui et al. (2012) showed that PLGA-poly(ethylene oxide) nanoparticles could be untaken and localized in human brain-derived endothelial cells, and induced oxidative stress generation and DNA-damaging effects in exposed human brain-derived endothelial cells.

3.4 Neurotoxicity of Polystyrene

Amine-modified polystyrene nanoparticles induced human brain astrocytoma cell death through the mitochondria damage-mediated apoptotic mechanism. Bexiga et al. (2011) found that cationic polystyrene nanoparticle triggered apoptosis followed by a secondary necrotic response in 1321N1 human brain astrocytoma cells. Western blot showed that apoptosis-related proteins such as caspase 3/7, caspase 9, and poly(ADP-ribose) polymerase (PARP)-1 were cleaved after cationic polystyrene nanoparticle treatment. Transmission electron microscopy analysis confirmed morphological changes of lysosomes and mitochondria, which was consistent with rapid increase in the formation of ROS.

3.5 Neurotoxicity of PAMAM Dendrimers

Wang et al. (2014) investigated neurotoxicity caused by PAMAM dendrimers in U87MG, U251MG, U118, and A172 glioma cells in vitro. Obvious morphology changes including reduced cell size, rounded shape, and loss of neuritis occurred and cell viability decreased in a dose-dependent manner after generation 5 PAMAM dendrimers treatment for 24h. Mouse embryonic hippocampal cells (mHippoE-18) were also used to explore neurotoxicity of cationic amino-terminated PAMAM dendrimers and modified PAMAM-pyrrolidone dendrimers. Lactate dehydrogenase (LDH) release and 3-(4,5-dimethyl-2-thiazolyl)-2,5-diphenyltetrazolium bromide assay showed that unmodified cationic PAMAM dendrimers induced obvious LDH release and cell growth inhibition. Cationic PAMAM dendrimers could significantly induce shrinkage of mouse embryonic hippocampal cells, shortening of axonal branches, and cell death, whereas pyrrolidone modification significantly diminished neurotoxicity of PAMAM dendrimers (Janaszewska et al., 2015).

Besides, PAMAM dendrimers-induced neuron cell death was tightly associated with ROS production, lysosomal activity, DNA damage, apoptosis, and activation of CNS resident immune system microglia (Li et al., 2014). PAMAM dendrimers could make a hole in the plasma membrane, which may trigger membrane disruption. Li et al. (2015b) found that generation 5 PAMAM dendrimers induced mitochondrial oxidative fluorescence and autophagic fluorescence accumulation in neuronal cells. Nyitrai et al. (2013) reported that the interaction of PAMAM dendrimers and plasma membrane led to robust activation of neurons and astroglial cells, leading to mitochondrial depolarization, and impaired mitochondrial oxidation in brain tissue: enhanced mitochondrial depolarization followed by impaired oxidative metabolism of neurons may be the primary basis of neurotoxicity. Immune competent cells of the CNS could be moderately activated in vivo by the presence of generation 4 PAMAM dendrimers. A single dose of PAMAM dendrimers was intranasally administered to BALB/c mice, and a microarray analysis showed the alterations of the gene expression level related to pluripotent network, serotonin-anxiety pathway, transforming growth factor (TGF)-beta receptor signaling, prostaglandin synthesis regulation, complement coagulation cascades, and chemokine signaling pathway and nonodorant G protein-coupled receptor signaling pathway in brain tissues (Win-Shwe et al., 2014).

4. MECHANISMS OF CATIONIC POLYMERS-INDUCED NEUROTOXICITY

The physicochemical and biochemical properties of cationic polymers are two main factors mediating interactions between cationic polymers and neuronal cells during cationic polymers-induced neurotoxicity (Bhattacharya et al., 2013). Among them, particle surface properties, sizes, shapes, dosage, mode of administration, and exposure times of cationic polymers are common factors affecting the severity of neurotoxicity.

4.1 Physicochemical Mechanisms of Cationic Polymers-Induced Neurotoxicity

4.1.1 Charge

The charge of cationic polymers significantly affected their cellular uptake in vitro (Hsu et al., 2014). As there is electrostatic attraction between cell membranes and positively charged cationic polymers, cationic polymers with positively charged surfaces could be more easily attached on the cell membranes and accumulated inside the cells (Yang et al., 2016). Although positively charged cationic polymers have potential interfering with cell membranes, they could also lead to Ca^{2+} influx into cells and inhibit cell growth (Zhu and Mahato, 2010).

Previous reports have indicated that the toxicity of PAMAM dendrimers is significantly influenced by size and surface charge. Khaled Greish reported that surface charge and functional group–dependent toxicity with amine-terminated dendrimers (both G4 and G7) were safely only at doses lower than 10 mg/kg. On the contrary, carboxyl (G3.5 and G6.5) and hydroxyl (G4 and G7) terminated dendrimers were safely administered at 50-fold or higher doses. Employment of amine-terminated dendrimers in human applications could be fatal unless amine-terminated surface properties were modified to reduce their toxic effects (Madaan et al., 2014). Albertazzi et al. (2013) examined cell internalization properties and diffusion of G4 and G4-C12-modified PAMAM dendrimers in primary neuronal cultures and in the CNS of live animals. Hundred nanomoles of G4-C12 PAMAM dendrimers could induce dramatic apoptotic cell death of neurons in vitro, whereas G4 PAMAM dendrimers did not induce apoptotic neural cell death in the submicromolar range of concentration.

The charge of nanoparticles could also influence the oral toxicity of PAMAM dendrimers. Thiagarajan et al. (2013) found that G7-OH PAMAM dendrimers treated at doses of 300 mg/kg proved to be toxic, whereas G7-NH_2 dendrimers with a dose of 30 mg/kg was at least 10 times less tolerated than G7-OH dendrimers.

The high positive charge on the surface of PEI led to electrostatic interaction with the negatively charged components of the cell membranes and caused membrane disruption and damage. Different methods were employed to reduce PEI-induced cytotoxicity by decreasing the positive charge density of the polymer such as acetylation, carboxyalkylation of primary amines, and conjugation of succinate, butanoate, and hexanoate to PEI (Rezvani Amin et al., 2013; Zhao et al., 2012).

Besides, the toxic properties of dendrimers also depend on their structural components including the core, interior monomers, and surface groups, of which amino surface groups are known to be the most toxic. Ziemba et al. (2011) investigated the toxicity of unmodified generation 4 PPI dendrimers and generation 4 PPI dendrimers with 25% and 100% of the surface amino groups substituted with maltotriose residues. A higher dose of unmodified PPI dendrimers induced remarkable toxicity, but surface modification nearly abolished these toxic effects.

4.1.2 Size

The molecular weights as well as the cationic charge density of the polymers were confirmed as key parameters for the interaction with the cell membranes and cell damage (Fischer et al., 2003, 2004).

Huang et al. (2004) evaluated the effects of molecular weight and degree of deacetylation on the cellular uptake and in vitro cytotoxicity of chitosan molecules and nanoparticles. Transformation of chitosan into nanoparticles did not change the cytotoxicity profile of the polymers but facilitated their endocytosis. Molecular weight affected the uptake efficiency by modulating the zeta potential of chitosan nanoparticles. The binding affinity and uptake capacity for the nanoparticles decreased with decreasing chitosan molecular weight.

Xiong et al. (2013) explored the size influence of PLGA on their potential cytotoxicity. Although biomedically used PLGA nanoparticles did not exhibit obvious cytotoxic effects, smaller (60 nm) PLGA nanoparticles triggered the release of TNF, whereas 200-nm PLGA nanoparticles did not trigger any negative response from cells. The size-dependent cytotoxicity of PLGA nanoparticles could be because smaller particles with larger specific surface area could adsorb more biomolecules, and the ability to adsorb proteins could be an important paradigm to predict the in vitro cytotoxicity of less toxic PLGA nanoparticles.

Florea et al. (2002) evaluated PEI of different molecular weights (600–1000, 60, and 25 KDa) as a DNA complexing agent for its efficiency and cytotoxicity, and found transfection efficacy was strongly correlated to PEI-induced cytotoxicity.

PLL polymers are commonly used for DNA compaction and cell transfection. Symonds et al. (2005) compared the cytotoxicity of PLL polymers of low molecular weights (2.9 kDa) and high molecular weights (27.4 kDa) through detecting mitochondrial membrane potential, cytochrome C release, and activation of caspase. Compared with high-molecular-weight PLL polymers, the action of low-molecular-weight PLL polymers on mitochondria did not result in cytochrome C release or changes in mitochondrial function. Berit L. Strand also found that toxicity of PLL polymers was tightly associated with their molecular weight. PLL polymers with molecular weight of 29.8 kDa seemed to be more toxic than PLL polymers with molecular weight of 4.0 and 7.5 kDa.

Cerda-Cristerna et al. (2011) evaluated the hemocompatibility of free-form PDMAEMA homopolymers differing in molecular weight (10, 2.64, and 40K) with or without a poly(ethylene glycol) sequence in the form of a palm tree-like structure, and found that molecular weight strongly affected the hemocompatibility of the polymers evaluated. A molecular weight of 40 KDa PDMAEMA polymers has been highlighted that could induce hemolysis and hemagglutination when injected in the bloodstream.

4.1.3 Shape

Cytotoxicity is affected by the architecture of the polymers. Polymers with increased toxicity usually have larger molecular weight (size) and more branching of the macromolecule (shape) (Hunter, 2006).

In vitro cytotoxicity of branched PEI (Br-PEI) (molecular weight 800 and 25 kDa) and linear PEI (L-PEI) (molecular weight 20 kDa) were compared. Branched PEI 25 kDa has been shown as the golden standard but its transfection efficiency is limited by high toxicity to different cells. Linear PEI has been shown to be less toxic to cells but less efficient in delivering DNA to cells when compared with their branched counterparts (Wiseman et al., 2003).

Kafil and Omidi (2011) also explored cytotoxic impacts of 25-kDa linear and branched PEI, and found that branched PEI could result in greater internalization and induce greater cytotoxicity than linear PEI. Annexin V assay and DNA fragmentation analyses also confirmed early and late apoptosis was induced by branched PEI. Cai et al. (2011) investigated the effect of chain length on the cytotoxicity and endocytosis of rhodamine B end-labeled cation linear poly(2-(N,N-dimethylamino)ethyl methacrylate) (*RhB*-PDMAEMA) polymers. Cai et al. (2011) found that for short PDMAEMA chains, cytotoxicity, membrane disruption, and apoptosis were very low, independent of the chain length; for the medium-length PDMAEMA chains, the cytotoxicity increased with the chain length and polymer concentration, mainly due to the cooperative effect of membrane disruption and apoptosis; for long-chain PDMAEMA polymers, of which cytotoxicity is independent of the PDMAEMA chain length, they were very disruptive to the cellular membrane, proapoptotic, and able to enter the cytoplasm and nucleus faster than short chains.

4.2 Biochemical Mechanisms of Cationic Polymers-Induced Neurotoxicity

4.2.1 Apoptosis

Apoptosis is the process of programmed cell death that occurs in multicellular organisms (Elmore, 2007). It could be triggered both by the extrinsic death receptors and by the intrinsic mitochondrial pathway (Fulda and Debatin, 2006; Ouyang et al., 2012). Mitochondrion is a major regulator of cell death, especially for cationic polymers-induced apoptotic cell death. Mitochondrial dysfunction associated with apoptosis is usually characterized as the loss of mitochondrial membrane potential, release of cytochrome C, and activation of the caspases, respectively (Baines, 2010; Li et al., 2014) (Fig. 12.2).

PEI has been identified as an apoptotic agent. High-molecular-weight PLLs acted by inducing cytosolic cytochrome C release from isolated mitochondria and dramatically altered respiration and membrane potential in the

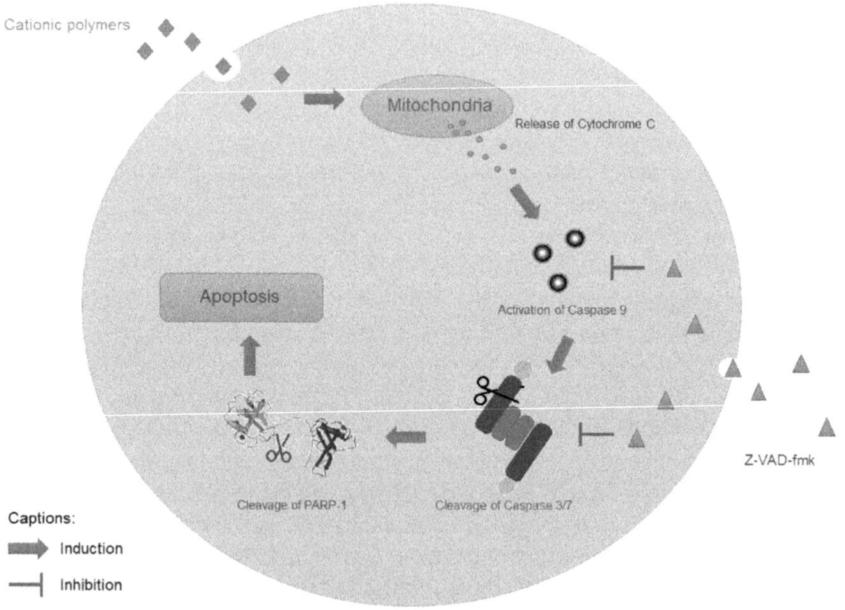

FIGURE 12.2 Apoptosis triggered by cationic polymers. Mitochondrion is a major regulator of cell death, especially for cationic polymers–induced apoptotic cell death. Mitochondrial dysfunction associated with apoptosis is usually characterized as the loss of mitochondrial membrane potential, release of cytochrome C, and activation of the caspases. *PARP-1*, poly(ADP-ribose) polymerase 1.

absence of swelling (Hunter, 2006). In whole cells, cytochrome C release from mitochondria was reduced by 50% following inhibition of Bax channel forming activity. Besides loss of mitochondrial membrane potential, PEI and PEI-DNA complexes treatment also elevated other markers of apoptosis in a time-dependent manner, including activation of caspase-3 and caspase-9 (Symonds et al., 2005).

The mechanism underlying mitochondria-mediated branched and linear PEI and PEI-DNA complex induction of apoptosis was demonstrated in both whole cells and isolated mitochondria. PEI-mediated membrane initial plasma membrane damage was confirmed by detection of high levels of LDH in whole cells. Redistribution of phosphatidylserine (a primary indicator of apoptosis) occurred after 20 μg/mL of PEI incubating for 30–60 min. Mechanistically active caspase-3 inhibiting aminophospholipid translocase resulted in phosphatidylserine externalization (Di Gioia and Conese, 2009; Hunter, 2006).

Omidi et al. (2005) assessed the effects of PPI dendrimers on cellular gene expression, and found that the upregulation of TGF beta and B-cell lymphoma 2 (BCL2) alpha induced by PPI dendrimers may impact on the apoptosis process. DNA damage within cells treated with PPI dendrimers alone or combined copolymers with DNA also confirmed the apoptotic effects triggered by PPI dendrimers.

Xia et al. (2008) compared the toxic effects of cationic polystyrene nanomaterials in five different cell lines, and found that the toxicity of cationic polystyrene nanomaterials was tightly dependent on cell-specific endocytic and mitochondrial injury pathways. Bexiga et al. (2011) applied biochemical and imaging techniques to study the mechanism of amine-modified polystyrene nanoparticles–induced cell death in human brain astrocytoma cell lines. Cells exposed to cationic nanoparticles revealed an increase in cell membrane permeability of YoPro-1 and propidium iodide (PI), indicating that cationic polystyrene induced an apoptotic followed by a secondary necrotic response. Cationic polystyrene also induced activation of caspases 3/7, caspase 9, and cleavage of PARP-1, but Wang et al. (2013a) found that inhibition of caspases using pan-caspase inhibitor Z-VAD fmk could not protect human brain astrocytoma cells from the damage induced by polystyrene nanoparticles. Besides, Lunov et al. (2011) found that amino-functionalized polystyrene nanomaterials induced proton accumulation in lysosomes triggering lysosomal destabilization, release of cathepsin B, and damage of the mitochondrial membrane in a time-dependent manner.

Jones et al. (2004) found that exposure to PDMAEMA increased the proportion of cells staining with annexin V fluorescein isothiocyanate, excluding PI in a dose-dependent manner. A gradual increase in the proportion of cells with depolarized mitochondria following exposure to PDMAEMA was also detected, which confirmed that PDMAEMA could induce apoptosis including the lipid phosphatidylserine exposure and mitochondrial depolarization.

Albertazzi et al. (2013) evaluated apoptotic toxicity induced by PAMAM dendrimers in primary cortical neurons in vitro and in vivo. Nuclei stain showed that most of the G4-C12 PAMAM dendrimers–treated neurons died by apoptosis (51% of the total cell death); PI staining and TdT-mediated dUTP (deoxyuridine triphosphate) nick end labeling (TUNEL) assay of mouse brain slices were also performed to verify PAMAM dendrimers–induced apoptotic effects in vivo; apoptotic detection with TUNEL assay showed the presence of programmed cell death reaction in G4-C12 PAMAM dendrimers–treated samples. Apoptotic cell signal concurred with PAMAM dendrimers accumulation in the brain parenchyma, and it indicated that higher concentration of dendrimers could be neurotoxic.

4.2.2 Necrosis

Traditionally, necrosis has been described as a passive and uncontrolled process, initiated by external factors like ischemia-reperfusion, toxins, infections, and cellular stress. It is characterized by a rapid breakdown of the cell membrane, resulting in the release of intracellular compounds into the extracellular space (Negroni et al., 2015).

Strand et al. found that PLL polymers at concentrations higher than 10 mg/mL induced massive cell death. Compared with cells exposed to agonistic antibody against Fas, which is known to cause apoptotic cell death, cells stained with annexin V and PI showed that PLL polymers induced necrosis but not apoptosis, as cells exposed to PLL nanoparticles have the ability to incorporate PI. Caspase 1 inhibitor could inhibit Fas-induced apoptosis, but could not inhibit PLL nanoparticles–induced cytotoxicity (Strand et al., 2001).

Moghimi et al. (2005) investigated that PEI-mediated cytotoxicity could be divided into two stages: necrosis for the first stage and apoptosis for the second stage. For the first phase, PEI induced toxicity as early necrotic-like changes resulting from compromised membrane integrity, assessed by considerable LDH release and phosphatidylserine translocation from the inner plasma membrane to the outer cell surface. For the second phage, cytotoxicity depended on the activation of mitochondria-mediated apoptotic program, resulting from PEI-induced channel formation in the outer mitochondrial membrane. This led to the release of proapoptotic cytochrome C, activation of caspase 3, and loss of mitochondrial membrane potential as a result of caspase translocation into the mitochondria.

Cationic polymers with different molecular weights present different potential for the interaction with the cell membranes and the cell damage (Song et al., 2011). Branched PEI 800 Da has been shown to cause unacceptable massive necrosis when compared with linear 25 kDa, branched PEI 1.8 kDa, PEI 2 kDa, and PEI 11 kDa, which showed more acceptable toxicity (Di Gioia and Conese, 2009). Fischer et al. (2003) also compared in vitro cytotoxicity of different cationic polymers by detecting cell viability and release of the cytosolic enzyme LDH, and found PEI treatment made cell nuclei retain their size; chromatin was homogenously distributed; and cell membranes lost their integrity rapidly. All these showed necrotic cell reaction.

4.2.3 Autophagy

Macroautophagy (hereafter referred to as autophagy) is an evolutionarily conserved self-eating process that can be triggered by a series of conditions of stress such as oxidative stress and nutrient deprivation (Xu et al., 2012). Activation of autophagy requires formation of autophagosomes in which eukaryotes degrade long-lived protein and dysfunctional organelles including mitochondria and Golgi apparatus by lysosomal enzymatic content (Piras et al., 2011). Characterized by the accumulation of autophagosomes and autophagolysosomes in the cytoplasm, autophagy has been classified as a new morphological form of programmed cell death (Tasdemir et al., 2008). Cationic polymers-mediated autophagy might be an adaptive cellular response aiding in the degradation and clearance of nanoparticles, and might also induce harmful cellular dysfunction (Wang et al., 2014) (Fig. 12.3).

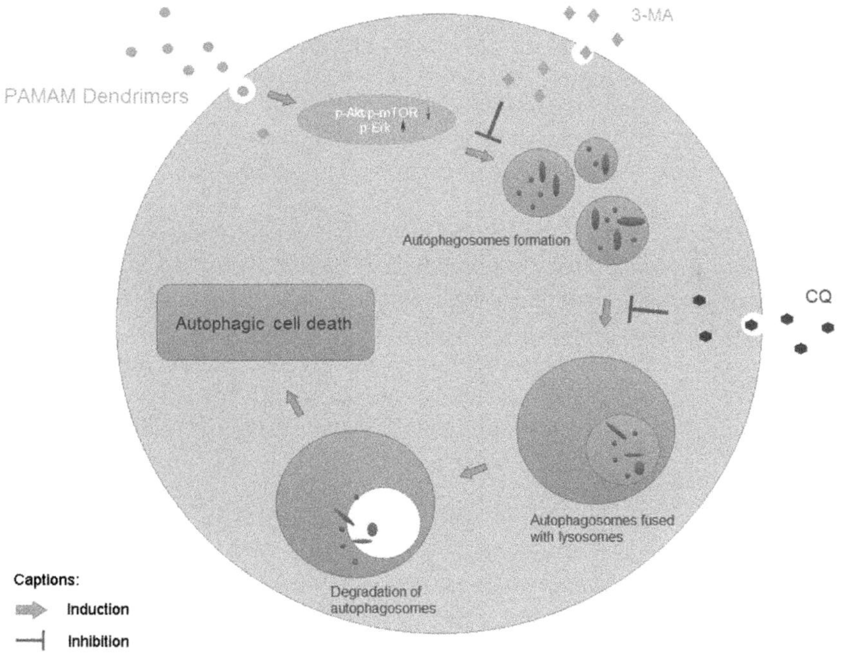

FIGURE 12.3 The role of autophagy in poly(amidoamine) (PAMAM) dendrimers–induced neurotoxicity. Cationic PAMAM dendrimers induced both cytotoxicity and autophagic flux in human glioma cells. Suppression of autophagy by autophagic inhibitors 3-methyladenine (3-MA) and chloroquine (CQ) significantly reversed glioma cell death caused by PAMAM dendrimers, indicating the cytotoxic role of autophagy in neurotoxicity caused by PAMAM dendrimers. *mTOR*, mechanistic target of rapamycin.

Man et al. (2010) found that cationic lipids induced genuine autophagy in mammalian cells, but not neutral lipids. Cationic lipids increased formation of autophagosomes in a mechanistic target of rapamycin–independent manner. Roberts et al. (2013) also showed that cationic liposome and polyplex carriers used for gene delivery activated autophagy and generated tubulovesicular autophagosomes. Activation of autophagy and capture within tubulovesicular autophagosomes provided a new cellular barrier against efficient gene delivery.

Gao et al. (2011) reported that autophagy was tightly associated with PEI-induced cytotoxicity, and autophagy contributed to lysosome injury and mitochondrial damage in different stage. Suppression of autophagy could significantly reduce PEI-induced cytotoxicity, whereas promotion of autophagy aggravated cell injury. Lin et al. (2012) investigated the role of autophagy in branched PEI and PLL polymers-induced cell death, and found that both PEI and PLL polymers could induce fluorescent puncta formation of green fluorescent protein-microtubule-associated protein 1 light chain 3(LC3) fusion protein, conversions of LC3-I to LC3-II, degradation of p62 in cells, whereas the proportion of cell death was higher in cells treated with PEI or PLL polymers plus autophagy inhibitors.

Halamoda Kenzaoui et al. (2012) found that PLGA-PEO (poly(ethylene oxide)) polymeric nanomaterials could induce autophagic process when taken up by human brain-derived endothelial cells.

Wang et al. (2014) reported that cationic PAMAM dendrimers induced both cytotoxicity and autophagic flux in human glioma cells. Suppression of autophagy by autophagic inhibitors 3-methyladenine and chloroquine significantly reversed glioma cell death caused by PAMAM dendrimers, indicating the cytotoxic role of autophagy in neurotoxicity caused by PAMAM dendrimers. Besides, intracellular oxidative stress generation was also involved in PAMAM dendrimers–induced autophagy in human glioma cells: inhibition of oxidative stress impaired expression of autophagic fluorescence and autophagic-related protein triggered by PAMAM dendrimers (Li et al., 2015b).

4.2.4 Oxidative Stress

ROS are highly reactive molecules containing an oxygen atom, and are mainly produced by the mitochondrial and cytoplasmic oxidation processes under physiological conditions (Buonocore et al., 2010). ROS are capable of inducing membrane damage that could affect the structure and function of inner proteins, lipid denaturation, and structural alteration of DNA (Marchi et al., 2012). An increasing amount of evidence indicated that cationic polymers could induce cytotoxicity due to their large surface area and positive charges, which is capable of inducing ROS (Jain et al., 2010) (Fig. 12.4).

The neuronal system is vulnerable to oxidative damage because of its high content of easily peroxidizable unsaturated fatty acids, high oxygen consumption rate, and relative paucity of antioxidant enzymes compared with other

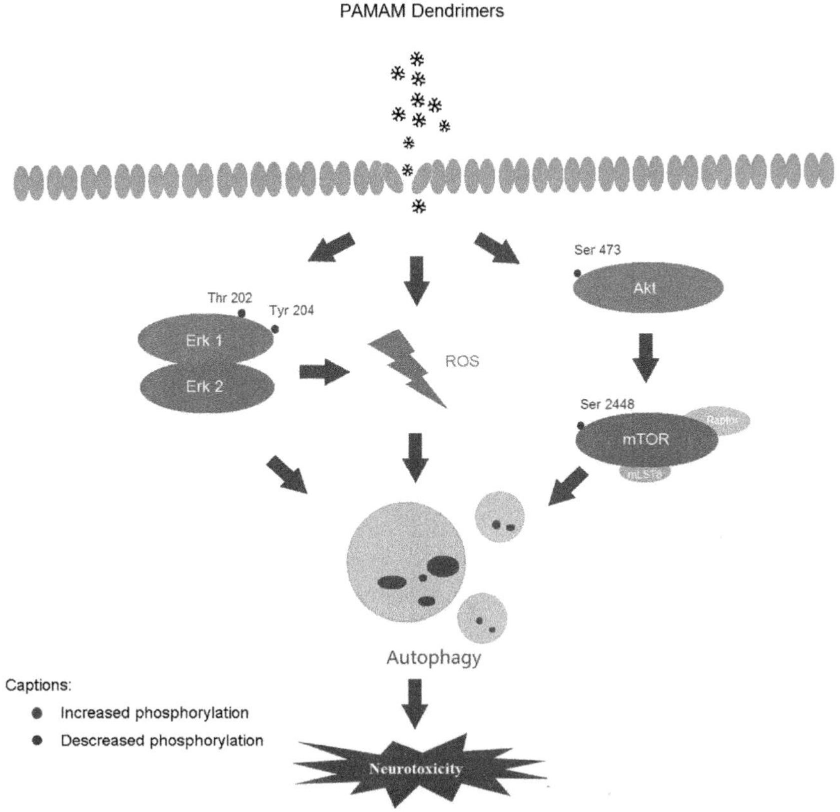

FIGURE 12.4 Interplay between oxidative stress and autophagy in poly(amidoamine) (PAMAM) dendrimers–induced neurotoxicity. Suppression of oxidative stress reduced PAMAM dendrimers–induced autophagic effects and cell death in neuronal cells, whereas inhibition of autophagy protected against PAMAM dendrimers–induced neurotoxicity. Extracellular signal-regulated kinase (ERK) 1/2 signaling pathway was activated and played a key role in the interplay between oxidative stress and autophagy induced by PAMAM dendrimers. *mTOR*, mechanistic target of rapamycin.

organs (Nunomura et al., 2006). Oxidative stress may result in inflammation and consequent damage to DNA and proteins, which are involved in neurotoxicity triggered by cationic polymers (Guan et al., 2012).

Membrane potential and membrane integrity are considered to play crucial roles in proper signaling in neuronal cells (Kohanski et al., 2008). A disturbance in membrane potential and membrane integrity could be induced by cationic polymers, which showed a significant effect on the normal function of the CNS, and became one of the most major causes of neurotoxicity (Verdurmen and Brock, 2011).

PAMAM dendrimers has been shown to evoke lasting neuronal depolarization and cell death in a concentration-dependent manner. Nyitrai et al. (2013)

explored the early progression of PAMAM dendrimers action on neuronal and astroglial cells, and found that interplay between PAMAM dendrimers and plasma membrane lead to robust activation of neurons and astroglial cells, resulting in mitochondrial depolarization followed by impaired oxidative metabolism of neurons. The oxidative stress may be the primary basis of neurotoxicity triggered by PAMAM dendrimers. Janaszewska et al. (2013) investigated the influence of pyrrolidone dendrimers on the mitochondrial membrane potential and intracellular ROS level. Pyrrolidone-modified dendrimers lacked the influence on the mitochondrial membrane potential and ROS generation, whereas unmodified PAMAM evoked toxic effects through the formation of ROS and damage of mitochondria. Li et al. (2015b) confirmed that ROS and autophagy were triggered by PAMAM dendrimers in neuronal cell death through detecting mitochondrial oxidative fluorescence and autophagic fluorescence, and explored the interplay of oxidative stress and autophagy in PAMAM dendrimers–induced neurotoxicity: suppression of oxidative stress reduced PAMAM dendrimers–induced autophagic effects and cell death in neuronal cells, whereas inhibition of autophagy protected against PAMAM dendrimers–induced neurotoxicity. ERK 1/2 signaling pathway was activated and played a key role in the interplay between oxidative stress and autophagy induced by PAMAM dendrimers in PC-12 cells.

Calarco et al. (2013) showed that PEI-PLGA copolymers triggered oxidative stress on cells by inducing the production of ROS, which cause DNA damage without apparently affecting cell viability. Bexiga et al. (2011) confirmed that cationic polystyrene nanoparticles induced morphological changes in lysosomes and mitochondria using transmission electron microscopy, and rapidly increased the formation of ROS in human brain astrocytoma cell lines.

4.2.5 Inflammation and Inflammasome

4.2.5.1 Inflammation

The neurotoxicity of cationic polymers was tightly associated with the expression level of inflammation-related cytokines in the brain. Cationic polymers triggered the release of proinflammatory cytokines, which might contribute to the neuroinjury.

Strand et al. (2011) found that PLL polymers could provoke the proinflammatory cytokine TNF, and promote inflammatory responses through its necrosis-inducing properties directly or indirectly. The ability of PLL polymers inducing TNF was also explored with different molecular weights: PLL polymers with molecular weight of 29.8 kDa were more potent than those with molecular weight of 4.0 kDa and 7.5 kDa, confirming that the ability to induce TNF of PLL polymers increased with larger molecular weight of PLL. Ayyappan et al. (2014) also investigated the immunocompatibility and toxicity behavior of PLL polymers in vivo. The levels of immunomodulatory genes such as monocyte chemotactic protein-1, TNF-alpha, intercellular adhesion molecule-1, and interleukin-6 showed least changes when compared with control. The protein level of major inflammatory markers such as cyclooxygenase-2, lipo-oxygenase-15, and nitric oxide synthase also had least expression showing the immunocompatibility of PLL nanomaterials.

Kawakami et al. (2006) evaluated the cytokine response induced by linear and branched PEI/plasmid DNA complex and lipoplex. Intravenously injected lipoplex induced the production of proinflammatory cytokines such as TNF-alpha, interferon-gamma, and interleukin-12. Among these cytokines, TNF-alpha is the primary source of toxicity because it induced septic shock in animals at high concentrations. The response to an intravenous administration of a preparation of PEI polyplex varying in nanoparticle/plasmid ratio and the structure and molecular weight of PEI was investigated. After the injection of lipoplex, a significantly high TNF-alpha level was observed in serum, when compared with PEI polyplex. The proinflammatory cytokines interferon-gamma and interleukin-12 were significantly induced by the administration of lipoplex.

Dendrimers are able to cross the cell membrane of primary neurons, and diffuse to the CNS tissues. When PAMAM dendrimers reached the brain, it affected the immune system and induced an inflammatory response. Gene expression levels of the potent inflammatory mediator *Cox2*, the oxidative stress marker *Ho1*, the immune cell cytokine *Il-1beta*, and the transcription factor *NfkB* in the olfactory bulb, hippocampus, and cerebral cortex were highly elevated after dendrimers treatment (Win-Shwe et al., 2014).

Inflammatory responses in the brain are also associated with the activation of CNS resident immune system, namely microglial cells (Rivest, 2009). Chronic microglial activation endangers neuronal survival through the secretion of various proinflammatory mediators, neurotoxic factors, and generation of ROS and peroxynitrites (Nam et al., 2010). The activated microglial cells were not normally found in the brain, but infection or inflammation can activate microglial cells and cause them to migrate to the brain where they damage normal neurocells (Luo et al., 2010).

Dai et al. (2010) explored the unusual neuroinflammation targeting of dendrimers (with no targeting ligands) in the brain, with significant consequences for nanoscale materials in medicine. Albertazzi et al. (2013) investigated brain immune resident cell activation on PAMAM dendrimers administration. PAMAM dendrimers enhanced GFAP (a hallmark of the activation of astrocytes) and ionized calcium binding adaptor molecule-1 expression in the injection site owing to the response to the brain tissue damage induced by needle penetration, and induced microglial cells activation during its induced toxicity in CNS. Newland et al. (2013) explored the neurotoxicity of PEI in the brain of adult Sprague-Dawley rats, and found that PEI induced the volume of the gross tissue loss on brain tissues. GFAP and OX-42 (microglia marker) immunohistochemical analyses were used to detect the reactivity of astrocytes and microglial recruitment. Compared with negative control, PEI triggered the accumulation of astrocytes and microglia during its induced neurotoxicity.

4.2.5.2 Inflammasome

Inflammasome is an intracellular multiprotein complex assembled by NOD-like receptors (NLRs) and DNA sensors. It can be triggered by series of dangerous signals evoked by the presence of pathogens and cellular stress (Le and Harton, 2013). The recognition component of the inflammasome is the receptor protein. These pattern recognition receptors include NLR family, pyrin domain-containing 3 (NLRP3), NLR family CARD domain-containing 4, or absent in melanoma 2, adaptor protein apoptosis-associated speck-like protein containing a CARD and procaspase-1 (Ting et al., 2008). Assembly of inflammasome leads to the autocleavage and activation of caspase-1. The activated caspase-1 triggers the processing and release of proinflammatory cytokines interleukin-1 beta and interleukin-18. Both interleukins are important for the shaping of adaptive immune responses (Satoh et al., 2015). Activation of caspase-1 triggers a form of cell death termed pyroptosis, and the release of interleukin-1 beta is a hallmark of inflammasome activation (Fernandes-Alnemri et al., 2007). Studies have shown that a large set of nanomaterials could trigger the activation of inflammasome, which is tightly associated with nanotoxicity (Sun et al., 2013) (Fig. 12.5).

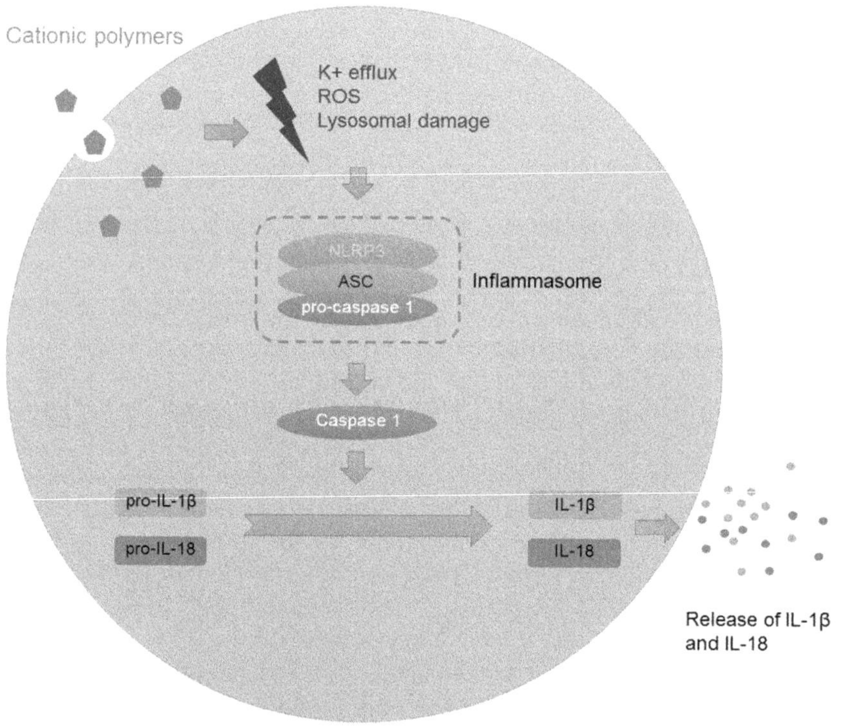

FIGURE 12.5 NLRP3 inflammasomes activated by cationic polymers. Cationic polymers activated NLRP3 inflammasome, induced caspase-1 cleavage, and subsequently triggered the release of proinflammatory interleukin (IL)-1 beta. *ASC*, apoptosis-associated speck-like protein containing a CARD; *ROS*, reactive oxygen species.

Bueter et al. (2011) detected the capacity of stimulating murine macrophages releasing the inflammasome-associated cytokine interleukin-1 beta triggered by chitosan, and confirmed that highly purified chitosan was a potent NLRP3 inflammasome activator. The size of the chitosan nanomaterials played an important role, as low-molecular-weight chitosan nanomaterials provoked the greatest activity on NLRP3 inflammasome activation. Suppression of phagocytosis using cytochalasin D eliminated the interleukin-1 beta activity of chitosan, which explained why the high-molecular-weight chitosan particles lack activity on inflammasome activation.

The reported lysosomal disruption by PEI in the gene transduction activity suggested that PEI might activate the NLRP3 inflammasome. Wegmann et al. (2012) found that the expression of interleukin-1 beta was elevated after intraperitoneal PEI administration. Compared with untreated cells, both PEI alone and PEI combined with LPS induced caspase-1 cleavage, suggesting that PEI was a potential activator of inflammasome.

Vaine et al. (2013) examined the role of polymeric microparticles surface texturing on uptake efficiency and the subsequent immune cell activation of the inflammasome. The textured particles were much more readily phagocytosed than smooth particles and induced more lipid raft recruitment to the phagosome. Textured particles also induced stronger interleukin-1 beta secretion than smooth particles through activation of the NALP3 inflammasome. Compared with carboxyl- or nonfunctionalized polystyrene nanomaterials, amino-functionalized nanomaterials with 100 nm diameter effectively activated NLRP3 inflammasome and subsequently triggered the release of proinflammatory interleukin-1 beta in human macrophages (Frohlich, 2012).

5. CONCLUSION

As cationic polymers possess unique physical structures and surface properties, various kinds of cationic polymers have been developed over the past few decades for a wide spectrum of nanomedical applications in CNS. Although cationic polymers could be successfully used for gene transfer, drug delivery, and diagnostic imaging, after entering into the CNS, they may cause neurotoxicity and induce CNS damage, which seriously limits their applications. Unfortunately, until now, the evaluation of neurotoxicity triggered by cationic polymers is limited, and our understanding of the role of cationic polymers on the CNS is still poor. Another question is, the neurotoxic effects of cationic polymers on CNS are mostly studied in mice, and have not been examined in detail. Thus, there is an instant need for scientists to address the problems about the neurotoxicity of cationic polymers, and this study might provide a foundation for the further design and development of the cationic polymers.

While evaluating the neurotoxicity of cationic polymers, the surface charge, surface area, coating, size, shape, and the basic materials that cationic polymers are made up of are expected to show important roles, and should be carefully considered. Besides, the research initiatives highlighted in the chapter further show biochemical mechanisms of cationic polymers–induced neurotoxicity, including apoptosis, necrosis, autophagy, oxidative stress, inflammation, and inflammasome; which are expected to be the most important problems in the evaluation of cationic polymers-induced neurotoxicity. These physicochemical and biochemical mechanisms of neurotoxicity triggered by cationic polymers might be helpful for establishing a guideline for the evaluation of the nanoneurotoxicity, designing, and developing new cationic polymers in the future.

REFERENCES

Abbasi, E., Aval, S.F., Akbarzadeh, A., Milani, M., Nasrabadi, H.T., Joo, S.W., Hanifehpour, Y., Nejati-Koshki, K., Pashaei-Asl, R., 2014. Dendrimers: synthesis, applications, and properties. Nanoscale Res. Lett. 9, 247.

Acharya, S., Sahoo, S.K., 2011. PLGA nanoparticles containing various anticancer agents and tumour delivery by EPR effect. Adv. Drug Deliv. Rev. 63, 170–183.

Aillon, K.L., Xie, Y., El-Gendy, N., Berkland, C.J., Forrest, M.L., 2009. Effects of nanomaterial physicochemical properties on in vivo toxicity. Adv. Drug Deliv. Rev. 61, 457–466.

Albertazzi, L., Gherardini, L., Brondi, M., Sulis Sato, S., Bifone, A., Pizzorusso, T., Ratto, G.M., Bardi, G., 2013. In vivo distribution and toxicity of PAMAM dendrimers in the central nervous system depend on their surface chemistry. Mol. Pharm. 10, 249–260.

Alhaddad, A., Durieu, C., Dantelle, G., Le Cam, E., Malvy, C., Treussart, F., Bertrand, J.R., 2012. Influence of the internalization pathway on the efficacy of siRNA delivery by cationic fluorescent nanodiamonds in the Ewing sarcoma cell model. PLoS One 7, e52207.

Alvarez-Paino, M., Munoz-Bonilla, A., Lopez-Fabal, F., Gomez-Garces, J.L., Heuts, J.P., Fernandez-Garcia, M., 2015. Effect of glycounits on the antimicrobial properties and toxicity behavior of polymers based on quaternized DMAEMA. Biomacromolecules 16, 295–303.

Arote, R., Kim, T.H., Kim, Y.K., Hwang, S.K., Jiang, H.L., Song, H.H., Nah, J.W., Cho, M.H., Cho, C.S., 2007. A biodegradable poly(ester amine) based on polycaprolactone and polyethylenimine as a gene carrier. Biomaterials 28, 735–744.

Arote, R.B., Jere, D., Jiang, H.L., Kim, Y.K., Choi, Y.J., Cho, C.S., 2009. Biodegradable poly(ester amine)s for gene delivery applications. Biomed. Mater. 4, 044102.

Arshad, A., Yang, B., Bienemann, A.S., Barua, N.U., Wyatt, M.J., Woolley, M., Johnson, D.E., Edler, K.J., Gill, S.S., 2015. Convection-enhanced delivery of carboplatin PLGA nanoparticles for the treatment of glioblastoma. PLoS One 10, e0132266.

Aslan, S., Loebick, C.Z., Kang, S., Elimelech, M., Pfefferle, L.D., Van Tassel, P.R., 2010. Antimicrobial biomaterials based on carbon nanotubes dispersed in poly(lactic-co-glycolic acid). Nanoscale 2, 1789–1794.

Ayyappan, J.P., Sami, H., Rajalekshmi, D.C., Sivakumar, S., Abraham, A., 2014. Immunocompatibility and toxicity studies of poly-L-lysine nanocapsules in Sprague-Dawley rats for drug-delivery applications. Chem. Biol. Drug Des. 84, 292–299.

Azevedo, M.M., Ramalho, P., Silva, A.P., Teixeira-Santos, R., Pina-Vaz, C., Rodrigues, A.G., 2014. Polyethyleneimine and polyethyleneimine-based nanoparticles: novel bacterial and yeast biofilm inhibitors. J. Med. Microbiol. 63, 1167–1173.

Baines, C.P., 2010. Role of the mitochondrion in programmed necrosis. Front. Physiol. 1, 156.

Bansal, R., Gupta, K.C., Kumar, P., 2015. Biodegradable and versatile polyethylenimine derivatives efficiently transfer DNA and siRNA into mammalian cells. Colloids Surf. B 135, 661–668.

Betancourt, T., Brown, B., Brannon-Peppas, L., 2007. Doxorubicin-loaded PLGA nanoparticles by nanoprecipitation: preparation, characterization and in vitro evaluation. Nanomedicine (Lond.) 2, 219–232.

Bexiga, M.G., Varela, J.A., Wang, F., Fenaroli, F., Salvati, A., Lynch, I., Simpson, J.C., Dawson, K.A., 2011. Cationic nanoparticles induce caspase 3-, 7- and 9-mediated cytotoxicity in a human astrocytoma cell line. Nanotoxicology 5, 557–567.

Bhattacharya, P., Geitner, N.K., Sarupria, S., Ke, P.C., 2013. Exploiting the physicochemical properties of dendritic polymers for environmental and biological applications. Phys. Chem. Chem. Phys. 15, 4477–4490.

Bivas-Benita, M., Romeijn, S., Junginger, H.E., Borchard, G., 2004. PLGA-PEI nanoparticles for gene delivery to pulmonary epithelium. Eur. J. Pharm. Biopharm. 58, 1–6.

Borchard, G., 2001. Chitosans for gene delivery. Adv. Drug Deliv. Rev. 52, 145–150.

Bromberg, L., Deshmukh, S., Temchenko, M., Iourtchenko, L., Alakhov, V., Alvarez-Lorenzo, C., Barreiro-Iglesias, R., Concheiro, A., Hatton, T.A., 2005. Polycationic block copolymers of poly(ethylene oxide) and poly(propylene oxide) for cell transfection. Bioconjug. Chem. 16, 626–633.

Bueter, C.L., Lee, C.K., Rathinam, V.A., Healy, G.J., Taron, C.H., Specht, C.A., Levitz, S.M., 2011. Chitosan but not chitin activates the inflammasome by a mechanism dependent upon phagocytosis. J. Biol. Chem. 286, 35447–35455.

Buonocore, G., Perrone, S., Tataranno, M.L., 2010. Oxygen toxicity: chemistry and biology of reactive oxygen species. Semin. Fetal Neonatal. Med. 15, 186–190.

Cadete, A., Figueiredo, L., Lopes, R., Calado, C.C., Almeida, A.J., Goncalves, L.M., 2012. Development and characterization of a new plasmid delivery system based on chitosan-sodium deoxycholate nanoparticles. Eur. J. Pharm. Sci. 45, 451–458.

Cai, J., Yue, Y., Rui, D., Zhang, Y., Liu, S., Wu, C., 2011. Effect of chain length on cytotoxicity and endocytosis of cationic polymers. Macromolecules 44, 2050–2057.

Calarco, A., Bosetti, M., Margarucci, S., Fusaro, L., Nicoli, E., Petillo, O., Cannas, M., Galderisi, U., Peluso, G., 2013. The genotoxicity of PEI-based nanoparticles is reduced by acetylation of polyethylenimine amines in human primary cells. Toxicol. Lett. 218, 10–17.

Cerda-Cristerna, B.I., Flores, H., Pozos-Guillen, A., Perez, E., Sevrin, C., Grandfils, C., 2011. Hemocompatibility assessment of poly(2-dimethylamino ethylmethacrylate) (PDMAEMA)-based polymers. J. Control Release 153, 269–277.

Chen, C.Z., Cooper, S.L., 2002. Interactions between dendrimer biocides and bacterial membranes. Biomaterials 23, 3359–3368.

Chen, C.Z., Beck-Tan, N.C., Dhurjati, P., van Dyk, T.K., LaRossa, R.A., Cooper, S.L., 2000. Quaternary ammonium functionalized poly(propylene imine) dendrimers as effective antimicrobials: structure-activity studies. Biomacromolecules 1, 473–480.

Cheng, Y., Li, M., Xu, T., 2008. Potential of poly(amidoamine) dendrimers as drug carriers of camptothecin based on encapsulation studies. Eur. J. Med. Chem. 43, 1791–1795.

Cheng, H., Deng, H., Zhou, L., Su, Y., Yu, S., Zhu, X., Zhou, Y., Yan, D., 2011. Improvement of gene delivery of PDMAEMA by incorporating a hyperbranched polymer core. J. Control Release 152 (Suppl. 1), e187–188.

Choi, J.S., Nam, K., Park, J.Y., Kim, J.B., Lee, J.K., Park, J.S., 2004. Enhanced transfection efficiency of PAMAM dendrimer by surface modification with L-arginine. J. Control Release 99, 445–456.

Costa, D., Valente, A.J., Queiroz, J., 2015. Stimuli-responsive polyamine-DNA blend nanogels for co-delivery in cancer therapy. Colloids Surf. B 132, 194–201.

Dai, H., Navath, R.S., Balakrishnan, B., Guru, B.R., Mishra, M.K., Romero, R., Kannan, R.M., Kannan, S., 2010. Intrinsic targeting of inflammatory cells in the brain by polyamidoamine dendrimers upon subarachnoid administration. Nanomedicine (Lond.) 5, 1317–1329.

Di Gioia, S., Conese, M., 2009. Polyethylenimine-mediated gene delivery to the lung and therapeutic applications. Drug Des. Dev. Ther. 2, 163–188.

Doane, T., Burda, C., 2013. Nanoparticle mediated non-covalent drug delivery. Adv. Drug Deliv. Rev. 65, 607–621.

Eichman, J.D., Bielinska, A.U., Kukowska-Latallo, J.F., Baker Jr., J.R., 2000. The use of PAMAM dendrimers in the efficient transfer of genetic material into cells. Pharm. Sci. Technol. Today 3, 232–245.

Elmore, S., 2007. Apoptosis: a review of programmed cell death. Toxicol. Pathol. 35, 495–516.

El-Sharif, A.A., Hussain, M.H., 2011. Chitosan-EDTA new combination is a promising candidate for treatment of bacterial and fungal infections. Curr. Microbiol. 62, 739–745.

Fernandes-Alnemri, T., Wu, J., Yu, J.W., Datta, P., Miller, B., Jankowski, W., Rosenberg, S., Zhang, J., Alnemri, E.S., 2007. The pyroptosome: a supramolecular assembly of ASC dimers mediating inflammatory cell death via caspase-1 activation. Cell Death Differ. 14, 1590–1604.

Fischer, D., Li, Y., Ahlemeyer, B., Krieglstein, J., Kissel, T., 2003. In vitro cytotoxicity testing of polycations: influence of polymer structure on cell viability and hemolysis. Biomaterials 24, 1121–1131.

Fischer, D., Dautzenberg, H., Kunath, K., Kissel, T., 2004. Poly(diallyldimethylammonium chlorides) and their N-methyl-N-vinylacetamide copolymer-based DNA-polyplexes: role of molecular weight and charge density in complex formation, stability, and in vitro activity. Int. J. Pharm. 280, 253–269.

Florea, B.I., Meaney, C., Junginger, H.E., Borchard, G., 2002. Transfection efficiency and toxicity of polyethylenimine in differentiated Calu-3 and nondifferentiated COS-1 cell cultures. AAPS PharmSci. 4, E12.

Frohlich, E., 2012. The role of surface charge in cellular uptake and cytotoxicity of medical nanoparticles. Int. J. Nanomed. 7, 5577–5591.

Fulda, S., Debatin, K.M., 2006. Extrinsic versus intrinsic apoptosis pathways in anticancer chemotherapy. Oncogene 25, 4798–4811.

Ganguly, K., Kulkarni, A.R., Aminabhavi, T.M., 2015. In vitro cytotoxicity and in vivo efficacy of 5-fluorouracil-loaded enteric-coated PEG-cross-linked chitosan microspheres in colorectal cancer therapy in rats. Drug Deliv. 1–14.

Gao, X., Yao, L., Song, Q., Zhu, L., Xia, Z., Xia, H., Jiang, X., Chen, J., Chen, H., 2011. The association of autophagy with polyethylenimine-induced cytotoxicity in nephritic and hepatic cell lines. Biomaterials 32, 8613–8625.

Gao, S., Tian, H., Guo, Y., Li, Y., Guo, Z., Zhu, X., Chen, X., 2015. miRNA oligonucleotide and sponge for miRNA-21 inhibition mediated by PEI-PLL in breast cancer therapy. Acta Biomater. 25, 184–193.

Geornaras, I., Yoon, Y., Belk, K.E., Smith, G.C., Sofos, J.N., 2007. Antimicrobial activity of epsilon-polylysine against *Escherichia coli* O157:H7, *Salmonella typhimurium*, and *Listeria monocytogenes* in various food extracts. J. Food Sci. 72, M330–M334.

Germano, I.M., Binello, E., 2009. Gene therapy as an adjuvant treatment for malignant gliomas: from bench to bedside. J. Neurooncol 93, 79–87.

Gomez-Valades, A.G., Molas, M., Vidal-Alabro, A., Bermudez, J., Bartrons, R., Perales, J.C., 2005. Copolymers of poly-L-lysine with serine and tryptophan form stable DNA vectors: implications for receptor-mediated gene transfer. J. Control Release 102, 277–291.

Goyal, R., Tripathi, S.K., Tyagi, S., Sharma, A., Ram, K.R., Chowdhuri, D.K., Shukla, Y., Kumar, P., Gupta, K.C., 2012. Linear PEI nanoparticles: efficient pDNA/siRNA carriers in vitro and in vivo. Nanomedicine 8, 167–175.

Gu, J., Fang, X., Hao, J., Sha, X., 2015. Reversal of P-glycoprotein-mediated multidrug resistance by CD44 antibody-targeted nanocomplexes for short hairpin RNA-encoding plasmid DNA delivery. Biomaterials 45, 99–114.

Guan, R., Kang, T., Lu, F., Zhang, Z., Shen, H., Liu, M., 2012. Cytotoxicity, oxidative stress, and genotoxicity in human hepatocyte and embryonic kidney cells exposed to ZnO nanoparticles. Nanoscale Res. Lett. 7, 602.

Guo, Z., Chen, R., Xing, R., Liu, S., Yu, H., Wang, P., Li, C., Li, P., 2006. Novel derivatives of chitosan and their antifungal activities in vitro. Carbohydr. Res. 341, 351–354.

Gupta, R., Mehra, N.K., Jain, N.K., 2014. Development and characterization of sulfasalazine loaded fucosylated PPI dendrimer for the treatment of cytokine-induced liver damage. Eur. J. Pharm. Biopharm. 86, 449–458.

Halamoda Kenzaoui, B., Chapuis Bernasconi, C., Guney-Ayra, S., Juillerat-Jeanneret, L., 2012. Induction of oxidative stress, lysosome activation and autophagy by nanoparticles in human brain-derived endothelial cells. Biochem. J. 441, 813–821.

Han, L., Li, J., Huang, S., Huang, R., Liu, S., Hu, X., Yi, P., Shan, D., Wang, X., Lei, H., Jiang, C., 2011. Peptide-conjugated polyamidoamine dendrimer as a nanoscale tumor-targeted T1 magnetic resonance imaging contrast agent. Biomaterials 32, 2989–2998.

Han, B., Wang, W., Wu, H., Fang, F., Wang, N., Zhang, X., Xu, S., 2012. Polyethyleneimine modified fluorescent carbon dots and their application in cell labeling. Colloids Surf. B 100, 209–214.

Hayder, M., Poupot, M., Baron, M., Nigon, D., Turrin, C.O., Caminade, A.M., Majoral, J.P., Eisenberg, R.A., Fournie, J.J., Cantagrel, A., Poupot, R., Davignon, J.L., 2011. A phosphorus-based dendrimer targets inflammation and osteoclastogenesis in experimental arthritis. Sci. Transl. Med. 3, 81ra35.

Hill, L.E., Taylor, T.M., Gomes, C., 2013. Antimicrobial efficacy of poly (DL-lactide-co-glycolide) (PLGA) nanoparticles with entrapped cinnamon bark extract against *Listeria monocytogenes* and *Salmonella typhimurium*. J. Food Sci. 78, N626–N632.

Hollins, A.J., Benboubetra, M., Omidi, Y., Zinselmeyer, B.H., Schatzlein, A.G., Uchegbu, I.F., Akhtar, S., 2004. Evaluation of generation 2 and 3 poly(propylenimine) dendrimers for the potential cellular delivery of antisense oligonucleotides targeting the epidermal growth factor receptor. Pharm. Res. 21, 458–466.

Hsu, E.W., Liu, S., Shrivats, A.R., Watt, A.C., McBride, S., Averick, S.E., Cho, H.Y., Matyjaszewski, K., Hollinger, J.O., 2014. Cationic nanostructured polymers for siRNA delivery in murine calvarial pre-osteoblasts. J. Biomed. Nanotechnol. 10, 1130–1136.

Hu, Y.L., Qi, W., Han, F., Shao, J.Z., Gao, J.Q., 2011. Toxicity evaluation of biodegradable chitosan nanoparticles using a zebrafish embryo model. Int. J. Nanomed. 6, 3351–3359.

Huang, M., Khor, E., Lim, L.Y., 2004. Uptake and cytotoxicity of chitosan molecules and nanoparticles: effects of molecular weight and degree of deacetylation. Pharm. Res. 21, 344–353.

Huang, F.W., Wang, H.Y., Li, C., Wang, H.F., Sun, Y.X., Feng, J., Zhang, X.Z., Zhuo, R.X., 2010. PEGylated PEI-based biodegradable polymers as non-viral gene vectors. Acta Biomater. 6, 4285–4295.

Hunter, A.C., 2006. Molecular hurdles in polyfectin design and mechanistic background to polycation induced cytotoxicity. Adv. Drug Deliv. Rev. 58, 1523–1531.

Imamura, M., Kodama, Y., Higuchi, N., Kanda, K., Nakagawa, H., Muro, T., Nakamura, T., Kitahara, T., Sasaki, H., 2014. Ternary complex of plasmid DNA electrostatically assembled with polyamidoamine dendrimer and chondroitin sulfate for effective and secure gene delivery. Biol. Pharm. Bull. 37, 552–559.

Jain, K., Kesharwani, P., Gupta, U., Jain, N.K., 2010. Dendrimer toxicity: let's meet the challenge. Int. J. Pharm. 394, 122–142.

Janaszewska, A., Ciolkowski, M., Wrobel, D., Petersen, J.F., Ficker, M., Christensen, J.B., Bryszewska, M., Klajnert, B., 2013. Modified PAMAM dendrimer with 4-carbomethoxypyrrolidone surface groups reveals negligible toxicity against three rodent cell-lines. Nanomedicine 9, 461–464.

Janaszewska, A., Studzian, M., Petersen, J.F., Ficker, M., Christensen, J.B., Klajnert-Maculewicz, B., 2015. PAMAM dendrimer with 4-carbomethoxypyrrolidone–in vitro assessment of neurotoxicity. Nanomedicine 11, 409–411.

Jere, D., Xu, C.X., Arote, R., Yun, C.H., Cho, M.H., Cho, C.S., 2008. Poly(beta-amino ester) as a carrier for si/shRNA delivery in lung cancer cells. Biomaterials 29, 2535–2547.

Jones, R.A., Poniris, M.H., Wilson, M.R., 2004. pDMAEMA is internalised by endocytosis but does not physically disrupt endosomes. J. Control Release 96, 379–391.

Jumaa, M., Furkert, F.H., Muller, B.W., 2002. A new lipid emulsion formulation with high antimicrobial efficacy using chitosan. Eur. J. Pharm. Biopharm. 53, 115–123.

Kafil, V., Omidi, Y., 2011. Cytotoxic impacts of linear and branched polyethylenimine nanostructures in a431 cells. Bioimpacts 1, 23–30.

Kang, L., Gao, Z., Huang, W., Jin, M., Wang, Q., 2015. Nanocarrier-mediated co-delivery of chemotherapeutic drugs and gene agents for cancer treatment. Acta Pharm. Sin. B 5, 169–175.

Katare, Y.K., Daya, R.P., Sookram Gray, C., Luckham, R.E., Bhandari, J., Chauhan, A.S., Mishra, R.K., 2015. Brain targeting of a water insoluble antipsychotic drug haloperidol via the intranasal route using PAMAM dendrimer. Mol. Pharm. 12, 3380–3388.

Kato, Y., Onishi, H., Machida, Y., 2003. Application of chitin and chitosan derivatives in the pharmaceutical field. Curr. Pharm. Biotechnol. 4, 303–309.

Kawakami, S., Ito, Y., Charoensit, P., Yamashita, F., Hashida, M., 2006. Evaluation of proinflammatory cytokine production induced by linear and branched polyethylenimine/plasmid DNA complexes in mice. J. Pharmacol. Exp. Ther. 317, 1382–1390.

Kean, T., Thanou, M., 2010. Biodegradation, biodistribution and toxicity of chitosan. Adv. Drug Deliv. Rev. 62, 3–11.

Kim, B.K., Kwon, S.Y., Lee, C.H., Choe, W.H., Choi, H.M., Koo, H.W., 2009. Analysis of the cost-effectiveness of antiviral therapies in chronic hepatitis B patients in Korea. Korean J. Hepatol. 15, 25–41.

Kim, Y.K., Cho, C.S., Cho, M.H., Jiang, H.L., 2013. Poly(ester amine) synthesized from trimethylolpropane triacrylate and spermine as an efficient siRNA carrier. J. Nanosci. Nanotechnol. 13, 5692–5697.

Kodama, Y., Nakamura, T., Kurosaki, T., Egashira, K., Mine, T., Nakagawa, H., Muro, T., Kitahara, T., Higuchi, N., Sasaki, H., 2014. Biodegradable nanoparticles composed of dendrigraft poly-L-lysine for gene delivery. Eur. J. Pharm. Biopharm. 87, 472–479.

Kohanski, M.A., Dwyer, D.J., Wierzbowski, J., Cottarel, G., Collins, J.J., 2008. Mistranslation of membrane proteins and two-component system activation trigger antibiotic-mediated cell death. Cell 135, 679–690.

Kojima, C., Turkbey, B., Ogawa, M., Bernardo, M., Regino, C.A., Bryant Jr., L.H., Choyke, P.L., Kono, K., Kobayashi, H., 2011. Dendrimer-based MRI contrast agents: the effects of PEGylation on relaxivity and pharmacokinetics. Nanomedicine 7, 1001–1008.

Kumar, M.N., Muzzarelli, R.A., Muzzarelli, C., Sashiwa, H., Domb, A.J., 2004. Chitosan chemistry and pharmaceutical perspectives. Chem. Rev. 104, 6017–6084.

Labena, A., Kabel, K.I., Farag, R.K., 2016. One-pot synthesize of dendritic hyperbranched PAMAM and assessment as a broad spectrum antimicrobial agent and anti-biofilm. Mater. Sci. Eng. C Mater. Biol. Appl. 58, 1150–1159.

Le, H.T., Harton, J.A., 2013. Pyrin- and CARD-only proteins as regulators of NLR functions. Front. Immunol. 4, 275.

Lesniak, W.G., Mishra, M.K., Jyoti, A., Balakrishnan, B., Zhang, F., Nance, E., Romero, R., Kannan, S., Kannan, R.M., 2013. Biodistribution of fluorescently labeled PAMAM dendrimers in neonatal rabbits: effect of neuroinflammation. Mol. Pharm. 10, 4560–4571.

Li, J., Fan, C., Pei, H., Shi, J., Huang, Q., 2013. Smart drug delivery nanocarriers with self-assembled DNA nanostructures. Adv. Mater. 25, 4386–4396.

Li, Y., Wang, S., Wang, Z., Qian, X., Fan, J., Zeng, X., Sun, Y., Song, P., Feng, M., Ju, D., 2014. Cationic poly(amidoamine) dendrimers induced cyto-protective autophagy in hepatocellular carcinoma cells. Nanotechnology 25, 365101.

Li, M., Li, Y., Huang, X., Lu, X., 2015a. Captopril-polyethyleneimine conjugate modified gold nanoparticles for co-delivery of drug and gene in anti-angiogenesis breast cancer therapy. J. Biomater. Sci. Polym. Ed. 26, 813–827.

Li, Y., Zhu, H., Wang, S., Qian, X., Fan, J., Wang, Z., Song, P., Zhang, X., Lu, W., Ju, D., 2015b. Interplay of oxidative stress and autophagy in PAMAM dendrimers-induced neuronal cell death. Theranostics 5, 1363–1377.

Liechty, W.B., Kryscio, D.R., Slaughter, B.V., Peppas, N.A., 2010. Polymers for drug delivery systems. Annu. Rev. Chem. Biomol. Eng. 1, 149–173.

Lim, E.K., Kim, T., Paik, S., Haam, S., Huh, Y.M., Lee, K., 2015. Nanomaterials for theranostics: recent advances and future challenges. Chem. Rev. 115, 327–394.

Lin, C.W., Jan, M.S., Kuo, J.H., Hsu, L.J., Lin, Y.S., 2012. Protective role of autophagy in branched polyethylenimine (25K)- and poly(L-lysine) (30-70K)-induced cell death. Eur. J. Pharm. Sci. 47, 865–874.

Liu, C., Zhang, P., Zhai, X., Tian, F., Li, W., Yang, J., Liu, Y., Wang, H., Wang, W., Liu, W., 2012. Nano-carrier for gene delivery and bioimaging based on carbon dots with PEI-passivation enhanced fluorescence. Biomaterials 33, 3604–3613.

Lopez, A.I., Reins, R.Y., McDermott, A.M., Trautner, B.W., Cai, C., 2009. Antibacterial activity and cytotoxicity of PEGylated poly(amidoamine) dendrimers. Mol. Biosyst. 5, 1148–1156.

Lunov, O., Syrovets, T., Loos, C., Beil, J., Delacher, M., Tron, K., Nienhaus, G.U., Musyanovych, A., Mailander, V., Landfester, K., Simmet, T., 2011. Differential uptake of functionalized polystyrene nanoparticles by human macrophages and a monocytic cell line. ACS Nano 5, 1657–1669.

Luo, Y., Wang, Q., 2014. Recent development of chitosan-based polyelectrolyte complexes with natural polysaccharides for drug delivery. Int. J. Biol. Macromol. 64, 353–367.

Luo, X.G., Ding, J.Q., Chen, S.D., 2010. Microglia in the aging brain: relevance to neurodegeneration. Mol. Neurodegener. 5, 12.

Lv, H., Zhang, S., Wang, B., Cui, S., Yan, J., 2006. Toxicity of cationic lipids and cationic polymers in gene delivery. J. Control Release 114, 100–109.

Madaan, K., Kumar, S., Poonia, N., Lather, V., Pandita, D., 2014. Dendrimers in drug delivery and targeting: drug-dendrimer interactions and toxicity issues. J. Pharm. Bioallied Sci. 6, 139–150.

Makadia, H.K., Siegel, S.J., 2011. Poly lactic-co-glycolic acid (PLGA) as biodegradable controlled drug delivery carrier. Polymers (Basel) 3, 1377–1397.

Man, N., Chen, Y., Zheng, F., Zhou, W., Wen, L.P., 2010. Induction of genuine autophagy by cationic lipids in mammalian cells. Autophagy 6, 449–454.

Mansur, A.A., de Carvalho, S.M., Mansur, H.S., 2015. Bioengineered quantum dot/chitosan-tripeptide nanoconjugates for targeting the receptors of cancer cells. Int. J. Biol. Macromol. 82, 780–789.

Marchi, S., Giorgi, C., Suski, J.M., Agnoletto, C., Bononi, A., Bonora, M., De Marchi, E., Missiroli, S., Patergnani, S., Poletti, F., Rimessi, A., Duszynski, J., Wieckowski, M.R., Pinton, P., 2012. Mitochondria-ros crosstalk in the control of cell death and aging. J. Signal Transduct. 2012, 329635.

Marin, E., Briceno, M.I., Caballero-George, C., 2013. Critical evaluation of biodegradable polymers used in nanodrugs. Int. J. Nanomed. 8, 3071–3090.

Mei, L., Lu, Z., Zhang, X., Li, C., Jia, Y., 2014. Polymer-Ag nanocomposites with enhanced antimicrobial activity against bacterial infection. ACS Appl. Mater. Interfaces 6, 15813–15821.

Mieszawska, A.J., Gianella, A., Cormode, D.P., Zhao, Y., Meijerink, A., Langer, R., Farokhzad, O.C., Fayad, Z.A., Mulder, W.J., 2012. Engineering of lipid-coated PLGA nanoparticles with a tunable payload of diagnostically active nanocrystals for medical imaging. Chem. Commun. (Camb.) 48, 5835–5837.

Mishra, V., Jain, N.K., 2014. Acetazolamide encapsulated dendritic nano-architectures for effective glaucoma management in rabbits. Int. J. Pharm. 461, 380–390.

Moghimi, S.M., Symonds, P., Murray, J.C., Hunter, A.C., Debska, G., Szewczyk, A., 2005. A two-stage poly(ethylenimine)-mediated cytotoxicity: implications for gene transfer/therapy. Mol. Ther. 11, 990–995.

Mori, Y., Hirano, T., Notomi, T., 2006. Sequence specific visual detection of LAMP reactions by addition of cationic polymers. BMC Biotechnol. 6, 3.

Moschonas, G., Geornaras, I., Stopforth, J.D., Wach, D., Woerner, D.R., Belk, K.E., Smith, G.C., Sofos, J.N., 2012. Activity of caprylic acid, carvacrol, epsilon-polylysine and their combinations against Salmonella in not-ready-to-eat surface-browned, frozen, breaded chicken products. J. Food Sci. 77, M405–M411.

Murugan, E., Geetha Rani, D.P., Srinivasan, K., Muthumary, J., 2013. New surface hydroxylated and internally quaternised poly(propylene imine) dendrimers as efficient biocompatible drug carriers of norfloxacin. Expert Opin. Drug Deliv. 10, 1319–1334.

Nam, K.N., Park, Y.M., Jung, H.J., Lee, J.Y., Min, B.D., Park, S.U., Jung, W.S., Cho, K.H., Park, J.H., Kang, I., Hong, J.W., Lee, E.H., 2010. Anti-inflammatory effects of crocin and crocetin in rat brain microglial cells. Eur. J. Pharmacol. 648, 110–116.

Nam, J.P., Park, S.C., Kim, T.H., Jang, J.Y., Choi, C., Jang, M.K., Nah, J.W., 2013. Encapsulation of paclitaxel into lauric acid-O-carboxymethyl chitosan-transferrin micelles for hydrophobic drug delivery and site-specific targeted delivery. Int. J. Pharm. 457, 124–135.

Nascimento, A.V., Singh, A., Bousbaa, H., Ferreira, D., Sarmento, B., Amiji, M.M., 2015. Combinatorial-designed epidermal growth factor receptor-targeted chitosan nanoparticles for encapsulation and delivery of lipid-modified platinum derivatives in wild-type and resistant non-small-cell lung Cancer cells. Mol. Pharm. 12 (12), 4466–4477.

Navath, R.S., Menjoge, A.R., Dai, H., Romero, R., Kannan, S., Kannan, R.M., 2011. Injectable PAMAM dendrimer-PEG hydrogels for the treatment of genital infections: formulation and in vitro and in vivo evaluation. Mol. Pharm. 8, 1209–1223.

Negroni, A., Cucchiara, S., Stronati, L., 2015. Apoptosis, necrosis, and necroptosis in the gut and intestinal homeostasis. Mediators Inflamm. 2015, 250762.

Newland, B., Moloney, T.C., Fontana, G., Browne, S., Abu-Rub, M.T., Dowd, E., Pandit, A.S., 2013. The neurotoxicity of gene vectors and its amelioration by packaging with collagen hollow spheres. Biomaterials 34, 2130–2141.

Niidome, T., Gokuden, R., Watanabe, K., Mori, T., Naganuma, T., Utsumi, H., Ichikawa, K., Katayama, Y., 2014. Nitroxyl radicals-modified dendritic poly(l-lysine) as a contrast agent for Overhauser-enhanced MRI. J. Biomater. Sci. Polym. Ed. 25, 1425–1439.

Nunomura, A., Honda, K., Takeda, A., Hirai, K., Zhu, X., Smith, M.A., Perry, G., 2006. Oxidative damage to RNA in neurodegenerative diseases. J. Biomed. Biotechnol. 2006, 82323.
Nurhasni, H., Cao, J., Choi, M., Kim, I., Lee, B.L., Jung, Y., Yoo, J.W., 2015. Nitric oxide-releasing poly(lactic-co-glycolic acid)-polyethylenimine nanoparticles for prolonged nitric oxide release, antibacterial efficacy, and in vivo wound healing activity. Int. J. Nanomed. 10, 3065–3080.
Nyitrai, G., Heja, L., Jablonkai, I., Pal, I., Visy, J., Kardos, J., 2013. Polyamidoamine dendrimer impairs mitochondrial oxidation in brain tissue. J. Nanobiotechnology 11, 9.
Omidi, Y., Hollins, A.J., Drayton, R.M., Akhtar, S., 2005. Polypropylenimine dendrimer-induced gene expression changes: the effect of complexation with DNA, dendrimer generation and cell type. J. Drug Target. 13, 431–443.
Ouyang, L., Shi, Z., Zhao, S., Wang, F.T., Zhou, T.T., Liu, B., Bao, J.K., 2012. Programmed cell death pathways in cancer: a review of apoptosis, autophagy and programmed necrosis. Cell Prolif. 45, 487–498.
Pandey, C.M., Sharma, A., Sumana, G., Tiwari, I., Malhotra, B.D., 2013. Cationic poly(lactic-co-glycolic acid) iron oxide microspheres for nucleic acid detection. Nanoscale 5, 3800–3807.
Parveen, S., Sahoo, S.K., 2011. Long circulating chitosan/PEG blended PLGA nanoparticle for tumor drug delivery. Eur. J. Pharmacol. 670, 372–383.
Patel, J., Garala, K., Basu, B., Raval, M., Dharamsi, A., 2011. Solubility of aceclofenac in polyamidoamine dendrimer solutions. Int. J. Pharm. Investig. 1, 135–138.
Perez, A.P., Romero, E.L., Morilla, M.J., 2009. Ethylendiamine core PAMAM dendrimers/siRNA complexes as in vitro silencing agents. Int. J. Pharm. 380, 189–200.
Piras, A., Gianetto, D., Conte, D., Bosone, A., Vercelli, A., 2011. Activation of autophagy in a rat model of retinal ischemia following high intraocular pressure. PLoS One 6, e22514.
Prados, J., Alvarez, P.J., Melguizo, C., Rodriguez-Serrano, F., Carrillo, E., Boulaiz, H., Velez, C., Marchal, J.A., Caba, O., Ortiz, R., Rama, A., Aranega, A., 2012. How is gene transfection able to improve current chemotherapy? the role of combined therapy in cancer treatment. Curr. Med. Chem. 19, 1870–1888.
Pramanik, S., Bharali, P., Konwar, B.K., Karak, N., 2014. Antimicrobial hyperbranched poly(ester amide)/polyaniline nanofiber modified montmorillonite nanocomposites. Mater. Sci. Eng. C Mater. Biol. Appl. 35, 61–69.
Qian, Y., Zha, Y., Feng, B., Pang, Z., Zhang, B., Sun, X., Ren, J., Zhang, C., Shao, X., Zhang, Q., Jiang, X., 2013. PEGylated poly(2-(dimethylamino) ethyl methacrylate)/DNA polyplex micelles decorated with phage-displayed TGN peptide for brain-targeted gene delivery. Biomaterials 34, 2117–2129.
Qian, X., Long, L., Shi, Z., Liu, C., Qiu, M., Sheng, J., Pu, P., Yuan, X., Ren, Y., Kang, C., 2014. Star-branched amphiphilic PLA-b-PDMAEMA copolymers for co-delivery of miR-21 inhibitor and doxorubicin to treat glioma. Biomaterials 35, 2322–2335.
Ravi Kumar, M.N., Bakowsky, U., Lehr, C.M., 2004. Preparation and characterization of cationic PLGA nanospheres as DNA carriers. Biomaterials 25, 1771–1777.
Rawlinson, L.A., O'Brien, P.J., Brayden, D.J., 2010. High content analysis of cytotoxic effects of pDMAEMA on human intestinal epithelial and monocyte cultures. J. Control Release 146, 84–92.
Rawlinson, L.A., O'Gara, J.P., Jones, D.S., Brayden, D.J., 2011. Resistance of *Staphylococcus aureus* to the cationic antimicrobial agent poly(2-(dimethylamino ethyl)methacrylate) (pDMAEMA) is influenced by cell-surface charge and hydrophobicity. J. Med. Microbiol. 60, 968–976.
Rezvani Amin, Z., Rahimizadeh, M., Eshghi, H., Dehshahri, A., Ramezani, M., 2013. The effect of cationic charge density change on transfection efficiency of polyethylenimine. Iran. J. Basic Med. Sci. 16, 150–156.

Rivest, S., 2009. Regulation of innate immune responses in the brain. Nat. Rev. Immunol. 9, 429–439.

Roberts, R., Al-Jamal, W.T., Whelband, M., Thomas, P., Jefferson, M., van den Bossche, J., Powell, P.P., Kostarelos, K., Wileman, T., 2013. Autophagy and formation of tubulovesicular autophagosomes provide a barrier against nonviral gene delivery. Autophagy 9, 667–682.

Russ, V., Gunther, M., Halama, A., Ogris, M., Wagner, E., 2008. Oligoethylenimine-grafted polypropylenimine dendrimers as degradable and biocompatible synthetic vectors for gene delivery. J. Control Release 132, 131–140.

Sarmento, B., Ribeiro, A., Veiga, F., Sampaio, P., Neufeld, R., Ferreira, D., 2007. Alginate/chitosan nanoparticles are effective for oral insulin delivery. Pharm. Res. 24, 2198–2206.

Satoh, T., Otsuka, A., Contassot, E., French, L.E., 2015. The inflammasome and IL-1beta: implications for the treatment of inflammatory diseases. Immunotherapy 7, 243–254.

Schafer, J., Hobel, S., Bakowsky, U., Aigner, A., 2010. Liposome-polyethylenimine complexes for enhanced DNA and siRNA delivery. Biomaterials 31, 6892–6900.

Serramia, M.J., Alvarez, S., Fuentes-Paniagua, E., Clemente, M.I., Sanchez-Nieves, J., Gomez, R., de la Mata, J., Munoz-Fernandez, M.A., 2015. In vivo delivery of siRNA to the brain by carbosilane dendrimer. J. Control Release 200, 60–70.

Shao, N., Su, Y., Hu, J., Zhang, J., Zhang, H., Cheng, Y., 2011. Comparison of generation 3 polyamidoamine dendrimer and generation 4 polypropylenimine dendrimer on drug loading, complex structure, release behavior, and cytotoxicity. Int. J. Nanomedicine 6, 3361–3372.

Shih, I.L., Shen, M.H., Van, Y.T., 2006. Microbial synthesis of poly(epsilon-lysine) and its various applications. Bioresour. Technol. 97, 1148–1159.

Shim, M.S., Kwon, Y.J., 2012. Stimuli-responsive polymers and nanomaterials for gene delivery and imaging applications. Adv. Drug Deliv. Rev. 64, 1046–1059.

Shukla, S.K., Mishra, A.K., Mamba, B.B., Arotiba, O.A., 2014. Zirconia-poly(propylene imine) dendrimer nanocomposite based electrochemical urea biosensor. Enzym. Microb. Technol. 66, 48–55.

Singhal, A., Morris, V.B., Labhasetwar, V., Ghorpade, A., 2013. Nanoparticle-mediated catalase delivery protects human neurons from oxidative stress. Cell Death Dis. 4, e903.

Song, A., Walker, S.G., Parker, K.A., Sampson, N.S., 2011. Antibacterial studies of cationic polymers with alternating, random, and uniform backbones. ACS Chem. Biol. 6, 590–599.

Strand, B.L., Ryan, T.L., In't Veld, P., Kulseng, B., Rokstad, A.M., Skjak-Brek, G., Espevik, T., 2001. Poly-L-Lysine induces fibrosis on alginate microcapsules via the induction of cytokines. Cell Transplant. 10, 263–275.

Sun, B., Slomberg, D.L., Chudasama, S.L., Lu, Y., Schoenfisch, M.H., 2012. Nitric oxide-releasing dendrimers as antibacterial agents. Biomacromolecules 13, 3343–3354.

Sun, B., Wang, X., Ji, Z., Li, R., Xia, T., 2013. NLRP3 inflammasome activation induced by engineered nanomaterials. Small 9, 1595–1607.

Sun, Y., Guo, F., Zou, Z., Li, C., Hong, X., Zhao, Y., Wang, C., Wang, H., Liu, H., Yang, P., Han, Z., Liu, K., Kuba, K., Song, B., Gao, J., Mo, Z., Li, D., Li, B., Li, Q., Zhong, N., Wang, C., Penninger, J.M., Jiang, C., 2015. Cationic nanoparticles directly bind angiotensin-converting enzyme 2 and induce acute lung injury in mice. Part Fibre Toxicol. 12, 4.

Symonds, P., Murray, J.C., Hunter, A.C., Debska, G., Szewczyk, A., Moghimi, S.M., 2005. Low and high molecular weight poly(L-lysine)s/poly(L-lysine)-DNA complexes initiate mitochondrial-mediated apoptosis differently. FEBS Lett. 579, 6191–6198.

Tasdemir, E., Galluzzi, L., Maiuri, M.C., Criollo, A., Vitale, I., Hangen, E., Modjtahedi, N., Kroemer, G., 2008. Methods for assessing autophagy and autophagic cell death. Methods Mol. Biol. 445, 29–76.

Thiagarajan, G., Greish, K., Ghandehari, H., 2013. Charge affects the oral toxicity of poly(amidoamine) dendrimers. Eur. J. Pharm. Biopharm. 84, 330–334.

Thoma, L.M., Boles, B.R., Kuroda, K., 2014. Cationic methacrylate polymers as topical antimicrobial agents against *Staphylococcus aureus* nasal colonization. Biomacromolecules 15, 2933–2943.

Ting, J.P., Willingham, S.B., Bergstralh, D.T., 2008. NLRs at the intersection of cell death and immunity. Nat. Rev. Immunol. 8, 372–379.

Torchilin, V.P., 1996. Liposomes as delivery agents for medical imaging. Mol. Med. Today 2, 242–249.

Vaine, C.A., Patel, M.K., Zhu, J., Lee, E., Finberg, R.W., Hayward, R.C., Kurt-Jones, E.A., 2013. Tuning innate immune activation by surface texturing of polymer microparticles: the role of shape in inflammasome activation. J. Immunol. 190, 3525–3532.

Verdurmen, W.P., Brock, R., 2011. Biological responses towards cationic peptides and drug carriers. Trends Pharmacol. Sci. 32, 116–124.

Vignesh, G., Arunachalam, S., Vignesh, S., James, R.A., 2012. BSA binding and antimicrobial studies of branched polyethyleneimine-copper(II)bipyridine/phenanthroline complexes. Spectrochim. Acta A Mol. Biomol. Spectrosc. 96, 108–116.

von Harpe, A., Petersen, H., Li, Y., Kissel, T., 2000. Characterization of commercially available and synthesized polyethylenimines for gene delivery. J. Control Release 69, 309–322.

Wagner, E., Ogris, M., Zauner, W., 1998. Polylysine-based transfection systems utilizing receptor-mediated delivery. Adv. Drug Deliv. Rev. 30, 97–113.

Wan, K., Li, J., Li, D., Ge, J., Wang, Y., Li, X., Guo, Y., Guo, J., Leng, M., Wang, P., An, Y., 2015. Novel hydroxybutyl chitosan nanoparticles for siRNA delivery targeting tissue factor inhibits proliferation and induces apoptosis in human vascular smooth muscle cells. Mol. Med. Rep. 12, 7957–7962.

Wang, B., Navath, R.S., Menjoge, A.R., Balakrishnan, B., Bellair, R., Dai, H., Romero, R., Kannan, S., Kannan, R.M., 2010a. Inhibition of bacterial growth and intramniotic infection in a guinea pig model of chorioamnionitis using PAMAM dendrimers. Int. J. Pharm. 395, 298–308.

Wang, H., Zhao, P., Su, W., Wang, S., Liao, Z., Niu, R., Chang, J., 2010b. PLGA/polymeric liposome for targeted drug and gene co-delivery. Biomaterials 31, 8741–8748.

Wang, F., Cai, X., Su, Y., Hu, J., Wu, Q., Zhang, H., Xiao, J., Cheng, Y., 2012. Reducing cytotoxicity while improving anti-cancer drug loading capacity of polypropylenimine dendrimers by surface acetylation. Acta Biomater. 8, 4304–4313.

Wang, F., Bexiga, M.G., Anguissola, S., Boya, P., Simpson, J.C., Salvati, A., Dawson, K.A., 2013a. Time resolved study of cell death mechanisms induced by amine-modified polystyrene nanoparticles. Nanoscale 5, 10868–10876.

Wang, H., Zheng, L., Peng, C., Shen, M., Shi, X., Zhang, G., 2013b. Folic acid-modified dendrimer-entrapped gold nanoparticles as nanoprobes for targeted CT imaging of human lung adenocarcinoma. Biomaterials 34, 470–480.

Wang, S., Li, Y., Fan, J., Wang, Z., Zeng, X., Sun, Y., Song, P., Ju, D., 2014. The role of autophagy in the neurotoxicity of cationic PAMAM dendrimers. Biomaterials 35, 7588–7597.

Wang, C., Li, M., Yang, T., Ding, X., Bao, X., Ding, Y., Xiong, H., Wu, Y., Wang, W., Zhou, J., 2015a. A self-assembled system for tumor-targeted co-delivery of drug and gene. Mater. Sci. Eng. C Mater. Biol. Appl. 56, 280–285.

Wang, C.W., Yang, S.P., Hu, H., Du, J., Li, F.H., 2015b. Synthesis, characterization and in vitro and in vivo investigation of C(3)F(8)-filled poly(lactic-co-glycolic acid) nanoparticles as an ultrasound contrast agent. Mol. Med. Rep. 11, 1885–1890.

Wang, K., Kievit, F.M., Jeon, M., Silber, J.R., Ellenbogen, R.G., Zhang, M., 2015c. Nanoparticle-mediated target delivery of TRAIL as gene therapy for glioblastoma. Adv. Healthc. Mater. 4 (17), 2719–2726.

Wegmann, F., Gartlan, K.H., Harandi, A.M., Brinckmann, S.A., Coccia, M., Hillson, W.R., Kok, W.L., Cole, S., Ho, L.P., Lambe, T., Puthia, M., Svanborg, C., Scherer, E.M., Krashias, G., Williams, A., Blattman, J.N., Greenberg, P.D., Flavell, R.A., Moghaddam, A.E., Sheppard, N.C., Sattentau, Q.J., 2012. Polyethyleneimine is a potent mucosal adjuvant for viral glycoprotein antigens. Nat. Biotechnol. 30, 883–888.

Winnicka, K., Wroblewska, M., Wieczorek, P., Sacha, P.T., Tryniszewska, E.A., 2013. The effect of PAMAM dendrimers on the antibacterial activity of antibiotics with different water solubility. Molecules 18, 8607–8617.

Win-Shwe, T.T., Sone, H., Kurokawa, Y., Zeng, Y., Zeng, Q., Nitta, H., Hirano, S., 2014. Effects of PAMAM dendrimers in the mouse brain after a single intranasal instillation. Toxicol. Lett. 228, 207–215.

Wiseman, J.W., Goddard, C.A., McLelland, D., Colledge, W.H., 2003. A comparison of linear and branched polyethylenimine (PEI) with DCChol/DOPE liposomes for gene delivery to epithelial cells in vitro and in vivo. Gene Ther. 10, 1654–1662.

Wu, Y., Yang, W., Wang, C., Hu, J., Fu, S., 2005. Chitosan nanoparticles as a novel delivery system for ammonium glycyrrhizinate. Int. J. Pharm. 295, 235–245.

Wu, J., Huang, W., He, Z., 2013. Dendrimers as carriers for siRNA delivery and gene silencing: a review. ScientificWorldJournal 2013, 630654.

Xia, T., Kovochich, M., Liong, M., Zink, J.I., Nel, A.E., 2008. Cationic polystyrene nanosphere toxicity depends on cell-specific endocytic and mitochondrial injury pathways. ACS Nano 2, 85–96.

Xiong, S., George, S., Yu, H., Damoiseaux, R., France, B., Ng, K.W., Loo, J.S., 2013. Size influences the cytotoxicity of poly (lactic-co-glycolic acid) (PLGA) and titanium dioxide (TiO(2)) nanoparticles. Arch. Toxicol. 87, 1075–1086.

Xu, P., Gullotti, E., Tong, L., Highley, C.B., Errabelli, D.R., Hasan, T., Cheng, J.X., Kohane, D.S., Yeo, Y., 2009. Intracellular drug delivery by poly(lactic-co-glycolic acid) nanoparticles, revisited. Mol. Pharm. 6, 190–201.

Xu, J., Wang, Y., Tan, X., Jing, H., 2012. MicroRNAs in autophagy and their emerging roles in crosstalk with apoptosis. Autophagy 8, 873–882.

Yang, Z.Y., Li, H., Zeng, Y.P., Hao, Y.H., Liu, C., Liu, J., Wang, W.D., Li, R., 2015. Photosensitizer-loaded branched polyethylenimine-PEGylated ceria nanoparticles for imaging-guided synchronous photochemotherapy. ACS Appl. Mater. Interfaces 7, 24218–24228.

Yang, X.N., Xue, D.D., Li, J.Y., Liu, M., Jia, S.R., Chu, L.Q., Wahid, F., Zhang, Y.M., Zhong, C., 2016. Improvement of antimicrobial activity of graphene oxide/bacterial cellulose nanocomposites through the electrostatic modification. Carbohydr. Polym. 136, 1152–1160.

Ye, R., Xu, H., Wan, C., Peng, S., Wang, L., Xu, H., Aguilar, Z.P., Xiong, Y., Zeng, Z., Wei, H., 2013. Antibacterial activity and mechanism of action of epsilon-poly-L-lysine. Biochem. Biophys. Res. Commun. 439, 148–153.

You, S., Cai, Q., Zheng, Y., He, B., Shen, J., Yang, W., Yin, M., 2014. Perylene-cored star-shaped polycations for fluorescent gene vectors and bioimaging. ACS Appl. Mater. Interfaces 6, 16327–16334.

Yuan, Z.Y., Hu, Y.L., Gao, J.Q., 2015. Brain localization and neurotoxicity evaluation of polysorbate 80-Modified chitosan nanoparticles in rats. PLoS One 10, e0134722.

Zamani, M., Prabhakaran, M.P., Ramakrishna, S., 2013. Advances in drug delivery via electrospun and electrosprayed nanomaterials. Int. J. Nanomedicine 8, 2997–3017.

Zarebkohan, A., Najafi, F., Moghimi, H.R., Hemmati, M., Deevband, M.R., Kazemi, B., 2015. Synthesis and characterization of a PAMAM dendrimer nanocarrier functionalized by SRL peptide for targeted gene delivery to the brain. Eur. J. Pharm. Sci. 78, 19–30.

Zhang, G., Zeng, X., Li, P., 2013. Nanomaterials in cancer-therapy drug delivery system. J. Biomed. Nanotechnol. 9, 741–750.

Zhao, Q.Q., Hu, Y.L., Zhou, Y., Li, N., Han, M., Tang, G.P., Qiu, F., Tabata, Y., Gao, J.Q., 2012. Gene-carried hepatoma targeting complex induced high gene transfection efficiency with low toxicity and significant antitumor activity. Int. J. Nanomedicine 7, 3191–3202.

Zhu, L., Mahato, R.I., 2010. Lipid and polymeric carrier-mediated nucleic acid delivery. Expert Opin. Drug Deliv. 7, 1209–1226.

Zhu, C., Zheng, M., Meng, F., Mickler, F.M., Ruthardt, N., Zhu, X., Zhong, Z., 2012. Reversibly shielded DNA polyplexes based on bioreducible PDMAEMA-SS-PEG-SS-PDMAEMA triblock copolymers mediate markedly enhanced nonviral gene transfection. Biomacromolecules 13, 769–778.

Ziemba, B., Janaszewska, A., Ciepluch, K., Krotewicz, M., Fogel, W.A., Appelhans, D., Voit, B., Bryszewska, M., Klajnert, B., 2011. In vivo toxicity of poly(propyleneimine) dendrimers. J. Biomed. Mater. Res. A 99, 261–268.

Zou, Q.C., Yan, Q.J., Song, G.W., Zhang, S.L., Wu, L.M., 2007. Detection of DNA using cationic polyhedral oligomeric silsesquioxane nanoparticles as the probe by resonance light scattering technique. Biosens. Bioelectron. 22, 1461–1465.

Chapter 13

Perspective on Strategies to Reduce the Neurotoxicity of Nanomaterials and Nanomedicines

H. Gao[1], X. Jiang[2]
[1]*Sichuan University, Chengdu, China;* [2]*Fudan University, Shanghai, China*

Chapter Outline

1. INTRODUCTION

The development of nanotechnology has provided various kinds of nanomaterials and nanomedicines that are greatly influencing our life. However, as reviewed in Chapters 4–12, the application of these nanomaterials and nanomedicines also leads to more or less neurotoxicity, which shadows the prospect of nanomaterials and nanomedicines. Therefore, reducing the neurotoxicity potential is an important topic that requires more attention.

Only limited studies have focused on the methods to reduce the neurotoxicity of nanomaterials. In this chapter, we provide a general perspective on the related strategies. Basically, we can categorize the strategies to reducing brain exposure and reducing the toxicity of nanomaterials.

Neurotoxicity of Nanomaterials and Nanomedicine. http://dx.doi.org/10.1016/B978-0-12-804598-5.00013-1

2. REDUCING BRAIN EXPOSURE

The central nervous system (CNS) is protected by the blood–brain barrier (BBB) to avoid the harm of toxins from the blood circulation. Because many nanomaterials are not intended to enter into brain, it is useful to reduce the potential brain exposure to minimize the neurotoxicity.

2.1 Reducing Inhalation of Nanomaterials

Compounds and nanomaterials could be directly transported from the nasal cavity to brain through the olfactory nerve pathway and olfactory mucosa epithelium pathway (Wu et al., 2008). Thus inhaled nanomaterials may enter into brain, leading to neurotoxicity. Intranasal administration and inhalation are important routes for nanomedicines. Thus when these nanomedicines are not designed to treat CNS disorders, particular attention should be paid to potential neurotoxicity. It would be useful to limit the contact time of nanomaterials with nose cavity endothelial cells. Apart from being used as medicines, nanomaterials may be inhaled by workers in factories involving nanomaterials. Therefore, protection face mask should be used to filter out the particle matters in the air.

2.2 Reducing Blood–Brain Barrier Penetration

Although the integrity of the BBB prevents almost 100% of large molecules and nanomaterials from entering into brain, there are several strategies to improve BBB penetration, including receptor-mediated endocytosis, transporter-mediated endocytosis, and adsorption-mediated endocytosis (Gao et al., 2013). But some nanomaterials, designed to target to peripheral tissues or cells may also penetrate through BBB using the above-mentioned strategies. For example, cell-penetrating peptides could interact with the negatively charged cell membrane and mediate the penetrating through cell membrane due to their positive charge. Therefore, many kinds of cell-penetrating peptide–modified nanomaterials were developed to deliver cargoes to peripheral tumors (Huang et al., 2013). However, the BBB is also negatively charged, and cell-penetrating peptide–modified liposomes showed higher brain concentration than PEGylated liposomes (Gao et al., 2014). In addition to adsorption-mediated endocytosis, there are several receptors and transporters overexpressed both on peripheral diseased cells and BBB. For example, transferrin receptor is overexpressed on most tumor cells and BBB (Gao et al., 2013). The corresponding ligand, transferrin, is widely used for targeting drug delivery. Some of the ligand-modified nanomaterials are even under clinical evaluation (van der Meel et al., 2013). MBP-426 (Mebiopharm) is a transferrin-modified liposome loaded with oxaliplatin undergoing phase I/II trials for treatment of second-line gastric, gastroesophageal, or esophageal adenocarcinomas in combination with other chemotherapeutics. SGT-53, SGT-94, and CALAA-01 are all nanoformulations under phase I trials that can target to transferrin receptors. These nanoformulations may also have higher brain distribution compared with the nanoformulations without ligand. Thus the neurotoxicity may increase, which should be taken into consideration when evaluating the toxicity of these nanoformulations. To avoid the intended neurotoxicity, the nanoformulations for peripheral disease targeting are suggested to use ligands that have low affinity to BBB.

In certain pathological conditions, such as neurodegenerative disease, inflammation, and infection, the integrity of the BBB is compromised and the permeability would increase. Using cerebrospinal fluid albumin as an index, a study showed the permeability of BBB in patients with Alzheimer's disease increased (Bowman et al., 2007). The application of nanomaterials in these conditions may have higher neurotoxicity risk compared with that in normal conditions. It is better to minimize the application of nanomaterials in such conditions.

2.3 Improving Cell Selectivity in Brain

Except to avoid nanomaterials unintended entering into brain, the distribution of nanomaterials in brain should be taken into consideration when using nanomaterials for brain disease diagnosis and treatment. Normally, the BBB is considered as the first and main barrier for brain targeting delivery, and many kinds of BBB penetration strategies were developed to deliver drugs and probes to brain. However, the diseased brain cells are only a small part of the whole brain. When the nanomaterials penetrate through the BBB, unselective distribution in brain may cause serious neurotoxicity to brain. To conquer this problem, dual targeting strategies were introduced to further improve the selectivity in brain (Gao et al., 2013). For example, we modified nanoparticles (NPs) with angiopep-2 for BBB penetration and EGF1 for neuroglial cell targeting (Gao et al., 2011). Results showed the dual targeting NPs could specifically colocalize with brain neuroglial cells rather than brain endothelial cells, suggesting that dual targeting could indeed improve the cell type selectivity in brain. Another study used TGN peptide for BBB penetration and AS1411 aptamer for brain tumor targeting. The dual targeting NPs considerably improved the tumor/normal brain drug concentration ratio (Gao et al., 2012). These studies demonstrated that the dual targeting delivery strategy could enhance cell selectivity in brain, which can minimize the distribution in normal brain, providing a useful strategy to enhance diagnosis and treatment outcome with minimized neurotoxicity.

3. REDUCING THE INHERENT TOXICITY OF NANOMATERIALS

The neurotoxicity of nanomaterials is partly due to their inherent toxicity, thus reducing the inherent toxicity of nanomaterials could also attenuate the potential neurotoxicity of nanomaterials. The toxicity of nanomaterials is influenced by many factors (Sharifi et al., 2012), which could be taken into consideration for minimizing nanomaterials toxicity.

3.1 Composition of Nanomaterials

The composition of nanomaterials considerably affects the toxicity. Especially, toxicity of many metal NPs ascribes to the metal ion. Silver ion shows antibacterial activity, which contributes to the toxicity of silver NPs. The toxicity of quantum dots also partly ascribes to the heavy metal in the dots. Developing nanomaterials without these toxic metals is a popular and useful method. For example, carbonaceous dots are demonstrated with similar fluorescent property compared with quantum dots, whereas the toxicity was significantly attenuated (Chen et al., 2015).

For the organic- or polymer-based NPs, the composition of constituted nanomaterials considerably attributes to the biocompatibility of NPs. NPs prepared from lipid, polyethylene glycol–polycaprolactone, and polyethylene glycol–poly-L-lactic acid are approved for human use because of their low toxicity. The low toxicity of these nanomaterials is because not only they are degradable but also the degraded fragments are not toxic. Thus the biodegradable nanomaterials attract more and more attention.

3.2 Size and Shape

Because the interaction of nanomaterials with organisms occurs at the surface of NPs and size determines the surface atom ratio of the NPs, size is a critical character that influences the biodistribution of nanomaterials, interaction with cells, and toxicity (Sharifi et al., 2012). Among different sizes (20, 50, 70, and 110 nm) of gold NPs, particles with 70 nm showed the highest ability to penetrate through the in vitro BBB model, resulting in maximum concentration in brain cells (Shilo et al., 2015). However, the influence of size on toxicity was various. Small silver NPs (10 nm) showed greater ability in inducing reactive oxygen species in MC3T3-E1 cells than large silver NPs (50 and 100 nm), whereas large silver NPs (100 nm) caused higher PC12 cell apoptosis than small NPs (10 and 50 nm) (Kim et al., 2012). Thus, for a particular kind of NPs, it is better to optimize the size before application.

Particle shape and aspect ratio are also key factors in determining the toxicity of nanomaterials. Commonly, the shape and aspect ratio could influence the endocytosis by cells and clearance by macrophages because the shape affects membrane wrapping during endocytosis (Canton and Battaglia, 2012; Sharifi et al., 2012). Among the CeO_2 NPs with different aspect ratios (1 to over 100), only the nanorods with higher aspect ratio (>100) induced significant interleukin 1β and transforming growth factor β1 production in the bronchoalveolar lavage fluid at 21 days after oropharyngeal instillation at 2 mg/kg (Lin et al., 2014). The higher aspect ratio (>100) of CeO_2 NPs also induced significant growth inhibition, a decrease in body weight, and delayed vertebral calcification, whereas NPs with other aspect ratios (1 and 22) displayed no effect. Thus reducing the aspect ratio of NPs may attenuate the toxicity of nanomaterials.

3.3 Surface Property of Nanomaterials

The surface properties include charge, functional groups, and surface coating, which also influence toxicity. Compared with neutral NPs and low concentration of anionic NPs, high concentration of anionic NPs and cationic NPs could disrupt the BBB determined by in situ rat brain perfusion, suggesting cationic NPs had higher neurotoxicity than neutral and anionic NPs (Lockman et al., 2004). Positively charged NPs also showed higher affinity to cells, resulting in higher cellular uptake and stronger toxicity potential.

Surface groups also influenced the interaction between NPs and cells. For example, silica NPs functionalized with amine, sulfonate, polyethylene glycol, and polyethylene imine showed different cellular uptake and cytotoxicity potential. Sulfonate-functionalized NPs were effectively taken up by cells and delivered more cargoes into cells, causing high cytotoxicity (Shahabi et al., 2015). Silver NPs coated with carbon showed lower cytotoxicity than uncoated silver NPs (Nishanth et al., 2011), and citrate-coated silver NPs showed lower cytotoxicity than polyvinylpyrrolidone-coated silver NPs (Kim and Ryu, 2013).

Because the surface properties affected BBB function and interaction with cells, leading to various levels of toxicity, they should be optimized for particular nanomaterials to balance the application efficiency (for example, targeting and being taken up by specific cells) and the unintended toxicity to brain.

4. CONCLUSION

Neurotoxicity is a critical concern of nanomaterials. To attenuate the neurotoxicity, we could reduce the brain exposure of nanomaterials and optimize the properties of nanomaterials. However, current strategies are based mostly on the general toxicity of nanomaterials, which may not be the same as neurotoxicity. Thus systemic evaluation of the influence of properties on neurotoxicity is still needed.

REFERENCES

Bowman, G.L., Kaye, J.A., Moore, M., Waichunas, D., Carlson, N.E., Quinn, J.F., 2007. Blood–brain barrier impairment in Alzheimer disease: stability and functional significance. Neurology 68, 1809–1814.

Canton, I., Battaglia, G., 2012. Endocytosis at the nanoscale. Chem. Soc. Rev. 41, 2718–2739.

Chen, J., Sun, H., Ruan, S., Wang, Y., Shen, S., Xu, W., et al., 2015. In vitro and in vivo toxicology of bare and PEGylated fluorescent carbonaceous nanodots in mice and zebrafish: the potential relation with autophagy. RSC Adv. 5, 38547–38557.

Gao, H., Pan, S., Yang, Z., Cao, S., Chen, C., Jiang, X., et al., 2011. A cascade targeting strategy for brain neuroglial cells employing nanoparticles modified with angiopep-2 peptide and EGFP-EGF1 protein. Biomaterials 32, 8669–8675.

Gao, H., Pang, Z., Jiang, X., 2013. Targeted delivery of nano-therapeutics for major disorders of the central nervous system. Pharm. Res. 30, 2485–2498.

Gao, H., Qian, J., Cao, S., Yang, Z., Pang, Z., Pan, S., et al., 2012. Precise glioma targeting of and penetration by aptamer and peptide dual-functioned nanoparticles. Biomaterials 33, 5115–5123.

Gao, H., Zhang, Q., Yu, Z., He, Q., 2014. Cell-penetrating peptide-based intelligent liposomal systems for enhanced drug delivery. Curr. Pharm. Biotechnol. 15, 210–219.

Huang, Y., Jiang, Y., Wang, H., Wang, J., Shin, M.C., Byun, Y., et al., 2013. Curb challenges of the "Trojan Horse" approach: smart strategies in achieving effective yet safe cell-penetrating peptide-based drug delivery. Adv. Drug Deliv. Rev. 65, 1299–1315.

Kim, S., Ryu, D.-Y., 2013. Silver nanoparticle-induced oxidative stress, genotoxicity and apoptosis in cultured cells and animal tissues. J. Appl. Toxicol. 33, 78–89.

Kim, T.H., Kim, M., Park, H.S., Shin, U.S., Gong, M.S., Kim, H.W., 2012. Size-dependent cellular toxicity of silver nanoparticles. J. Biomed. Mater. Res. A 100, 1033–1043.

Lin, S., Wang, X., Ji, Z., Chang, C.H., Dong, Y., Meng, H., et al., 2014. Aspect ratio plays a role in the hazard potential of CeO_2 nanoparticles in mouse lung and zebrafish gastrointestinal tract. ACS Nano 8, 4450–4464.

Lockman, P.R., Koziara, J.M., Mumper, R.J., Allen, D.D., 2004. Nanoparticle surface charges alter blood–brain barrier integrity and permeability. J. Drug Target. 12, 635–641.

van der Meel, R., Vehmeijer, L.J., Kok, R.J., Storm, G., van Gaal, E.V., 2013. Ligand-targeted particulate nanomedicines undergoing clinical evaluation: current status. Adv. Drug Deliv. Rev. 65, 1284–1298.

Nishanth, R.P., Jyotsna, R.G., Schlager, J.J., Hussain, S.M., Reddanna, P., 2011. Inflammatory responses of RAW 264.7 macrophages upon exposure to nanoparticles: role of ROS-NFkappaB signaling pathway. Nanotoxicology 5, 502–516.

Shahabi, S., Doscher, S., Bollhorst, T., Treccani, L., Maas, M., Dringen, R., et al., 2015. Enhancing cellular uptake and doxorubicin delivery of mesoporous silica nanoparticles via surface functionalization: effects of serum. ACS Appl. Mater. Interfaces 7, 26880–26891.

Sharifi, S., Behzadi, S., Laurent, S., Forrest, M.L., Stroeve, P., Mahmoudi, M., 2012. Toxicity of nanomaterials. Chem. Soc. Rev. 41, 2323–2343.

Shilo, M., Sharon, A., Baranes, K., Motiei, M., Lellouche, J.P., Popovtzer, R., 2015. The effect of nanoparticle size on the probability to cross the blood–brain barrier: an in-vitro endothelial cell model. J. Nanobiotechnol. 13, 19.

Wu, H., Hu, K., Jiang, X., 2008. From nose to brain: understanding transport capacity and transport rate of drugs. Expert Opin. Drug Deliv. 5, 1159–1168.

Index

'*Note*: Page numbers followed by "f" indicate figures and "t" indicate tables.'

D

E

F

G